W9-CFQ-458

Peninj

A Research Project on Human Origins 1995-2005

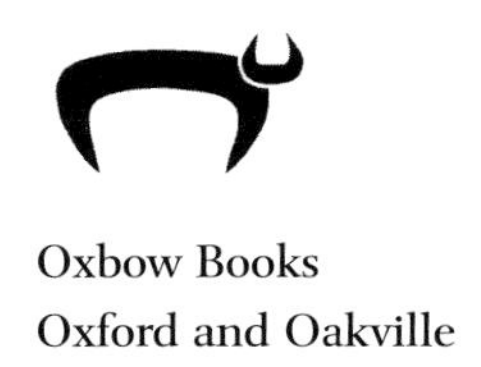

Oxbow Books
Oxford and Oakville

American School of Prehistoric Research Monograph Series

The American School of Prehistoric Research (ASPR) Monographs in Archaeology and Paleoanthropology present a series of documents covering a variety of subjects in the archaeology of the Old World (Eurasia, Africa, Australia, and Oceania). This series encompasses a broad range of subjects - from the early prehistory to the Neolithic Revolution in the Old World, and beyond including: hunter-gatherers to complex societies; the rise of agriculture; the emergence of urban societies; human physical morphology, evolution and adaptation, as well as; various technologies such as metallurgy, pottery production, tool making, and shelter construction. Additionally, the subjects of symbolism, religion, and art will be presented within the context of archaeological studies including mortuary practices and rock art. Volumes may be authored by one investigator, a team of investigators, or may be an edited collection of shorter articles by a number of different specialists working on related topics.

American School of Prehistoric Research (ASPR), Harvard University,
Peabody Museum, 11 Divinity Avenue, Cambridge, MA 02138, USA

The publication of this volume was made possible in part with the support of Fundación Conjunto Paleontológico Teruel-Dinópolis

Peninj

A Research Project on Human Origins 1995–2005

Edited by Manuel Domínguez-Rodrigo, Luis Alcalá, and Luis Luque

www.oxbowbooks.com

Published by Oxbow Books on behalf of the American School of Prehistoric Research.

ISBN 978-1-84217-382-4

Library of Congress Cataloging-in-Publication Data

Peninj : a research project on human origins, 1995-2005 / edited by Manuel Domínguez-Rodrigo, Luis Alcalá, and Luis Luque.
p. cm. -- (American School of Prehistoric Research monograph series)
Includes bibliographical references.
ISBN 978-1-84217-382-4
1. Peninj Site (Tanzania) 2. Acheulian culture--Tanzania. 3. Fossil hominids--Tanzania. 4. Human evolution--Tanzania. 5. Excavations (Archaeology)--Tanzania. 6. Paleontology--Tanzania. 7. Paleoecology--Tanzania. 8. Paleontology--Pleistocene. 9. Paleoecology--Pleistocene. I. Domínguez-Rodrigo, Manuel. II. Alcalá, Luis. III. Luque, Luis.
GN772.4.A53P46 2009
569.9--dc22

2009037184

TYPESET AND PRINTED IN THE UNITED STATES OF AMERICA

Contents

Preface

Manuel Domínguez-Rodrigo, Luis Alcalá, and Luis Luque

Preface

The Peninj archaeological area is situated to the west of Lake Natron in northern Tanzania (South 2°00′ to 2°50′, East 35°40′ to 36°20′) (Figure P.1). It lies at one of the lowest points of the Gregory Rift System. The Peninj region is geologically shaped by the southern extension of the Nguruman Escarpment in Kenya, which extends all the way from the Mau Hills in Central Kenya to the Lake Manyara region in the South. It has a semi-graben morphology, longest along the north-south axis, with a series of elevated platforms caused by the occurrence of fault escarpments along its western side.

Past and present research at Peninj has underscored the importance of archaeology in human evolutionary studies. The reconstruction of the human evolutionary process in Africa has traditionally relegated archaeology to a secondary role, trailing behind human paleontology. Yet prehistoric archaeology is much more than the descriptive study of stone tools and bones. The analytical tools developed in this interdisciplinary field over the past two decades have enabled archaeologists to ask new questions and answer old ones about human behavior. Archaeology often requires years of hard work in the field, slowly acquiring data until one can produce informative results, while in human paleontology, one hominid discovery can more easily produce results. But our understanding of human evolution would be incomplete if we failed to reconstruct that behavioral process we call "becoming human." The behavioral side of the human evolutionary process can only be approached at its fullest through careful archaeological work.

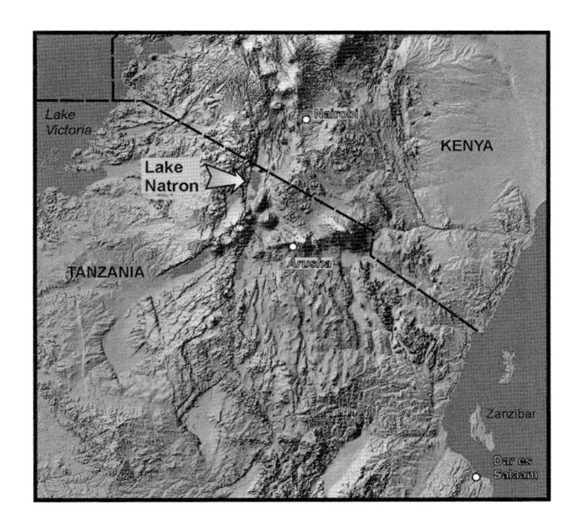

Figure P.1 Location of Lake Natron in northern Tanzania.

The authors of this monograph first became interested in working in the Peninj region in 1992, when one of us (M. Domínguez-Rodrigo) spent a few days in the Lake Natron area after some weeks of neo-taphonomic research in the Loliondo Game Reserve, close to the boundary with the Serengeti. He had been interested in Peninj since reading the preliminary interpretations offered by Isaac regarding the purported openness of the Peninj landscape during the Early Pleistocene. Given that modern studies on savanna ecology showed the difficulties of scavenging in open ecosystems, the idea of studying human adaptation to an open landscape seemed extremely appealing from the perspective of hominid strategies of carcass obtainment and early site function. The exposures with fossils encountered at Peninj in 1992 were as abundant and as promising as they were in 1964 when the

Peninj mandible, the only hominid find from the site, was discovered.

The complete record of our research in the Peninj area from 1995–2005 has been compiled in this volume. Some of our data have been previously published in various scientific journals, and Peninj has also been the subject of two major exhibits: one in Barcelona in 1996 and the other in Arusha in 2003, the latter officially inaugurated by the previous Spanish ambassador to Tanzania, José María Castroviejo.

When we began our work, access to the Peninj area was difficult. In his autobiography *One Life*, R. Leakey described how in the 1960s he had to clear out a road so that cars could go through the lake margin and the escarpment. The dirt road that links the Peninj valley to Sonjo and Malambo was also in very bad condition when we encountered it in the early 1990s. It took two days to reach Peninj from Arusha, a distance of barely 300 kilometers. The 12 kilometers that separated Sonjo from Peninj took more than three hours to drive through. In the beginning we did some work on the road, just enough to allow us to drive through. However in recent years, the deterioration of the road had made driving almost impossible. A new dirt track linking Loliondo to Mto wa Mbu was recently opened following the swampy lakeshore, saving at least three hours of driving.

Over the last decade of our research in the area, we have witnessed major changes in Peninj: a population that was isolated from nearly everything outside the Lake Natron basin has rapidly developed more intense relations with populations from other areas, partly due to the new dirt track mentioned above. Peninj now has a school and there is a dispensary in the nearby locality of Ngare Sero, in the south of the lake. Today the trip from Arusha to Peninj can be made in less than ten hours (if one does not get stuck in the swampy shores of Lake Natron). Through the years, the experience of working among the Maasai and Sonjo populations in the area has been enriching and enlightening. A Spanish physician offering medical care, who volunteered as a member of our team for several seasons, created a constant flow of people into and out of our camp and provided countless anecdotes around the bonfire.

It is our wish to dedicate this book to all those colleagues who have devoted their lives to helping us understand our origins. A special dedication goes to G. L. Isaac, who was in our opinion one of the best African archaeologists, and whose pioneering work in Koobi Fora and Peninj set the basis for very important ideas on early human behavior that remain tenable today. We are also indebted to the following people and institutions: José María Castroviejo, former Spanish ambassador to Tanzania; Agustin Zapata and Carlos Andradas, vice-rectors of research at Complutense University; Fernando Bouza, former Dean of Research at Complutense University; Barbara Isaac, who generously shared her husband's unpublished information with us; Ofer Bar Yosef, Richard Meadow, T. R. Pickering, N. Sikes, and S. Semaw, for their help and comments; Audax Mabulla, Head of the Archaeology Unit at the University of Dar es Salaam; Alfredo Pérez-González, Rafel González (MNCN), Dr. Pablo Silva, and Sergio Sánchez for their help. We thank N. Morán for the drawings of the lithic artifacts in Chapter 6.

The isotopic ratios reported in Chapter 5 were measured in the Stable Light isotope Facility of the Archaeology Department, University of Cape Town. N. van der Merwe thanks John Lanham and Ian Newton for technical laboratory assistance. Mariagrazia Galimberti produced the figures in Chapter 5. Funds were provided by the National Research Foundation of South Africa, the University of Cape Town, and the Landon T. Clay Fund of the Peabody Museum, Harvard University.

We thank Elsevier for permission to reproduce excerpts of a previous publication. We are grateful to the Spanish Ministry of Culture, and the Ministry of Science and Technology for their financial support through annual grants for archaeological projects abroad and through for Development and Technology (I+D) grants: BHA2000-0405, BHA2002-11667-E and BHA2003-00839.

We also thank Complutense University, Earthwatch, the Paleontological Network Foundation of Teruel and Dinopolis for special support, and Kibo safaris, especially to Julio Teigell, Nuria Panizo, and Willy Chambulo (for logistical support). Permission to conduct research was granted by the Tanzanian Commission for Science and Technology (COSTECH) and the Antiquities Department in Dar es-Salaam. We also thank Jose Luis Gómez Encinas, Fernando Diez Martín, Rebeca Barba Egido, Rafael Royo-Torres, Alberto Cobos, Victoria Medina, Jordi Serrallonga, Benigno Perez, Rafael Mora, Ignacio de la Torre, Dolores Soria and Pastory Bushozi for their participation in various years of fieldwork. We prematurely lost Dolores Soria due to a car accident but her memory remains with us.

MDR is thankful to Mary Prendergast for her editorial help, support, understanding and inspiration. He is also thankful to Travis Pickering for years of learning and sharing taphonomic experiences through his friendship.

We hope that with this work, we have contributed a little more information towards unravelling behavioral patterns in the beginning of the evolution of our genus. A little grain of sand may seem too small to signify anything, but when put together with others of its kind, it eventually creates a beach.

Contributor List

Alcalá, L.
Fundación Conjunto Paleontológico de Teruel-Dinópolis, Avda. Sagunto (Edificio Dinopolis) s/n, 44002 Teruel, Spain

Barba, R.
Department of Prehistory, Complutense University, 28040 Madrid, Spain

Bushozi, P.
Anthropology Department, University of Alberta, Canada

Diez-Martin, F.
Department of Prehistory and Archaeology, Valladolid University, Valladolid, Spain

Domínguez-Rodrigo, M.
Department of Prehistory, Complutense University, 28040 Madrid, Spain

Luque, L.
Fundación Conjunto Paleontológico de Teruel-Dinópolis, Avda. Sagunto (Edificio Dinopolis) s/n, 44002 Teruel, Spain

Mora, R.
Autonomous University of Barcelona, Departament d'Antropologia, Social i Prehistòria, 08193 Bellaterra, Barcelona, Spain

Serrallonga, J.
HOMINID - Grup d'Origens Humans, Parc Científic de Barcelona, 08028 Barcelona, Spain

de la Torre, I.,
Institute of Archaeology, University College London, 31-34 Gordon Square, London, WC1H 0PY, UK

van der Merwe, Nikolaas J.
Archaeology Department, University of Cape Town, South Africa

Figure List

Preface

Chapter 1

Chapter 2

Chapter 3

Chapter 4

Chapter 5

Chapter 6

Chapter 7

Chapter 8

Chapter 9

Chapter 10

Chapter 11

Appendix: Color Plates

Table List

Chapter 8

Chapter 9

Chapter 10

1

Structural and Referential Principles as Applied to Research Design at Peninj

Manuel Domínguez-Rodrigo and Luis Alcalá

Introduction: The evolution of paradigms

Paradigms (*sensu* Kuhn) or programs of scientific research (*sensu* Lakatos) condition how research is designed and carried out. When traditional Darwinian evolutionary frameworks were applied to the study of the archaeology of human evolution, special emphasis was given to the diachronic change of stone tools through time. In East Africa, this was clearly shown in the first monographic volume of research at Olduvai Gorge, which focused on the evolution of stone tool morphology along consecutive beds (Leakey 1951). Up to 11 types of Acheulian assemblages were distinguished, with the most crafted artifacts appearing in younger strata than the most rudimentary ones. Even when the relationships of artifacts in their archaeological contexts prompted extensive open-air excavation of sites, the evolution of the morphology of artifacts, understood as sets of stone industries, was particularly emphasized in subsequent research at Olduvai Gorge, as shown in the third monographic volume of the series (Leakey). Oldowan from Bed I evolved into Developed Oldowan A, B, and C in Bed II, followed by Acheulian, which itself evolved through Beds III, IVA, and Masek into more and more complex forms. Archaeologists studying human evolution were strongly committed to the idea of a firm link between hominid evolution and its reflection in stone tool evolution. This conditioned the way that field research was carried out. This way of conceptualizing archaeology and its field methods could be inserted into what Binford (1981) called the "artifact and assemblage" phase of archaeology.

Subsequently, when increasing awareness of ecological issues influenced society and science during the second half of the twentieth century, archaeologists began to introduce new questions into their research agenda. The New Archaeology and its systemic approach (with a strong economic component) prompted the study of archaeological sites in their ecological contexts to understand human adaptation to environments. Adaptation was seen as the interplay between energy investment and energy retrieval, as best exemplified in the Optimal Foraging Theory. The application of these New Archaeology schemes to the early Plio-Pleistocene archaeology of East Africa, predominantly by North American researchers, had a significant impact on field archaeology from the 1970s to the present. The Koobi Fora archaeological research project epitomizes these new conceptual frameworks. As clearly specified by G. L. Isaac (1997) in the archaeological monograph of Koobi Fora, this long-term project was aimed at reconstructing the diet of early hominids, their land-use patterns, their activity patterning as indicated by variation in site characteristics, aspects of their socio-economic organization, and the role played by technology and material culture in the evolutionary adaptation process (Isaac 1997).

In this new paradigm, stone tools were seen as adaptive functional items rather than strictly the end product of mental templates that reflected hominid skills. Mary Leakey's (1971) typologies of the Oldowan, reviewed by Toth (1982), gave way to Isaac's (1997) more aseptic morphological identification. All the core types of Leakey's typology were transformed into flaked tools and all the flake types were simply conceived of as detached pieces. Functional analysis was applied for the first time during the Koobi Fora project to prove that the "debris" might have been "more functional" than the core artifacts previously thought of as tools. The issue of functionality and adaptation was extended even to the concept of the site itself. Although according to the association of the materials contained in them, sites had been divided by Leakey (1971) into several distinct functional *loci* (living floors, butchering or kill sites and sites with diffuse material), it was during the Koobi Fora project that Isaac created a scientific framework to understand site formation and site behavioral function. Following his more sterile approach, Isaac created a typology of sites according to their structural contents. Type A sites are concentrations of artifacts without bone. Type B sites are clusters of bones from a single skeleton associated with stone tools. Type C sites are concentrations of stone tools and bones from a diversity of animals. Type M sites are concentrations of bones showing artifact-inflicted damage but lacking discarded artifacts (Bunn 1994). Assuming hominid authorship of any given archaeological site, these site types could be behaviorally meaningful. Type A sites could be stone knapping places. Type B sites could represent butchering *loci*. Type C could be the result of living floors or home bases. However, Issac's neutral terminology allowed classification of sites without a functional link, enabling archeologists to take taphonomic biases into consideration.

The behavioral meaning of all these types of sites is still unknown. The intervention of both hominids and hyenas in the formation of sites can be inferred by the presence of tooth-marks and cut-marks on the surface of several bones (Bunn 1981, 1982, 1983; Bunn and Kroll 1986; Potts and Shipman 1981). During the past two decades, the academic discourse stemming from neotaphonomic studies has focused on establishing the priority of both agents in the access to carcasses. A given site might be interpreted as: 1) the result of bone accumulations made by carnivores in which hominids might have participated secondarily (Binford 1981, 1985); or 2) the result of hominids' primary access to faunal remains and the later intervention of hyenas, who processed remains abandoned by hominids (Bunn 1982; Isaac 1983; Bunn and Kroll 1986; Blumenschine 1988; Marean et al. 1992).

After three decades of taphonomic research, archaeologists have reached a stage where there is enough information to grant that hominids were the primary agents of the accumulated faunal remains in some early sites. The wide range of studies undertaken during these years has substantially informed our understanding that although sites can be the result of a complex web of agents, hominids were probably the main importers of bone to some of them (Bunn and Kroll 1986; Blumenschine and Bunn 1987; Domínguez-Rodrigo and Barba 2006). There is no reason to argue otherwise. The assessment that some authors previously made (September 1992) that proof thereof has "so far eluded researchers" is a statement that can no longer be supported in the light of taphonomic research (Domínguez-Rodrigo 1994b; Domínguez-Rodrigo and Barba 2006). However, not all "classical" sites are anthropogenic (see extensive reanalysis of Olduvai Bed I sites in Domínguez-Rodrigo et al. 2007).

Recognizing that some sites are hominid-made does not necessarily mean that these sites were home bases. Although site integrity has been tested by three decades of taphonomic work, the behavioral meaning of these artifact-and-bone concentrations remains to be understood. In basic terms - and in the current state of the debate about the hominid role in site formation - there were two models that fit the taphonomic premises exposed above: the "stone cache" hypothesis (Potts 1982, 1984, 1988) - later re-named the "transport resource model" (Potts 1991) - and the "central-place foraging" model (Isaac 1983). Pott's model has failed to withstand taphonomic scrutiny whereas Isaac's model is still supported; after more than one decade of studies aimed at testing both models, the consensus is that the "central-place foraging" model possesses greater heuristic value.

Today, the bulk of the debate on early site formation and functionality revolves around the issue of whether hominids were transporting substantial amounts of food (which could be potentially shared), or instead small amounts that were transported with the intention of seeking refuge for individual consumption. This discussion is directly linked to the hunting-versus-scavenging debate. "Central-place foraging" models are only sustainable if hominids had been transporting high-yielding resources (i.e., fleshed carcasses) to sites. Therefore, in the past twenty years, the study of the order of access of hominids and carnivores to carcasses has been one of the most relevant topics in archaeological research and also one of the most innovative fields in archaeological taphonomy.

For this reason, one of our main goals in the Peninj research project was to study early human adaptation to open environments and its relevance to early site formation, site functionality and carcass obtainment strategies. Following modern savanna ecological standards, open ecosystems in East Africa grant few occasions for scavenging (Blumenschine 1986). Peninj revealed itself as a fairly open ecosystem where hominids left traces of carcass exploitation. How they managed to acquire animal resources and what they did with them was one of the *foci* of our research; to examine this problem, it was necessary to reconstruct the landscape and its trophic dynamics.

The landscape approach

As discussed above, the dense concentrations of faunal remains and stone tools accumulated in thin horizons at some Plio-Pleistocene sites in Olduvai (Tanzania) and Koobi Fora (Kenya) have been interpreted as the result of hominids selecting places which they repeatedly visited for carcass processing (Leakey 1971; Isaac 1978, 1983, 1984; Bunn 1982, 1991; Bunn and Kroll 1986; Potts 1982, 1988; Blumenschine 1988, 1991, 1995; Bunn and Ezzo 1993; Schick and Toth 1993; Rose and Marshall 1996; Oliver 1994; O'Connell 1997). One of the main characteristics of most of these "Type C" (Isaac 1978) sites is a high density of archaeological materials in spatially restricted concentrations. For several authors, these archaeological sites represent places where hominids might have stayed for long periods of time, very likely performing more activities than just stone tool manufacture and carcass processing (Leakey 1971; Isaac 1978, 1984; Bunn 1982; Stanley 1992; Oliver 1994; Domínguez-Rodrigo 1994a). For other researchers, such sites are the result of brief occupational episodes linked only to carcass-processing activities (Potts 1982, 1988; Blumenschine 1991; Blumenschine et al. 1994; O'Connell 1997). Some of the landscape archaeology projects carried out in East African Plio-Pleistocene archaeological areas in the past few years have attempted to evaluate if these concentrations of materials - usually referred to as "patches" (Isaac and Harris 1975, 1980) - are relevant in terms of hominid behavior rather than resulting from archaeological sampling bias (Stern 1991, 1993). This type of research, focused on areas away from archaeological sites, is also aimed at testing the reliability of smaller concentrations of materials – usually referred to as "scatters" (Isaac and Harris 1975, 1980) – as indicators of distinct hominid activities in the diverse paleohabitats sampled in each area (Isaac 1981; Isaac et al., 1981; Foley 1980, 1981; Bunn 1994; Blumenschine and Masao 1991, 1995; Potts 1994; Potts et al. 1999; Rogers 1997). One of the major shortcomings of these landscape archaeology approaches is the time-averaging character of the archaeological deposits. Current research at Olduvai is focusing on the lower Bed II (Lemuta member) with a deposit which could have been formed over a period of 80,000 years (Hay 1976; Blumenschine and Masao 1995; Manega 1993; Walter et al. 1991). At Olorgesailie, the landscape approach focused on the member 1 stratum, which could span as many as 15,000 years (Potts 1994), although recent estimates of sedimentation and pedogenesis have reduced the time span of the deposition of the artifact-bearing horizon to 1000 years (Potts et al. 1999). At Koobi Fora, Stern (1991, 1993) carried out a study on surface materials from the Lower Okote member, which could span 100,000 years. Rogers (1997) targeted the same stratum by analyzing the concentration of excavated materials to compare results with surface densities. All these studies show that archaeological patches and scatters could be contingent upon both hominid behaviors and biophysical processes that are not easily discernible today, since the time-depth of their formation cannot be ascertained. Despite these problems, most of these studies show that high-density clusters and low-density concentrations of materials exist and some seem to be linked primarily to early hominid behaviors (Rogers 1997; Potts et al. 1999).

Another shortcoming in the landscape approach is the resolution of the archaeological record itself. Most of the landscape archaeology projects mentioned above involve quantifying the distribution of materials vertically dispersed in deposits. High density of materials in this situation can represent either more intensive occupation of specific areas or less intensive but more repeated occupations of the same areas. This makes deposits rather unreliable for reconstructions of hominid land use, regardless of their well-bracketed chronology.

There are several constraints to the use of landscape approaches to archaeological research. A basic one is the fact that most of the "landscape" falls outside the sedimentary record. The first reason is because sedimentation occurs selectively in contexts regulated by hydraulic dynamics, such as around lakes and rivers. The array of environments thus preserved are mostly representative of alluvial habitats (river channels, deltas, alluvial fans, floodplains, swamps, proximal alluvial plains). The second reason for bias in the preservation of "archaeological" landscapes is that erosion of sedimentary units is very often restricted to specific areas, further reducing the spatial availability of outcrops suitable to be analyzed from a landscape perspective.

However, the sedimentary exposures displaying all these alluvial habitats can be very useful in determining trophic dynamics of the immediately surrounding "non-preserved" portion of landscape. Analyses of soil micromorphology, fossil pollen, soil carbonates, soil phytoliths, faunal isotopic indicators, macro- and micro-faunal taxonomic ranges, ecomorphology of bovid postcranial bones, fossil ostracods, paleobotanical remains, and landscape taphonomy can produce enough information to infer the array of potential habitats and the degree of trophic dynamics in each of them.

The landscape approach is very useful for the following research purposes:

1) Site variation/formation. As was described above, the widespread occurrence of Plio-Pleistocene sites in Africa, dated between 2 Ma and 1.5 Ma, composed of lithic artifacts and bone remains from several animals in single thin horizons, led evolutionary anthropologists to conclude that meat consumption had triggered the emergence of stone tool use among early hominids. This debated issue was mainly focused on the information obtained in a single area (Olduvai) and, more specifically, from a single archaeological site: the FLK Zinjanthropus site. However important the archeological information from the FLK Zinj has been to reconstruct hominid behavior, it has nonetheless limited archaeologists' views on early human adaptive patterns in two ways: by neglecting regional variability (clearly stressed in Potts 1994) and by deterring the development of explanatory frameworks to account for archaeological site diversity, both intra- and inter-regionally, in the same time periods. Although "Type C" sites have generated a fruitful discussion in the past two decades, researchers have hardly advanced in the interpretation of other sites which suggest alternative and complementary stone tool-using activities resulting in a different kind of archaeological record.

2) Landscape taphonomy, carnivore competition and hominid scavenging. The scavenging hypotheses are based on actualistic observations and studies about the ecology of scavenging (Blumenschine 1986), in which the key process analyzed is competition among carnivores in the diverse ecological settings of savanna ecosystems (Blumenschine et al. 1994). It has been observed that the extent of competition is highly variable according to the season (wet/dry) and habitat type (open/closed). This led Blumenschine (1986, 1988, 1989) to suggest that the most plausible scavenging scenario for early hominids could have been one determined by the time of year in which scavenging opportunities were greater (part of the dry season) and by the habitat showing the most reduced competition among carnivores (lacustrine/riverine woodlands). These conditions provide greater carcass availability and enhance hominid exploitation of carcasses, as observed in the Serengeti and Ngorongoro ecosystems of Tanzania today (Blumenschine 1986, 1988, 1989).

As a measure of competition among carnivores for fresh carcasses, Blumenschine (1989) carried out a landscape taphonomic study on bones scattered in modern East African savanna ecosystems, which was a continuation of his research on the processes of carcass consumption by predators and scavengers (Blumenschine 1986). When comparing both sets of data, Blumenschine concluded that fresh carcass availability could be predicted by three aspects of the bone scatters: the ratio of the minimum number of elements to the minimum number of individuals (MNE:MNI ratio), skeletal part representation, and the intensity of damage inflicted by bone-gnawing carnivores, expressed in terms of bone completeness (Blumenschine 1989:347). Thus, he found that in riparian woodlands, where fresh carcasses last longer due to reduced competition among carnivores, the number of bones per individual and the frequency of long bone completeness were higher than in other habitats (Blumenschine 1989).

However, this referential framework was challenged by Tappen's (1995) study of bone scatters in the savanna of the Virunga National Park of the Democratic Republic of Congo (formerly Zaire). Tappen observed that some open grasslands had a higher number of bone patches, with more individuals represented than in riparian woodland, and with a more significant number of bones per individual than in the latter habitat. She claimed, therefore, that the ecology of scavenging is different in Virunga, which she confirmed by the fact that most of the 15 recent kills made by lions that she observed during the course of her research were also located in the open plains (Tappen 1995).

Tappen's study appears to contradict the actualistic landscape studies conducted by Blumenschine. Even so, if we take a closer look at both studies, we notice that they have important points in common. For example, Blumenschine's use of landscape bone scatters as a measure of competition among carnivores for fresh carcasses still applies; his measures of competition, both qualitative (skeletal representation, bone damage) and quantitative (MNE:MNI ratio, MNI per patch), are not invalidated. Bones are differentially represented in certain areas, not because predation was more intense, but due to the fact that preservation processes, namely, relaxed competition among predators/scavengers, were operating there.

The ecology of the Serengeti and Virunga ecosystems accounts for the different landscape bone scatter patterns. In the Serengeti, lion and hyena territories overlap (Kruuk 1972; Schaller 1972). Therefore, competition is high because of their interaction. In Virunga, there is clear differential use of land by the two types of carnivores: lions are adapted to the grassland of the northern plateau while hyenas prefer the southern plateau (Tappen 1995). Therefore, competition is low. In the Serengeti, lions are more frequently found near rivers and bushy areas than in open plains, which are preferred by hyenas (Kruuk 1972). In Virunga, the situation is reversed: lions prefer the open grasslands and hyenas the bushy open spaces (Tappen 1995). Another feature that differentiates both ecosystems is their annual rainfall average: the Serengeti is a semi-dry savanna/steppe, whereas Virunga is a wet savanna. Thus, while there is a marked seasonality in the Serengeti, together with a large-scale migration, in Virunga, the lack of seasonal stress accounts for the year-round residence of fauna.

The good ecological conditions – food and water availability – in the wet savannas limit the mobility of the resident fauna. In turn, this creates different hunting opportunities for the resident carnivores, which show reduced inter-specific competition. Consequently, bones are better preserved in the zones or habitats where bone-crunching species are either absent or more marginal. In clear contrast, the lack or limited supply of food in dry seasonal savannas forces animals to migrate. In such a case, carnivores have to overlap their territories to exploit a changing herbivore biomass. The lack of water during the dry season also obliges some animals to congregate around the places where it is available, thus creating a seasonal movement towards rivers, lakes, ponds and/or waterholes. This concentrates predation, and therefore carcass availability, in a few fixed locations. Under these conditions, competition is high and carcasses are better conserved in places where hyenas are scarce (namely, riparian woodland).

Bearing in mind these differences between two types of ecosystems, the question is which one can be applied as a reference to Plio-Pleistocene savanna environments. What these studies suggest is that further research in other forms of savanna/steppe is needed before we can develop complete and applicable referential frameworks. New studies should not only aim at studying competition, but also at creating accurate and reliable signature criteria that can be applied to the fossil record (Domínguez-Rodrigo 1996). Tappen (1995) claims that variation in the ecology of scavenging is wider than previously thought. Can this variation be accounted for in terms of a two-patterned model (wet/dry savannas) or a multiple-patterned model (regional variability among wet/dry savannas themselves)? With this question in mind, Domínguez-Rodrigo (1996) carried out a landscape taphonomy study of the Galana and Kulalu (south Kenya) dry savanna/steppe ecosystems and observed the same duality of bone preservation as was documented by Blumenschine for the Serengeti. That is, riparian woodlands preserved a higher number of bones per individual, the number of complete bones was higher than in other habitats, and the number of smaller-sized animals represented was also higher as a result of reduced carnivore competition, compared with more open environments. Therefore, the duality seems to be expressed in terms of trophic dynamics in wet savannas and dry savannas. With this referential framework and the methodology described above, trophic dynamics can be reconstructed in past landscapes. This is an important part of the research that was undertaken and applied to the Peninj archaeological area.

3) Expansion of the adaptive capabilities of early hominids: adaptation to open habitats. African savannas are ecosystems containing a variety of habitats but with a predominance of open-vegetation environments. The habitats in these ecosystems vary from dense forest to open grasslands. Forests are conditioned by both humidity (i.e., water sources: gallery and lacustrine forest are fed by rivers and lakes, repectively) and altitude (montane forests). In between these two extremes, a diversity of habitats appears according to the interplay between humidity and altitude. The following types of savanna are defined by the percentage of canopy cover: a) grassy woodland. This also includes bushland and shrubland. Canopy cover may occupy as much as 80% of the landscape. The habitats contained in this type of ecosystem can also be identified by a soil isotopic component of C_4 $\leq$50%; b) wooded grassland. This also includes bush and shrubland but is characterized by a canopy cover of less than 20% of the landscape and a soil isotopic component of C_4 between 50% and 80%; c) grassland. Canopy cover occupies less than 2% and the soil isotopic component of C_4 is higher than 80%.

Faunal biomass distribution in African savannas occurs mostly in grassland habitats, given that most savanna fauna are herbivores (see summary in Potts 1988; Domínguez-Rodrigo 1994a). There is a progressive decrease in faunal density according to the degree of increasing canopy cover.

Most early Plio-Pleistocene sites occur in paleoecological contexts associated with riparian (either lacustrine or riverine) habitats, in which woodland areas must have been present. The archaeological sites of Bed I at Olduvai Gorge – used by most paleoanthropologists and their critics for making their behavioral and paleoecological inferences – were probably situated in the lacustrine plain of a shallow, slightly saline lake (Hay 1976) not far from the alluvial plain. Oxygen isotope analyses on carbonates suggest that the basin was wetter than today, with a mean annual precipitation >800 mm (Bonnefille 1984; Cerling and Hay 1986; Sikes 1994; Peters and Blumenschine 1995). Pollen analyses, fossil rhizomes and root casts indicate a mosaic of different landscape forms in which marshland must have been one of the predominant habitats near the lake margin (Hay 1976; Bonnefille 1984; Peters and Blumenschine 1995). This is further suggested by the presence of some murid rodents, especially those belonging to the genera Oenomys, Pelomys and Aethomys (Jaeger 1976). Moreover, the association of Oenomys and Grannomys in Middle Bed I suggests the presence of a riverine gallery forest or dense brush

(Jaeger 1976). The paleontological studies of the macrofauna, dominated by Reduncini and Bovini, also indicate a wet and closed vegetation environment nearby (Shipman and Harris 1988; Marean 1989; Plummer and Bishop 1994).

The indication of tree cover is further supported by Sikes' (1994) stable isotope results, which suggest a riverine or ground water forest in a 1 km^2 area covered by Blumenschine and Masao's (1991) initial landscape study area in 1989. It could be assumed that the C_3 results obtained by Sikes (1994) could be the result of either the presence of plants in marshy environments, like Typha, or the presence of tree cover on a floodplain. As shown in Sikes' (1994) Figure 1, the carbon isotope values for wetlands soils and sediments can be similar to those of nonwetland soils. Sikes used a nonwetland soil interpretation for the FLK Zinj paleosol carbon isotope values based on two reasons. First, Cerling et al. (1989) have shown that there is a systematic difference of 14–17‰ between the carbon values for coexisting soil carbonate and soil organic matter (SOM). At the bottom of Table 1.2, it shows there is a 14·1‰ difference between the Zinj SOM and $CaCO_3$, so this fits with theoretical expectations. Second, as also argued in Sikes' (1994) paper for the basal Bed II paleosols, the existence of the pedogenic carbonate and its oxygen isotope values supports a nonwetland terrestrial interpretation. However, whether one argues on the basis of organic or inorganic carbon or both, the isotopic data shows that there is still a significant woody C_3 component present in Bed I. Specific to Bed I, we can take a look at the carbonate isotope data in Cerling and Hay (1986). For seven pedogenic carbonates reported (three just above the basal lava in the western lake margin; four in the eastern lake margin – the Zinj carbonate between Tuffs IB and IC, one from FLK-N about a meter below Tuff IF, and two more just below IF) the carbon values range from 5–8 to 3–7‰. This represents around 45 to 60% C_4 plants. Four data points represent grassy woodland; three fall into an interpretational overlap zone of grassy woodland or wooded grassland. Because of isotopic effects during carbonate precipitation, there is an overlap of up to 2‰ between physiognomic categories. The Zinj carbonate – referenced in Sikes' (1994) Table 1.2 – with a value of 5·7‰ falls in the grassy woodland category.

Thus, the presence of woodland in the vicinity of the lacustrine plain, in relation to archeological sites, seems to be indicated by the studies of microfauna (Jaeger 1976; Fernández-Jalvo et al. 1998), macrofauna (Shipman and Harris 1988; Marean 1989; Plummer and Bishop 1994), fossil pollen (Bonnefille 1984) and carbonate isotopic analyses (Sikes 1994), as is shown in the landscape reconstruction developed by Peters and Blumenschine (1995; and Blumenschine and Peters 1998).

This paleoecological reconstruction provides a good reference for the general environment surrounding the locations where early archaeological sites appear. Sites could have been formed either in a woodland setting, on the open floodplain surrounded by grasssland or in other semi-open transitional areas. In all these cases, woodland seems to have been an important landscape feature near the sites. All these micro-habitats show similar ecological conditions, despite their diversity, with respect to carnivore competition and territory overlap, as is documented in the present study. Even if the actual locations of the sites might have been in open spots, their association with closed vegetation habitats (riparian woodlands) would have determined the degree of carnivore competition in such places. Therefore, paleoecological reconstructions are extremely useful to infer trophic dynamics in different areas, even if the degree of resolution regarding the actual location of sites is not very high. Since the knowledge of carnivore competition within these types of habitats is crucial for supporting or rejecting the aforementioned hominid behavioral models, only the study of trophic dynamics in modern African ecosystems can shed some light on this issue.

Although several previous ethological studies suggested that competition in riparian habitats is low, compared to more open habitats in which carnivores preferentially establish their ecological niche (Kruuk 1972; Schaller 1972; Sinclair 1979), it was only with the pioneering study of the ecology of scavenging in the Serengeti by Blumenschine (1986) that this assertion was clearly demonstrated. Blumenschine documented a longer survival of carcasses in riparian woodlands than in open vegetation settings away from perennial lakes and rivers. Whereas size 3 carcasses (e.g., zebra and wildebeest) were reported to last an average of 10–18 hours in open grasslands, they remained undiscovered by hyenas for an average of 71 hours in riparian environments. This low degree of competition was further reflected in the amount of bones, carcass size, type and degree of completeness of bones deposited in this habitat (Blumenschine 1989; Domínguez-Rodrigo 1996).

Competition in open habitats is high because most hunting episodes by predators occur there. This is due to the fact that carnivores are better adapted to these settings, because they contain the greatest part of the savanna herbivore biomass (Foster and Kerney 1967; Rudnai 1973; Western 1973; Sinclair 1979; Domínguez-Rodrigo 1994b). Competition is also enhanced by the fact that visibility in open spaces is better and kills are more easily spotted than in closed vegetation areas (Cavallo 1998). A third element to consider is that vultures are constantly present in open habitats, competing with smaller and/or solitary carnivores (e.g., cheetah, jackal) for their prey and carrion and also providing larger predators with clues to the locations of kill sites. Both hyenas and lions monitor vultures to locate scavengeable carcasses (Kruuk 1972; Schaller 1972; Cavallo 1998). Thus, it is not surprising that in his Serengeti study, Blumenschine (1986) documented the presence of other carnivores besides the first consumers at most of the kills in open habitats.

Blumenschine's (1986) study, based on the survival of carcasses according to habitat type, provided us with the first relational analogy for the opportunities of scavenging carcasses for hominids. Blumenschine also differentiated the main products that carcasses provide for their consumers (viscerae, flesh, brain, and bone marrow) and their differential preservation. A study by Domínguez-Rodrigo (2001) in riparian woodlands vs. open plain habitats further supported the conclusions reached by Blumenschine in his study of the Serengeti trophic dynamics and showed the contrast in carnivore competition between riparian habitats and open-vegetation habitats.

Riparian habitats are diverse (Riou 1995). They can comprise floodplains, marshlands, lacustrine/riverine woodlands, and transitional (semi-open) areas connecting to proximal alluvial plains. When referring to riparian habitat or riparian woodland, we refer to closed woodlands and their immediate surroundings: semi-closed transitional areas and the nearest open vegetation sections of floodplains and alluvial plains connected to them. Following the French tradition in ecological studies, floodplains can be defined as the areas near hydraulic sources periodically covered by water due to the transgressive processes undergone by lakes or to overflow by rivers (non-wetland soils), and also those areas where water is present on a permanent or semi-permanent basis, such as in swamps and marshes (Riou 1995). Floodplains are usually covered with short edaphic grasslands during regressive cycles. Floodplains can often be distinguished from alluvial plains in lacustrine environments because a riparian forest marks the transition between them. When this transition is not clearly marked, as Riou (1995) describes it, floodplains can be considered the most proximal ends of alluvial plains in their connection to water sources, showing a higher proportion of humidity, a change of lithology (with a higher amount of carbonates) and a smaller vegetation cover, where meadow grasses (Poa, Festuca) and some grasses of net bottomlands (Phragmites, Typha, Oryza) can be present (Riou 1995).

Competition can be studied by analyzing the duration of different carcass products according to their ecological context (Blumenschine 1986) or by observing the degree of interaction of carnivores with respect to carcass processing in different habitats (Domínguez-Rodrigo 2001). Competition, both intraspecific and interspecific, is usually expressed in the number of organisms overlapping in the same area to gain access to the same resources in a specific time period (Severtzov 1947; Margalef 1968, 1977). Regarding carnivore competition in African savannas, the number of predators and the diversity of predatory species in a particular geographical area and in the same time frame determines the degree of competition (Schaller 1972; Kruuk 1972). Highly competitive settings are those in which carnivores are more abundant, since this situation creates overlapping in the consumption of carcasses (Ayeni 1975; Blumenschine 1986; Kruuk 1972; Schaller 1972). This propitiates frequent encounters among different carnivores and makes them defend their prey from competitors. In low competition habitats, organisms tend to exploit limited resources sequentially at different times (Margalef 1977), very frequently uninterrupted by competitors, and the resources last longer than in highly competitive habitats (Sinclair 1979; Blumenschine 1986). Therefore, space, time and number of predators are the principal variables to measure competition in modern ecosystems. For this reason, the process of carcass consumption and the timing of the presence of predators/scavengers at kills are good indicators of the degree of carnivore pressure in different habitats.

Since a "central-place" behavior would imply several hominid activities – not strictly linked with carcass consumption – on the same spot for at least a few hours, time becomes a crucial variable in the analysis of carnivore competition. Time is also relevant in the "refuge" and "near-kill location" models. An early appearance of carnivores in the same spot would have prompted hominids to carry out carcass-processing activities somewhere else. If Plio-Pleistocene trophic networks were similar to those observed today, then we could reliably infer low and high carnivore pressure in different paleoenvironments, using modern ecosystems as a comparison.

As mentioned above, most archaeological sites appear located in riparian environments associated with wooded vegetation. If that was the case for Olduvai Bed I sites, it seems to be the same for Koobi Fora sites between 1.9 Ma. and 1.4 Ma. (Isaac et al. 1997). The analysis of the Koobi Fora paleontological faunas show that during the formation of the KBS member, woodland taxa were common. In the Okote member, the wet grassland taxa (reduncini) far outnumber the dry grassland and woodland species. Wet grassland bovids are predominant and antilopini and alcelaphini (especially the latter) are comparatively underrepresented (Harris 1991).

Although it has been claimed that by 1.5 Ma. hominids had successfully adapted to all savanna habitat and ecosystem types (Rogers et al. 1994), which prompted their expansion and early exit from the African continent, the truth is that what has been documented archaeologically at 1.5 Ma. both in Olduvai and in Koobi Fora is just a larger area of occurrence of sites situated distant from lake margins. Sites in the KBS member in Koobi Fora and in Bed I in Olduvai appear located in the vicinity of nearby paleolakes (Koobi Fora and Olduvai) or the major river Omo (Koobi Fora). In Olduvai Bed II and in the Okote member in Koobi Fora, archaeological sites are also distributed away from these areas. However, this may be the result of taphonomic/sediment exposure bias. Sediments dated around 1.5 Ma. in both areas are more widely eroded and exposed than in older strata. Therefore, the more widespread nature of the archaeological record in the younger strata may be more artificial than resulting from the increasing home range of hominids. Irrespective of this, in either area, sites appear repeatedly associated with areas where riparian environments and closed-vegetation habitats seem to have been well-represented. Therefore, no claim has yet been confidently made about hominids of this period being well-adapted to open ecosystems. This raises the question of whether or not hominids spread and left the African continent prior to their complete success in adapting to all the habitats of open savannas. If one could provide archaeological proof of hominid adaptation to open savannas during the Early Pleistocene, then the exit of the African continent could easily be explained as a result of successful adaptation to this type of ecosystem, probably due to technology. One of the main goals of our decade of research at Peninj, as described below, was to better understand this adaptation.

History of research at Peninj

The first Europeans to explore the Peninj area were J. Thompson and G. Fisher, between 1880-1890. An expedition led by M. Schuller in 1896-1897 took geological samples from the Oldonyo Lengai volcano in the south of Lake Natron (Gregory 1921); but the earliest substantial geological studies of the area were carried out in 1904 and 1910 by the German geographers and ethnographers C. Uhlig and F. Jaeger (1942), who studied the Magadi, Natron and Manyara lacustrine basins. In 1951, another expedition carried out a study on the hydrochemistry and economic potential of the saline deposits of the lake (Guest and Stevens 1951, unpublished). N. J. Guest further studied the geology and petrology of the Natron and Engaruka basins. He mapped the Natron basin and compared the saline deposits of Natron with those of Magadi, which were being exploited at the time (Guest 1953; Geological Survey of Tanganyika 1961; Mineral Resources Division, Tanzania 1966). The main geological contribution of Guest's study, besides mapping, was the division of the volcanism into two main phases, Older Extrusives and Younger Extrusives, which will be discussed further in the following chapter. A later Russian expedition provided the first description of the geomorphology of the area, and briefly described the sediments of the Peninj Group (Kapitsa 1968).

A few years after Guest's work in the area, B. H. Baker (1958, 1963) studied the Magadi basin and the northernmost part of the Natron basin. It was during these years that R. Leakey discovered the fossiliferous potential of the area while flying over the region from Nairobi to Olduvai Gorge in Tanzania. A 1963 expedition led by L.S.B. Leakey (formally), R. Leakey and G. Isaac discovered the first hominid remains: a complete mandible of a robust australopithecine. Between 1963-1964, field work resulted in the discovery of several fossils and two sites that, at the time, were considered the oldest Acheulian sites in Africa (Isaac 1967). During this time Isaac provided the first description of the stratigraphy of the Peninj Group (Isaac 1965). These preliminary field seasons did not extend to more systematic work in the area because the Leakeys and Isaac were committed to their respective projects in Olduvai, Omo, and eventually in Koobi Fora. Therefore, the area remained almost unexplored from paleontological and archaeological points of view until much later.

In 1981, an international team composed of French, North American, and Tanzanian researchers and led by M. Taieb, G. Isaac and M. Mturi resumed work in the area. The archaeological research led by Isaac and Mturi was carried out during 1981–1982. The geological work, led by M. Taieb, was carried out from 1981 to 1984 in successive field seasons. The prolific geological work since then has resulted in extensive analyses of the volcanology of the area (Dawson 1989; Wooley 1989; Brooker and Hamilton 1990), as well as sedimentology, lake-level change and hydrochemistry studies (Eugster 1986; Casanova 1986, 1987; Hillaire-Marcel 1987; Hillaire-Marcel and Casanova 1987; Hillaire-Marcel et al. 1987, 1996; Schubel and Simonson 1990; Vincens et al. 1991; Damnati 1993; Roberts et al. 1993); these are discussed further in the following chapter. In 1987, Taieb and Fritz published a monograph which mainly focused on their geological studies of the area.

However, archaeological research in the Peninj region remained in its early stages, due to the unexpected death of Isaac. Thus our project, begun in 1995 under the leadership of M. Domínguez-Rodrigo, is the first long-term project aimed at studying hominid adaptation to Lake Natron during Plio-Pleistocene times.

Summary of the inception of the Peninj archaeological project

The Peninj region comprises three different areas with Plio-Pleistocene sedimentary exposures situated in three different paleoecological settings: these are the Type Section, South Escarpment, and North Escarpment. In the chapters which follow, each of these three windows is presented sequentially: first, we discuss the geology, taphonomy, isotopic analysis and technology of the Humbu Formation sites in the Type Section, including the youngest site discovered at Peninj, within the Moinik Formation. This group of chapters is then followed by two chapters on the archaeology of the sites of the South Escarpment and North Escarpment, respectively. These three windows to the past offer an excellent opportunity to study human adaptation from an ecological perspective. One of these windows has preserved a paleosurface spanning about 1 km^2, in which archaeological materials on a thin paleosol can be horizontally traced along the preserved portion of the paleolandscape. This makes Peninj a unique place for the application of a landscape archaeological approach.

Isaac's preliminary information about the faunal materials discovered at Peninj showed a great abundance of bovids belonging to the antilopini and alcelaphini tribes, which prompted him to identify the Peninj area as a former "savanna grassland" (Isaac 1984). This emerged as a unique opportunity to test hominid adaptation to this now-widespread type of African savanna. Plants other than grasses are very scarce and are spatially separated in savanna grasslands. If modern trophic dynamics apply, these open habitats should have exhibited both a high density of faunal remains and a high degree of carnivore competition for animal resources. This would imply that scavenging opportunities for hominids might have been very limited. Peninj therefore becomes an ideal archaeological landscape to test hominid access to animal carcasses. Hominid adaptability to this highly competitive setting could have implied a different set of selective criteria in which the extent of hominids' intelligence and adaptive skills would have been tested more rigorously than in other landscapes.

Successful adaptation to open savannas at 1.5-1.0 Ma. would imply that hominids might have been actively involved in the acquisition of animal protein resources and might have had a high-quality diet. This could explain their rapid expansion and exit from the African continent. For this reason, a landscape approach in the archaeological project at Peninj was essential. The first step was to reconstruct as accurately as possible the degree of openness of the early Pleistocene savanna at Peninj. For this purpose, detailed geological analysis of the three areas with Plio-Pleistocene sedimentary exposures was carried out. Given the exceptional preservation of an extensive paleosurface in one of the areas (Type Section) in a non-deposit format, the whole paleolandscape has been carefully studied applying the following analyses: study of soil chemistry and micromorphology, isotopic analysis of the faunas, fossil pollen analysis, taxonomic and taphonomic analyses of faunal remains (Figure 1.1). Once the type of vegetation scenario was analyzed, a landscape taphonomic study was conducted on the portion of the paleolandscape preserved and was compared to several modern referential scenarios to analyze trophic dynamics.

Once the distribution of vegetation and the carnivore use of the space was examined, a study of the distribution of archaeological traces over the paleolandscape was carried out. We conducted our research in two phases. During phase 1, a method was developed to analyze the distribution of surface materials. During phase 2, systematic test trenches were excavated over the exposed area of the paleolandscape. Archaeological site analysis and distribution was later carried out with a thorough zooarchaeological and taphonomic study of the archaeofaunas and a technological study of the stone tools. Stone artifacts were analyzed from a techno-typological perspective. The results enabled us to compare the different functions of Oldowan and Acheulian sites and to understand hominid behavioral variability reflected in the paleoecological distribution of different types of archaeological sites. The results enabled the reconstruction of a 1.5-1.2 Ma. paleolandscape with enough paleoecological information to understand the contextual distribution of archaeological sites and hominid activities along the landscape. Sufficient information was also produced to understand hominid-carnivore interaction and the behavioral function of some of these sites.

Comparative research among these three areas informs our understanding of the differential distribution of contemporary Oldowan and Acheulian sites and their different behavioral meanings. Extensive open-air excavations were carried out in three sites (one per each sedimentary window): two Acheulian and one Oldowan. The landscape archaeological approaches applied to East Africa over the past 20 years have neglected the traditional excavation of large surfaces. Modern analytical techniques that could not previously be applied to such types of excavations are now available, and are very effective for understanding site formation and site behavioral meaning. The application of these techniques has allowed, for example, the improved understanding of hominid exploitation of carcasses from an archaeological perspective.

All this information, understood within its contextual framework, has enabled us to study human adaptation to a 1.5-1.2 million-year-old savanna landscape and to observe behavioral variability intra-regionally (via the three windows at Peninj) and inter-regionally (when comparing Peninj to Olduvai, for instance). This work, together with research projects carried out in other areas of East Africa, shows that hominid behavior in the Early Pleistocene exhibited a higher degree of variability than previously thought. This project is not a landscape archaeology project as is currently understood, but a project with a set of questions that mix traditional and more modern criteria, and which attempts to address these questions from an ecological and adaptive perspective.

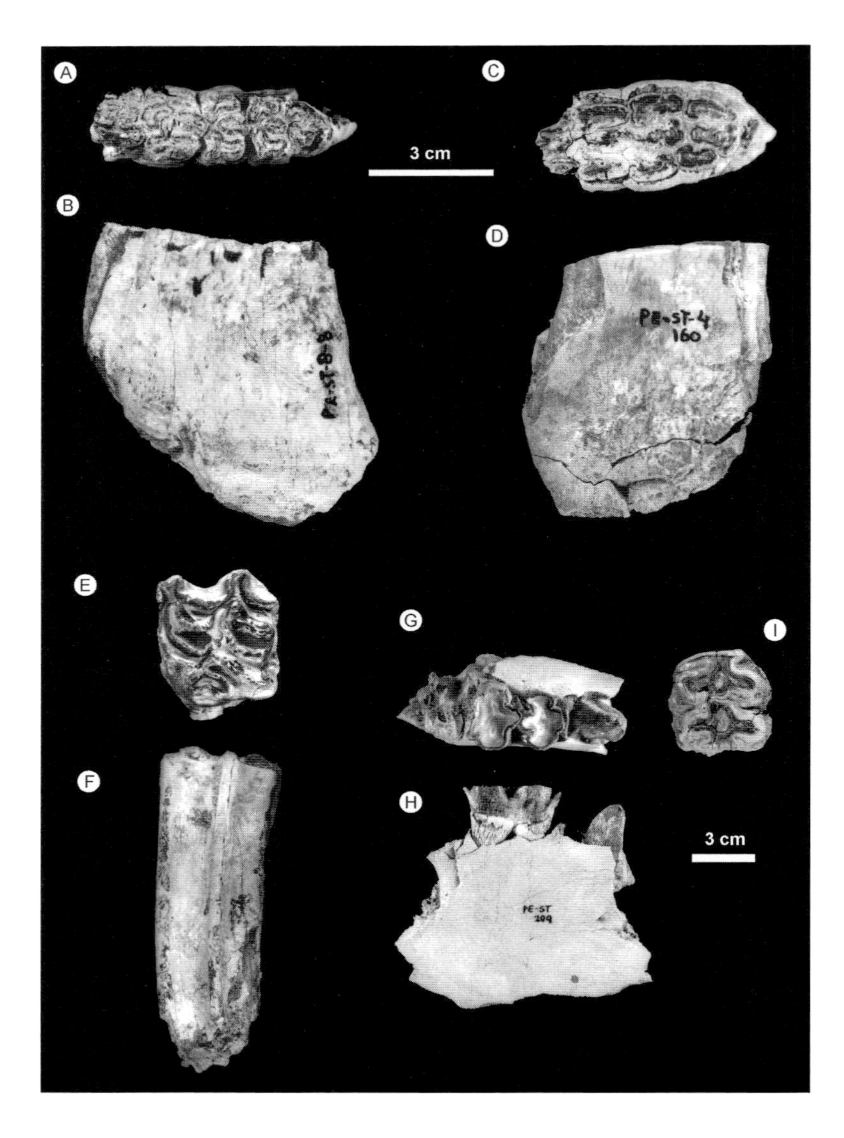

Figure 1.1 Sample of fossils from various strata and sites (ST4, ST8) in Humbu Formation found at Type Section (Maritanane): A-D) molar fragments (lateral and occlusal views) of Metridiochoerus compactus*; E-F)* Equus *sp.; G-I) dental remains from* Hippopotamus gorgops.

2

The Peninj Group: Tectonics, Volcanism, and Sedimentary Paleoenvironments During the Lower Pleistocene in the Lake Natron Basin (Tanzania)

Luis Luque, Luis Alcalá, and Manuel Domínguez-Rodrigo

Introduction

The sedimentary deposits of the Peninj Group are exposed along the western margin of the Lake Natron basin. This sedimentary unit, previously described by Isaac (1965, 1967), contains abundant archaeological and paleontological sites. Geological contextualization of this unit is crucial to understand the depositional environments, the facies distribution, the stratigraphical position and the age of these sites. Our research has focused much attention on stratigraphy and facies distribution. In this chapter, we discuss the main geological results obtained both by previous researchers in the area and by our own research.

As typically occurs in rifting areas, basin morphology, sedimentary processes and facies distribution are strongly conditioned by their tectonic-magmatic history. East African Rift Valley basins are exceptional sedimentary traps that enable excellent site preservation. The following factors favor preservation, making these areas into fossil reservoirs: tectonic escarpment shoulders and volcanics which produce large amounts of sediment, fast burial processes in alluvial environments, large lacustrine low-energy depositional environments, and volcanic ash layers that frequently coat and seal the landscape and contribute to bone preservation (e.g., Hay 1986; Pickford 1986). Landscape evolution in a basin can be traced through its sediments, which record lake-level changes, tectonic displacements, and volcano location, size and eruptive periods. Lake level regression is reflected in alluvial facies progradation, local erosional processes and channel incision, whereas during high lake levels we see evidence for shoreline expansion and lowland flooding. Stability is indicated by sedimentary infill and aggradation in central areas. Lake level changes can be deduced from several criteria: tectonics, rate of deposition/erosion, eruptive processes and climatic changes (i.e., variability in the ratio of precipitation to evapotranspiration). The relationship between these sedimentary facies and site distribution can shed light on hominid behavior in different landscapes.

The Peninj Group deposits have been influenced by all of these processes. The Peninj Group sediments only outcrop in the western margin of the Lake Natron basin, which was deeply affected by tectonic movements (Isaac 1965, 1967). Exposures are restricted to the foothill of the main tectonic escarpments of the basin and to a narrow belt along their rim. For this reason, the total extent of the sedimentary basin and the Natron paleolake during the deposition of the Peninj Group is unknown. The Natron sub-basin is part of a larger hydrological basin that includes both the Natron and Magadi lakes. The morphology of the Natron sub-basin is a typical half-graben that shows a stepped western flank and an eastern flexural margin that includes at least two successive generations of volcanoes, both inside and outside the main graben. The modern shape and size of the half-graben resulted from Middle Pleistocene tectonic movements that deepened the graben, changing the previously wider paleo-Natron basin

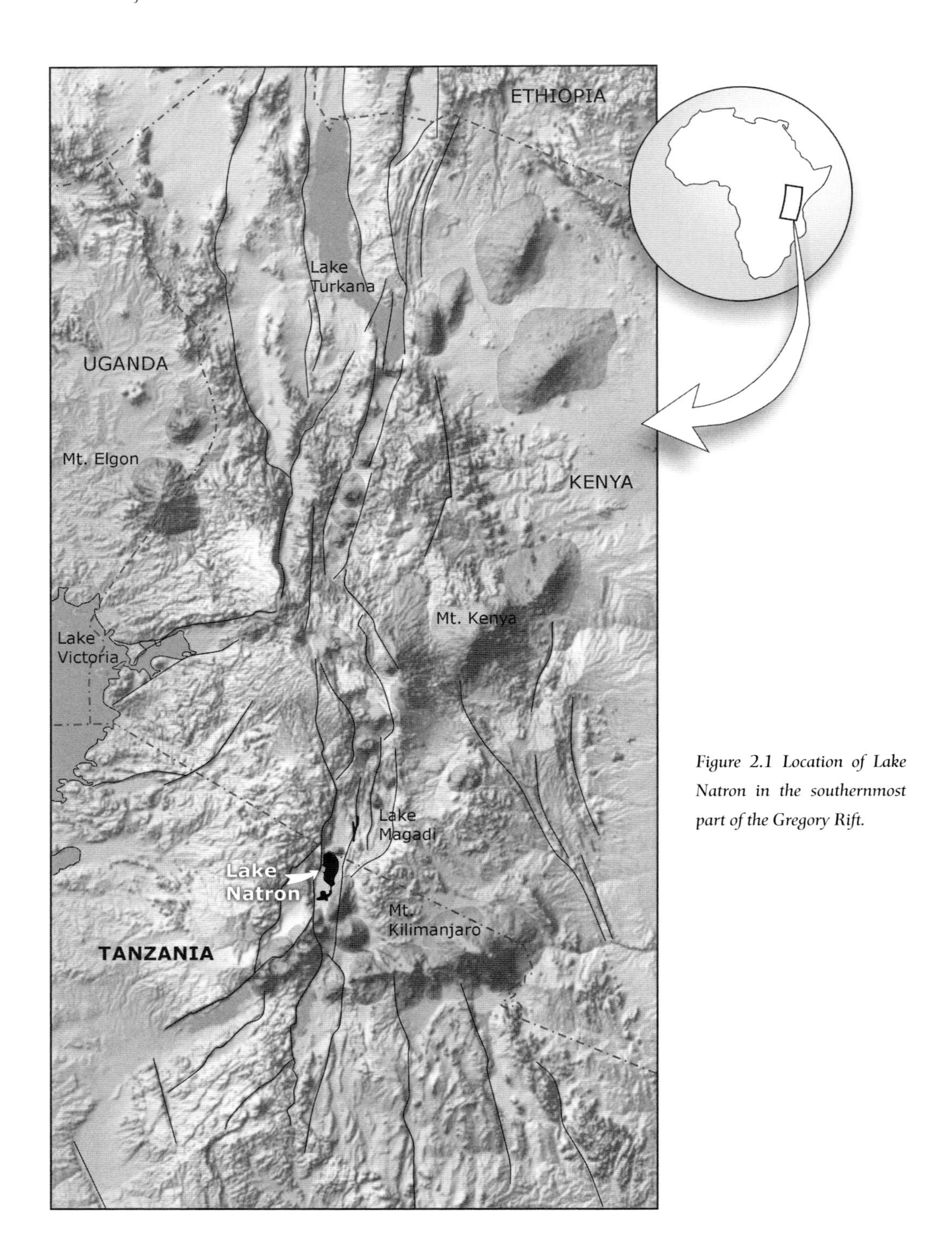

Figure 2.1 Location of Lake Natron in the southernmost part of the Gregory Rift.

(Isaac 1965, 1967; MacIntyre et al. 1974). These well-marked geological features allows us to conclude that tectonics and volcanism were active factors shaping sedimentary processes and exposure distribution from the deposition of the Peninj Group up to recent times.

Geological setting

The Lake Natron half-graben (East 35° to 37°, South 2° to 3°) forms part of the long chain of tectonic basins that configures the Gregory Rift, the southern part of the eastern branch of the Rift Valley (Figure 2.1). It is located in a complex zone a few kilometers north of the Northern Tanzania Divergence Zone (Dawson 1964; Dawson and Powell 1969), which gave rise to intense volcanism in the Crater Highlands area and the splitting of the eastern branch of the Rift Valley in three structural directions (NW-SE, N-S, and W-E). The active trona-carbonatite volcano called Oldoinyo Lengai (Dawson 1962) is a clear example of the continuing tectonics in the area.

The lower part of the Natron basin is occupied by Lake Natron, a shallow (less than 1 m deep), highly alkaline lake situated 610 m above sea level (Figure 2.2; see physiography in Vincens and Casanova 1987). The lake is fed by three permanent rivers and a number of hotsprings joining its shores. Lake waters are extremely saline and alkaline, creating sodium carbonate (trona) as well as a number of other evaporites (thermonatrite, halite, etc.) forming a thick crust on part of the lake surface (Fritz et al. 1987). The high concentration of these brines is caused by the high evapotranspiration rate, the shallow level of the lake, and the richness of ions supplied by volcanics surrounding the lake. The Natron sub-basin is almost completely situated in northern Tanzania, but the northern half of the greater basin (Magadi sub-basin and Ewaso Nyiro river) is in Kenya. The Shombole volcano (1,565 m) separates the Magadi and Natron sub-basins. During Lower Pleistocene times the barrier was most probably the Oldoinyo Sambu volcano (2,045 m), which today is deeply dissected by a rift fault, exposing a cliff higher than 1,400 m.

Apart from volcanoes, the main geographical feature of the basin is a large tectonic escarpment called Nguruman, more than 400 m high and extending from the north basin into the west Lake Natron margin. Its southward extension into the Natron basin consists of two large parallel escarpments called Sambu and Sanjan. The Sonjo Fault limits the basin in the west due to the uplift of Precambrian to Proterozoic metamorphic rocks (gneisses, schists and quartzite) from the craton substrate (Figure 2.3). On the East side, the Natron basin meets the flexural margin where intense grid faulting results in a dense network of horsts and grabens, oriented N-S, a typical feature of the rift axis in Kenya and Tanzania (Baker 1963, 1986). The large volcano Gelai (2,942 m) extends over fractured lavas of the substrate. A few kilometers away, the Precambrian substrate outcrops in the form of folded highlands. South of Natron lies the Engaruka sedimentary basin, separated from Natron by Oldoinyo Lengai (2,878 m), the Kerimasi volcano, and a horst created by oblique faults. On the uplifted western blocks, the Crater Highlands also limit the basin in the south.

Previous geological research in the Lake Natron area

Early geological work in the Lake Natron area has been described in Chapter 1, where the history of archaeological research is also discussed. The Natron basin has been far less studied than the Magadi basin, where salts and sodium carbonates have been exploited over the past 100 years, even though both are sub-basins of the same hydrographic basin. Research on rift valley

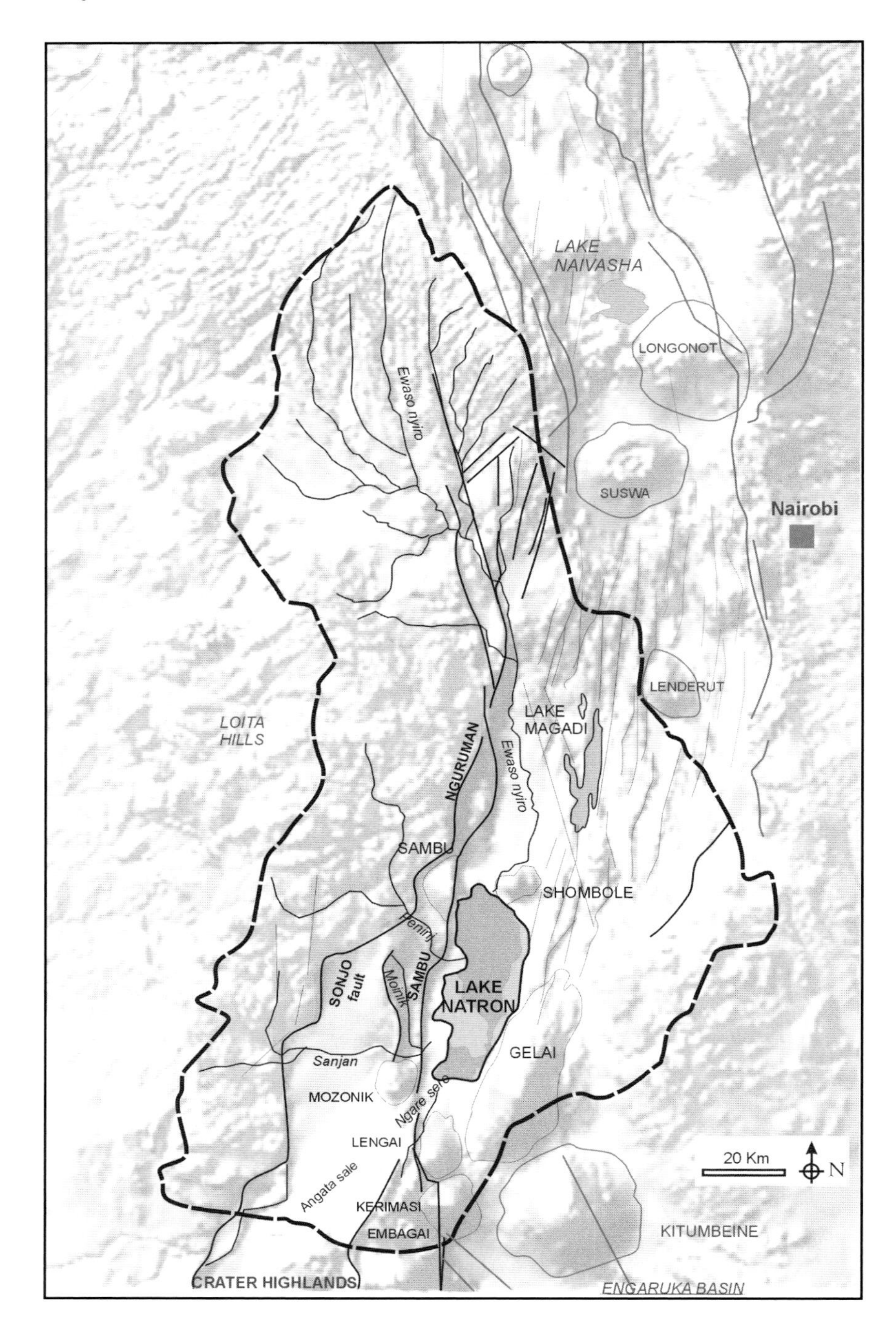

Figure 2.2 The hydrographic Natron-Magadi basin, showing a much larger surface than the Pleistocene Natron sub-basin under study.

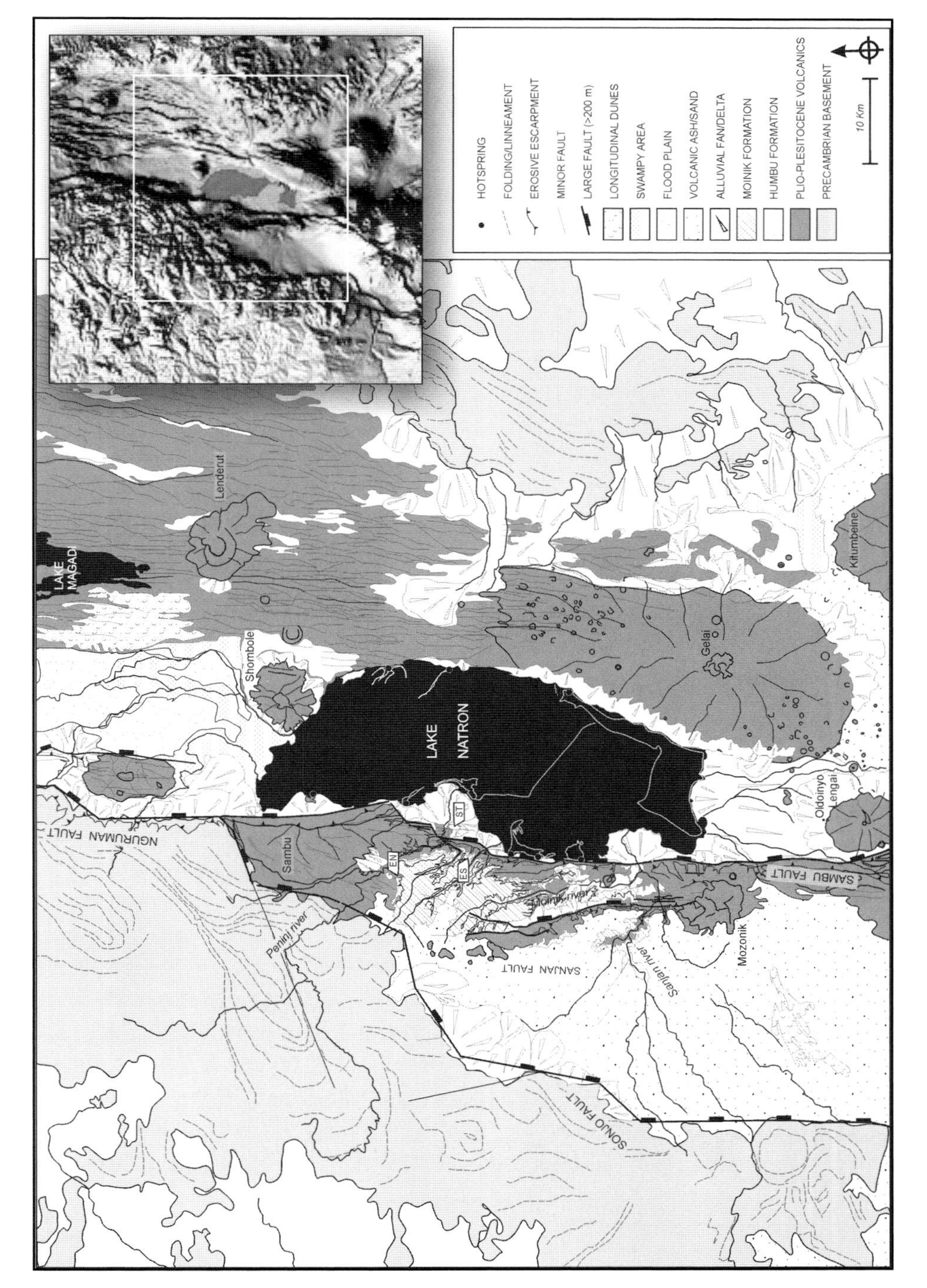

Figure 2.3 Geological map showing the main sedimentary fields as well as volcanic and tectonic features of the Lake Natron sub-basin. Satellite composite image from GoogleEarth for the Natron sub-basin.

processes has been very intense in the Magadi basin, especially because of the KRISP project, whose data can often be extrapolated to the Natron basin (e.g., Baker 1958, 1963; Grimaud et al. 1994; Byrne et al. 1997; Mechie et al. 1997; Novak et al. 1997; Simiyu and Keller 1998). There are also regional geological descriptions such as those of Baker et al. (1971, 1972), Baker (1986), Darracot et al. (1972), Fairhead (1980) and MacDonald et al. (1994).

Following Isaac's (1965, 1967) detailed work on the Peninj Group, new works appeared focusing on the chronology of the area and rift tectonic processes (Evernden and Curtis 1965; Curtis 1967; Howell 1972; Isaac and Curtis 1974; MacIntyre et al. 1974). In the early 1980s numerous geological works were published, mainly in association with a multidisciplinary French-Canadian-Tanzanian project (Taieb and Fritz 1987). These included studies of volcanism and its geodynamic significance (Lubala and Rafoni 1987), hydro-chemistry and lake sedimentation (Fritz et al. 1987; Hillaire-Marcel 1987; Manega and Bieda 1987), lake-level changes (Casanova 1987; Casanova and Hillaire-Marcel 1987), as well as paleomagnetism of the Peninj Group (Thouveny and Taieb 1987), mineralogy (Icole et al. 1987) and the paleontology of the fossiliferous sediments (Denys 1987; Geraads 1987).

Other geological studies of the basin have included the trona-carbonatitic volcanism of Oldonyo Lengai (e.g., Dawson 1962, 1964, 1989; Dawson and Powell 1969; Wooley 1989; Brooker and Hamilton 1990; Bailey 1993; Church and Jones 1994; Dawson et al. 1994; Bell and Keller 1995), the trona and silica deposits in the Natron and Magadi lakes (Eugster and Jones 1979; Eugster 1986; Schubel and Simonson 1990) and the record of recent climatic changes based on lake-level changes in the Natron-Magadi system (Casanova 1986; Roberts 1990; Damnati et al. 1992; Damnati 1993; Williamson et al. 1993). During the 1990s our team surveyed the region and carried out geological studies of the fossiliferous deposits (Luque 1996).

Geomorphology

As noted above, the Lake Natron sub-basin is a half-graben with two well-defined tectonic areas. The main tectonic escarpments appear in the West, consisting of three step-faults running N-S, but other important transforming faults follow a NE-SW direction. The first main fault, the Sonjo Fault, is the westernmost and the oldest; it splits the relief of the Mozambique mobile belt from that created by the infill of the Plio-Pleistocene basin beginning five million years ago. The second fault is the Sanjan escarpment, which begins at the foot of the Crater Highlands and rises up to 500 m high, becoming more diffuse to the north. The Sambu fault is parallel to Sanjan and generates a 400-600 m high relief in the north, limiting the area of modern alluvial and lacustrine sedimentation. This fault affects the Oldoinyo Sambu volcano by generating a vertical escarpment of 1,400 m. At the foot of the Sambu escarpment, the maximum subsidence of the basin can be observed.

The second tectonic area is the eastern low-relief region. This network of small basins affects the Older Extrusive lavas and ends up in contact with the westernmost Precambrian reliefs (Figure 2.3). This structure corresponds to the grid faulting defined by Baker (1958), which began in the axis of the rift system in the more recent phases, about 400,000 years ago. This Pliocene section of the rift is estimated to extend 5 km (Darracot et al. 1972). The volcanoes in this area, which constitute the southern and northern boundaries of the sub-basin, are conditioned by transforming oblique faults, as can be observed in other areas of the Gregory Rift.

Tectonic features and cortical structure below the Natron rift basin

The Lake Natron basin lies near the southernmost boundary of the Gregory Rift. In the basin, tectonic, volcanic and cortical characteristics are different from the northern areas. Volcanism is very intense in the southernmost area of the Gregory Rift, represented by the Crater Highlands and by an ultrabasic effusive volcanism that results in a number of minor tephra vents to the south (Dawson 1964; Dawson and Powell 1969; Fairhead 1980). These tectonic variations seem to be the result of regional differences both in cortical and sub-cortical structures.

In contrast with the Kenyan rift, where a number of geophysical studies have developed models of rift basin evolution (e.g. Swain et al. 1994), such studies are not very abundant in the Lake Natron basin. However, the proximity of Natron to the well-known Magadi basin allows extrapolation from nearby cortical data. Under the rift, the gravimetric data show a wide negative anomaly which contains a smaller anomaly. This anomaly suggests a cortical thinning and a deep mantle anomaly under the Tanzanian craton (Simiyu and Keller 1997:figure 10.5). The anomalies seem to be deeper in the south of the Western and Eastern rifts than in Kenya. In general, the structure of this southern part of the rift shows a thicker cortex together with a higher topography and narrower, better-defined boundaries compared with the north (Mechie et al. 1997). Between the Natron and the Magadi lakes, the asthenosphere deepens towards the south by at least 50 km (from 40 to 90 km), which can be related to the end of the dome in Kenya and to a smaller upwelling of the mantellic layer as well as of the temperature (Simiyu and Keller 1998). These data support classical tectonic models which suggest a migration of the rift towards the south, where the crust is thinner (20–4 km; Baker and Wohlenberg 1971; Darracot et al. 1972; Fairhead 1980).

Based on geomorphological studies, several morpho-structural regions within the basin itself can be defined (Figures 2.4, 2.6) as below.

Western tectonic escarpments. These are the main characteristics of the rift in the area. Geophysical studies show that the Nguruman escarpment, equivalent to the Sambu escarpment in Natron, is formed by two faults separated by an intermediate terrace. It shows a drastic decrease of the gravimetric values throughout the lower escarpment (Birt et al. 1997). However, in the western margin of Magadi, the peak-shaped anomaly is not documented, which could be due to an increase of the basement density (Birt et al. 1997). Near the Mozonik volcano, several tectonic steps can be observed affecting the lavas of this volcano. These blocks are due to gravitational faulting and complicate the relief of the rift in that area. The Sambu fault, which deeply splits the Oldoinyo Sambu volcano, also shows some smaller faults oblique to the mountain (Figure 2. 4), following a tectonic model similar to that proposed by van Wyk de Vries and Merle (1998) for strike-slip pull-apart structures.

Tectonic platforms. A wide plain extends between the Sambu and Sonjo faults, replicating the modern morphology of the bottom of the semi-graben between the Sambu fault and the Gelai volcano. This platform is called the Sanjan Plain (1,100–1,000 m above sea level). This platform seems relatively stable and has been covered by abundant sediments supplied from the Precambrian relief of the Loita Hills and ashes from eruptions of Kerimasi and Oldoinyo Lengai.

Grid faults at the bottom of the basin. The areas that have not been covered by recent lacustrine and

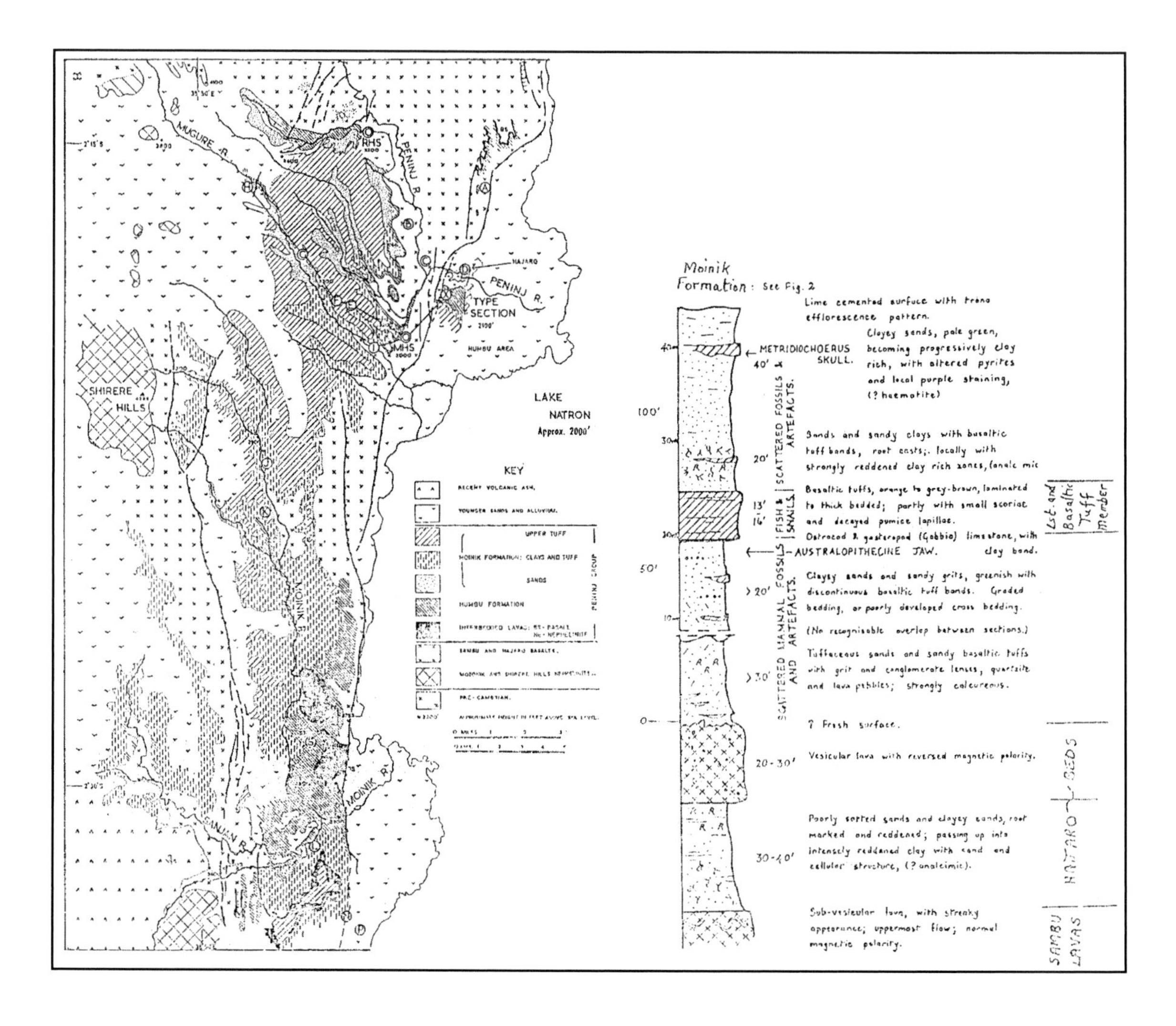

Figure 2.4 Geological map and stratigraphic section of the Peninj Group units from Isaac (1965).

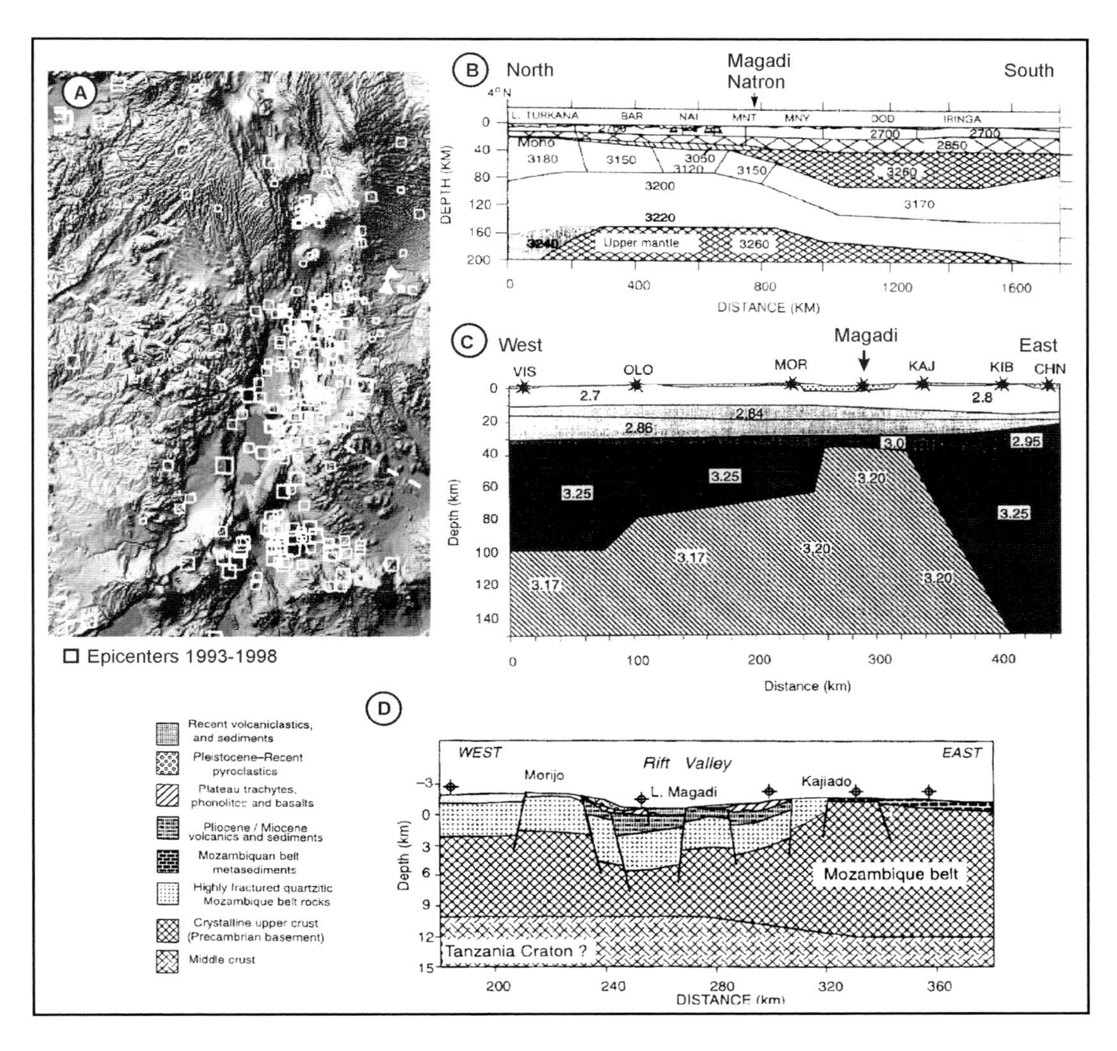

Figure 2.5 Cortical and structural data from different geophysical studies: a) epicenters recorded between 1993 and 1998 (Hollnack and Stangl 1998); b) regional model of gravity anomalies along the Kenyan Rift (Simiyu and Keller 1997); c) density model along the KRISP 94 line G (Birt et al. 1997); d) integrated analysis of seismic, gravity and geological data along the KRISP 94 seismic line G (Simiyu and Keller 1998).

alluvial sediments show distinct tectonic features, with a network of horsts and elongated grabens running N-S and NNE-SSW overlying the Older Extrusives lavas from Sambu, Gelai and Shombole. This geological feature was described by Baker (1958, 1963) and it is associated with a change in the tectonic dynamics of the main axis of the rift over the past 400,000 years. This intense faulting of the bottom of the rift can also be documented inside the half-graben in the sediments of the Peninj Group, but this area shows a higher diversity of tectonic directions.

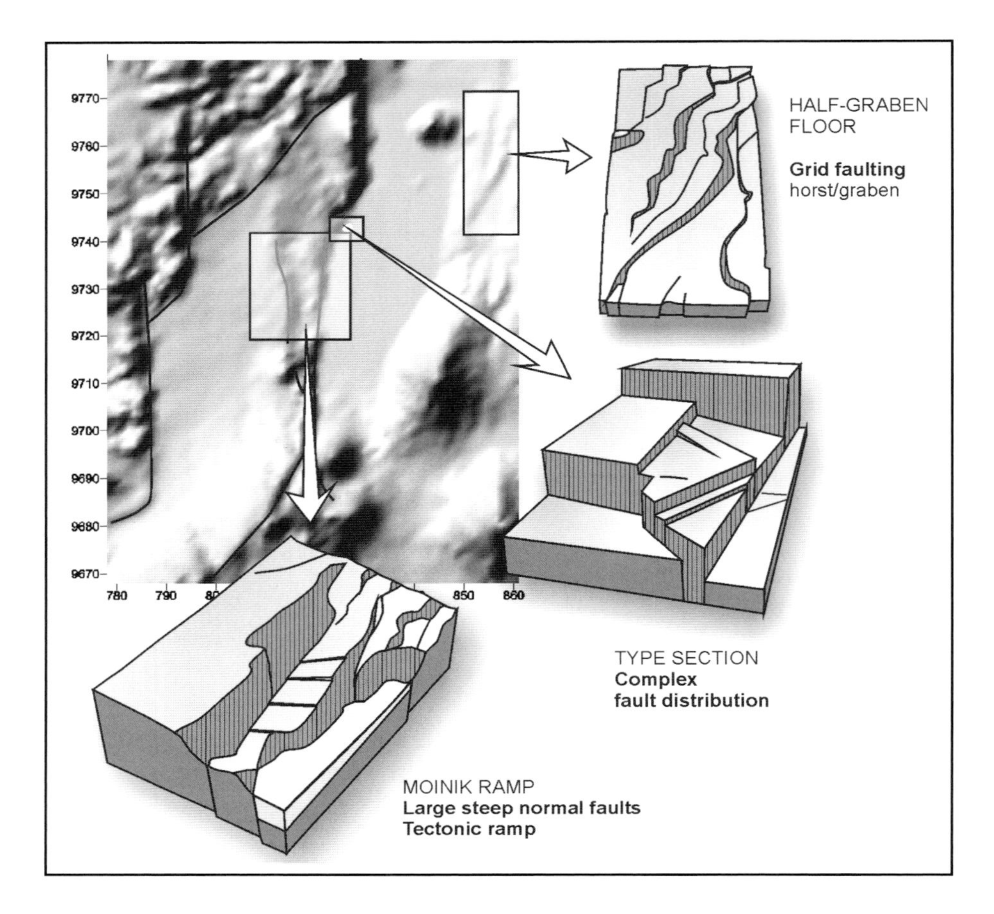

Figure 2.6 Different structural fields in Lake Natron: grid faulting (top right), complex multi-directional faulting (middle) and large steep faults and ramps (bottom).

Transform faults. These are mainly found at the north and south margins of the Natron sub-basin. They condition the northern limit of the Precambrian reliefs and the Sonjo fault, which shows the intersection of a north-south fault with another one at a 30° angle. This oblique direction can also be observed in the grid faulting of the rift axis. In the northern extreme of the Crater Highlands, other fractures running in a similar direction split the volcanic structures to the north of the Empakai caldera. In the south, within the semi-graben, these structures have been covered by recent volcanic sediments.

The Moinik tectonic ramp. This is a slope, 5-6 km wide and 18 km long, which deepens over 400 m from north to south (Figure 2.3). Along this ramp, between the Sanjan and Sambu escarpments, one finds the largest area of outcrops of Peninj Group sediments. This structure is similar to those of the strike-slip type that results from a horizontal north-south movement added to the dominant vertical east-west extensional movement. The contrasting fractures in the relay ramp are clearly reflected in the drainage network, although the fractures of the sedimentary materials within the ramp are fairly scarce. In the zone connecting the two main faults, there are oblique slip faults in contact with the higher blocks. The outcrop where the Type Section1 was defined at Peninj may be the highest part of another parallel ramp almost covered with recent sediments. The sloping strata of the Peninj Group and the direction of faults would suggest this process. In this area, the main fractures are N-S and W-E; that is, those following the rift trend and those perpendicular to it, formed by the stepping of the ramp during its subsidence. Other minor fractures follow an oblique trajectory. In the zone where the Peninj Group reaches its greatest thickness, a dominant system of N-S fractures can be observed together with normal faults that sink blocks towards the east, following the extensional direction.

Volcanism and eruptive phases during the Plio-Pleistocene

Volcanic features are typical products of rifting. In the Natron sub-basin, several volcanic structures are documented (Guest 1953; Figure 2.7). These structures show changes through time both in distribution through the basin as well as in their chemical components and eruptive processes. The Older Extrusives date from the Pliocene to the Middle Pleistocene and created large volcanic structures of basaltic lavas, trachytic basalts and basanites that are located on the rift axis (Figures 2.7-2.8). These lavas outcrop in a sequence that is more than 400 m thick in the Sambu escarpment. Within the Older Extrusives there are two groups: the old volcanoes, such as Mozonik or Shombole, which are more explosive and ultrabasic, including nephelinitic lava and ashes (Isaac 1967), and the more recent basaltic and trachytic-basaltic volcanoes, with basanites and ankaratrites (Lubala and Rafoni 1987) like Oldoinyo Sambu, Gelai or Kitumbeine. The latter were formed during the early Pliocene and lower Pleistocene.

Together with these volcanic structures, fissural lava flows fill a good part of the rift graben in this region. The Older Extrusives are partially contemporary with the Peninj Group and range between 4.8-2.0 Ma (in part of the Crater Highlands and Oldonyo Sambu) and 1.5-0.6 Ma (at Gelai or Kitumbeine; Baker and Wohlenberg 1971; MacIntyre et al. 1974; Fairhead 1980). The first generation is associated with the rift margin, whereas the second generation is produced more centrally around Lake Natron. Most of the substrate underlying the Peninj Group is composed of lavas from the Sambu volcano. The top of them is dated between 2.5 to 2.0 Ma (Isaac 1965, 1967; Kapitsa 1968; Howell 1972; Isaac and Curtis 1974; Crossley 1979). The Sambu lavas consist of a minimum of 40 layers corresponding to 40 separate flows in between stability episodes which allowed the erosion and formation of a paleorelief with sedimentary infilling. The Sambu lavas consist of olivine-basalts and basanites with plagioclase, olivine and clinopyroxene phenocrystals in a cryptocrystalline matrix (Lubala and Rafoni 1987).

Toward the south of the outcrops, a greater influence of the Mozonik volcano in the composition of lavas can be detected. Lavas appear enriched in Al_2O_3 and K_2O (plus nepheline, orthose, biotite and an increase of potassium in the plagioclase). The older volcanoes seem to be more similar in composition to the more recent volcanoes from the Younger Extrusives, indicating a change in magma composition. The relief of these volcanoes is also different. The oldest volcanic deposits (Mozonik and Shirere) show a high degree of erosion given the lack of compaction in their ash layers, giving rise to spiky forms and

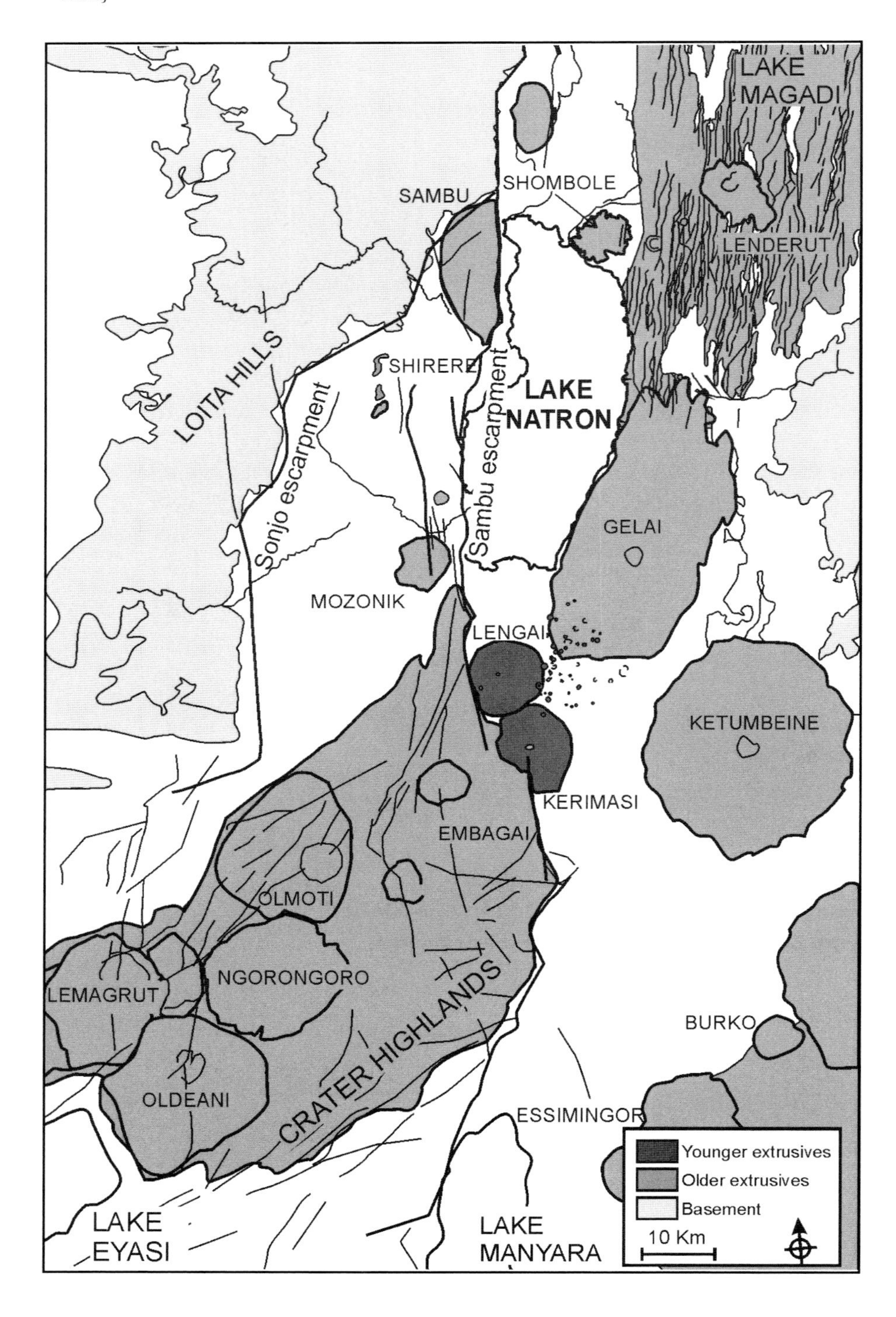

Figure 2.7 Volcanism in the Lake Natron area.

Figure 2.8 Geomorphological features of the Lake Natron area: Sambu Fault Escarpment and Lake Natron shore (top left); alluvial fan at the foot of the Sambu Escarpment (top right); stepped Sanjan and Sambu Faults with Mozonik Older Extrusive volcano at the top (middle left); Shombole volcano (Kenya) and Lake Natron from Type Section (middle right); Oldoinyo Lengai from the east (note white carbonatite lavas on top) (bottom left); and Oldoinyo Lengai active craters and lava flows (bottom right).

gullies (Figure 2.8). The following generation (Sambu, Gelai) is comprised of large volcanic structures with a radial drainage network. Slopes are steep but smooth due to the resistant lava flows.

The Younger Extrusives are situated on the rift axis and are composed of vents and volcanoes like Kerimasi (0.6-0.4 Ma) and Oldoinyo Lengai (0.37-present; Dawson 1962; Dawson and Powell 1969). The components are highly explosive and ultrabasic in composition, including nephelinite, phonolite and carbonatites. Their most active tectonic phase occurred during the Middle Pleistocene. Oldoinyo Lengai (Figure 2.8) in particular began to erupt when the tectonic activity of the rift axis together with grid faulting became more intense (Baker and Mitchell 1976; Baker et al. 1971, 1972; MacIntyre et al. 1974; Crossley 1979). The eruptions are also associated with the creation of a number of small ash vents around the main volcanoes (Dawson 1964; Baker 1986). A number of collapsed vents gave rise to a series of small calderas. The last generation of volcanoes (Oldoinyo Lengai, Kerimasi) are the product of a very effusive volcanism, with abundant pyroclasts and ashes. Cones are eroded in deep radial gullies associated with alluvial fans on their slopes. In many parts of the south of the basin, remains of ignimbrites produced by Oldonyo Lengai are abundant. The most recent unconsolidated ashes spread over the Sanjan platform as longitudinal dunes.

Alluvial deposits: Sedimentary response to volcano-tectonic forcing

The emergence of the modern drainage network on the flanks of the Rift Valley beginning 1.1 Ma created outcropping of large areas of the Peninj Group sediments. However, it also brought a substantial amount of alluvial deposits to the foot of the tectonic escarpments that partially mask these outcrops (Figure 2.9). Most of the water entering the basin is seasonal and arrives via steep streams carrying lava blocks and coarse-grained sands, which are deposited downslope as alluvial cones or coalescent fans. Permanent waters feeding the lake flow through four main rivers: Peninj, Ewaso Nyiro, Moinik and Ngare Sero, as well as several permanent hotsprings. The morphology of valleys cut by these rivers is tectonically conditioned by faults as well as by the prismatic structure of the underlying lavas. When channels cut softer sedimentary materials such as clayey or sandy sediments or volcanic ashes, the relief exposes a wide vertical wall of these materials, leaving a plain on the more resistant lavas. This increases the occurrence of outcrops containing Peninj Group sediments.

Alluvial fans

At the foot of the escarpment, there is a wide system of alluvial fans which rarely span surfaces greater than 1 km^2, with the exception of some which occasionally reach 3 km^2 (Figures 2.8–2.9). In many cases, these fans show a steep downward slope and are composed of coarse-grained materials, with abundant blocks and cobbles from the escarpment lavas and sands from the Precambrian relief, as well as Plio-Pleistocene reworked sediments and lavas. Gravity deposits, such as alluvial cones and colluvium on the slopes of the tectonic escarpments, are also frequent. The morphology of the fans developed over the past 4,000 years, when Lake Natron reached its modern level after the beginning of the modern arid phase (e.g. Kutzbach and Street-Perrot 1985; Casanova 1987; Casanova and Hillaire-Marcel 1992; Damnati 1993). The fan system developed both at the foot of the tectonic escarpment and on the slopes of the Sambu and Gelai volcanoes and basement relief (Figures 2.3, 2.9).

Deltas

The fan delta deposits created by the Ewaso Nyiro, Peninj and Moinik rivers are the most extensive alluvial deposits in the Natron basin. The Ewaso Nyiro delta grades into a swampy area since the river lies in a flat floodplain. The deltaic fan deposits of the Moinik and Peninj rivers are very similar to each other (Figure 2.9). The coalescent and prograding deltaic fans show

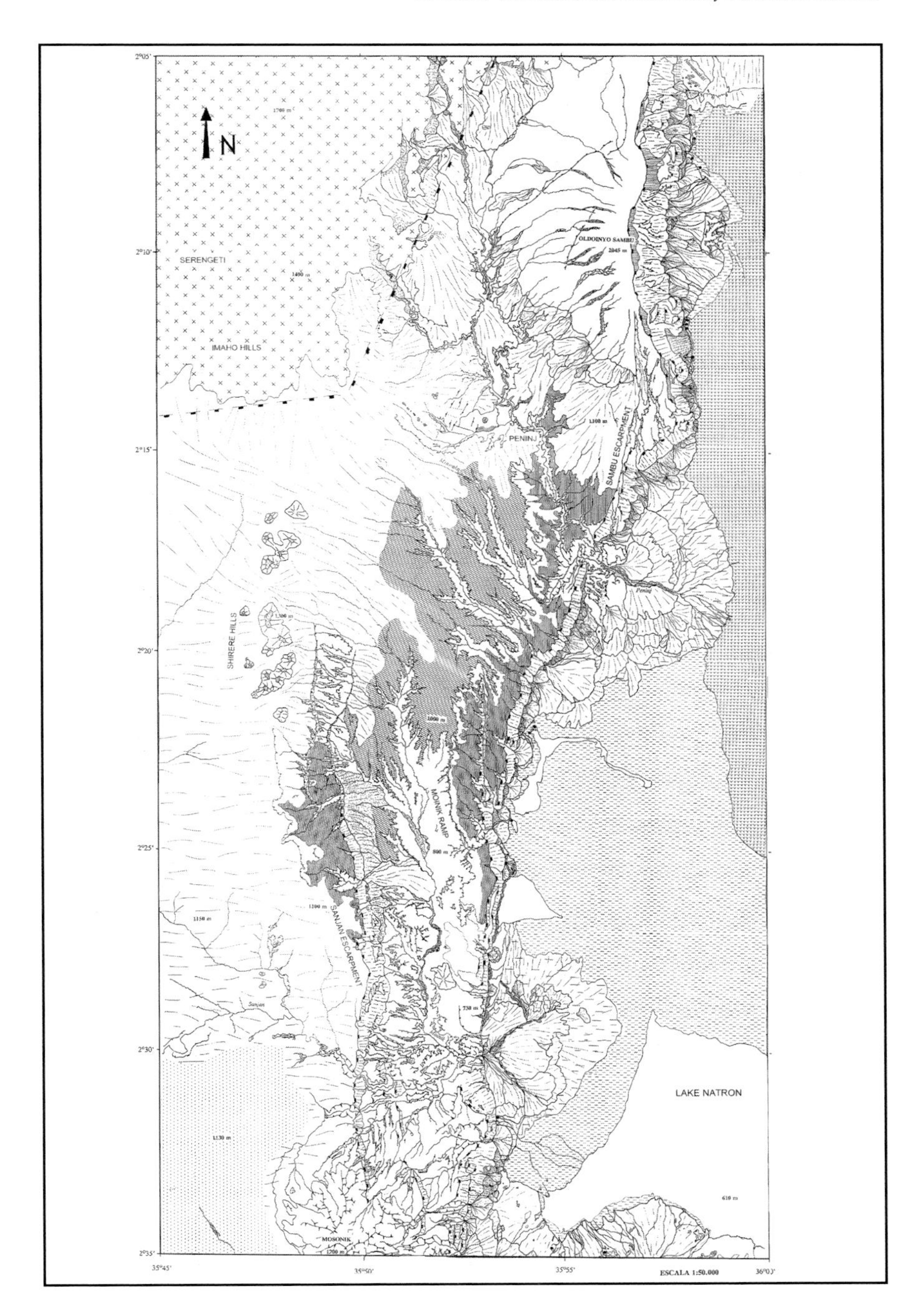

Figure 2.9 Geomorphological map of the West Lake Natron margin (from Luque 1996). Areas showing parallel lines correspond to structural surfaces developed on Sambu Lavas (vertical) and the Upper Moinik Tuff (oblique). Lake Natron mudflat and salt crust brackish water areas in the south and north-eastern areas.

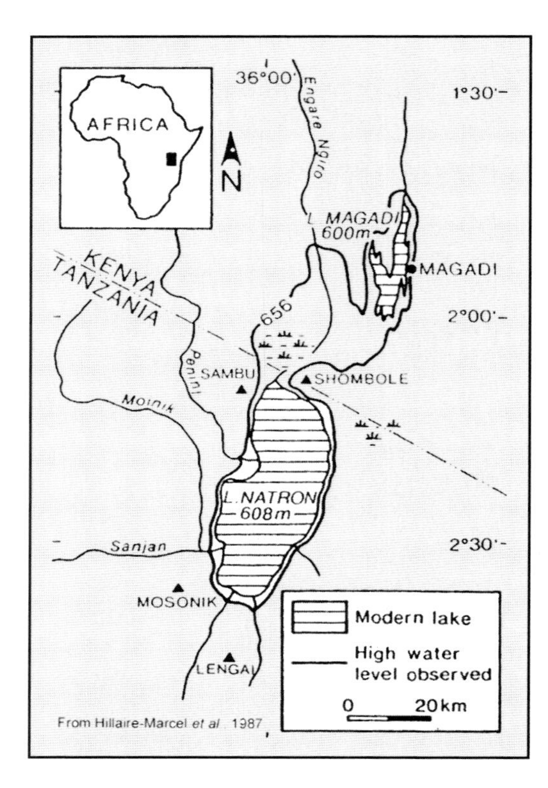

Figure 2.10 Holocene-Upper Pleistocene Natron-Magadi shoreline from Hillaire-Marcel et al. (1987).

the migration of the rivers' courses over time, although today both flow towards the north and northwest. Today, the Peninj river is embedded in braided gravel and cobble deposits incised into the deltaic fan, generating numerous boulder to sand bars. The distal parts of the deltas show a rich system of distributary channels which end in sandy lobules or swampy areas where salt crusts precipitate.

Lake Natron

The lacustrine system of Lake Natron is unique. As noted above, this extremely shallow lake is highly saline and alkaline, with a high content of sodium carbonate. The markedly endorrheic character of the basin and the high degree of evapotranspiration enhance the formation of these salts. The gradual slope and lack of precipitation favor the development of mudflats and saltpans during the dry season. The lake is very sensitive to climatic changes, and although lake-level changes have occurred throughout the Quaternary, most of them took place at the Pleistocene-Holocene transition (Casanova 1986, 1987; Casanova and Hillaire-Marcel 1987; Roberts 1990; Roberts et al. 1993; Damnati et al. 1992; Damnati 1993; Williamson et al. 1993). During the Younger Dryas event ca. 11,000 BP, there was a sudden drop in lake levels (Hillaire-Marcel and Casanova 1987a, 1987b). Higher lake levels during humid phases are evident from relict deltaic deposits and a multiple-generation stromatolite belt.

During maximum pluviosity, the Natron and Magadi sub-basins were joined (Figure 2.10).

Tecto-magmatic and sedimentary phases: Geological evolution of the Natron basin

The regional data obtained by various authors allow the reconstruction of the geological evolution of the Lake Natron basin (Figure 2.11). In the southern part of the rift, tectonic movements began about 12 to 7 Ma (Baker et al. 1988), substantially later than volcanism, which began some time between 20 and 16 Ma (Baker et al. 1971, 1972; Baker and Wohlenberg 1971; Baker and Mitchell 1976). The earliest faults are produced late compared to those of the northern rift and probably follow the structure of previous orogenies. The comparison of these data with those from basins in the north confirms the spreading of the rift from north to south (Baker et al. 1972; Fairhead 1980; Morley et al. 1992). The earliest fracture in the Natron basin is the Sonjo Fault, situated to the west, which emerged between 7 and 3.5 Ma, creating a 900 m high escarpment over

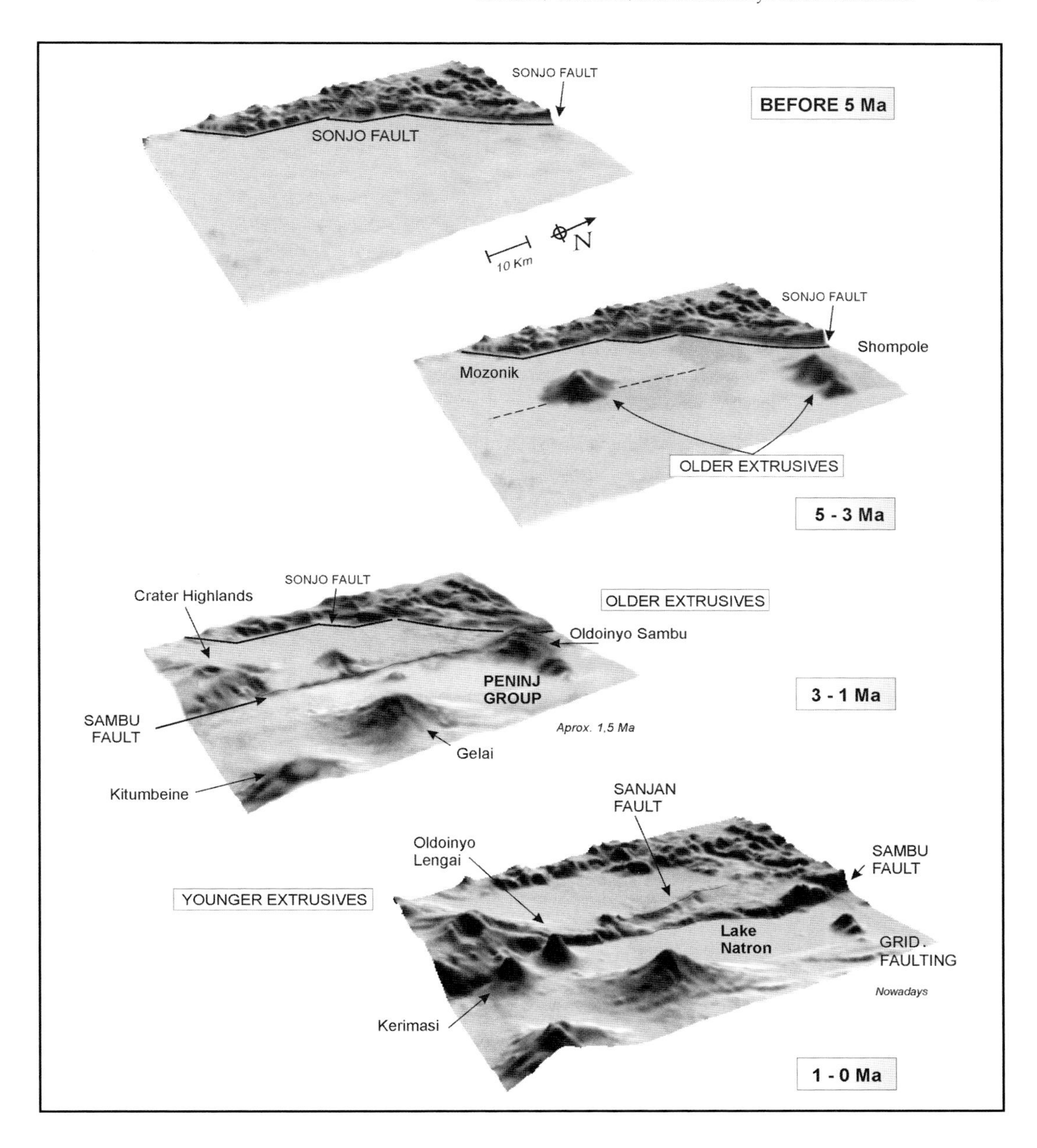

Figure 2.11 Tectonic evolution of the Natron basin from the Upper Miocene-Lower Pliocene to recent times (from top): Lower Pliocene Sonjo Fault in the west, starting of alluvial sedimentation into the half-graben (no known geological record) (top); first Older Extrusives eruptions (Ngorongoro, Mozonik, Shirere and Shombole volcanoes) (second from top); first Sambu Fault movements, second phase of Older Extrusives (Crater Highlands-Ngorongoro, Oldoinyo Sambu, Gelai, Kitumbeine and Shombole), Peninj Group deposition (second from bottom); modern landscape, large Sambu and Sanjan fault escarpments, narrow half-graben basin, Younger Extrusives (Oldoinyo Lengai, Kerimasi) (bottom).

time. The ultra-alkaline, nephelinitic and phonolitic Older Extrusives were produced afterwards, between 3 and 3.5 Ma (Isaac 1965, 1967; Curtis 1967; Kapitsa 1968; Isaac and Curtis 1974). The largest structure is the Mozonik volcano, formed by nephelinitic and phonolitic ash interbedded with lava.

In the north, the Shombole and Lenderut volcanoes are trachytic and basaltic. The former is associated with N-S and NNE-SSW fractures formed inside the rift and the latter appears on the rift axis itself, very likely associated with the oblique fractures that nowadays limit the Natron and Magadi basins. Other trachytic and basaltic lava flows were produced at the bottom and the margins of the rift basin at that time; the Shirere Hills are a relict of this activity (Isaac 1965, 1967; Curtis 1967; Kapitsa 1968; Isaac and Curtis 1974). During the upper Pliocene, this faulting is re-activated and the volcanic structures that currently appear in the basin can be recorded at 2.1-2 Ma both in southern Kenya and northern Tanzania (MacIntyre et al. 1974; Baker and Mitchell 1976). This activity allowed the intrusion of magma and produced a second phase within the Older Extrusives. The basalts, trachy-basalts and basanites configure the Crater Highlands and Kitumbeine beyond the basin, and the Oldoinyo Sambu and Gelai volcanoes within the basin (MacIntyre et al. 1974; Baker and Mitchell 1976; Lubala and Rafoni 1987).

The Sambu volcano was the main agent responsible for the filling of the basin, with deposits that span 400 m in depth and even 1,000 m in some places. Oldoinyo Sambu is a key player in the morphological and sedimentary evolution of the basin. The lavas span several episodes of magnetic polarity and new dates suggest an age of 2.5-2 Ma on the top of the Sambu lavas and 3.5 in the middle, although the latter date is not secure (Isaac 1965, 1967; Kapitsa 1968; Howell 1972; Isaac and Curtis 1974; Crossley 1979). The Sambu lavas have been correlated with the trachytic lavas of the central rift plateau in Kenya, which reaches the Magadi basin (Baker 1963; Kapitsa 1968). The period of erosion and fluvial encasement produced between the two eruptive episodes of Oldoinyo Sambu created a 20-meter sequence of detritic sediments and volcanic ashes (Isaac 1965, 1967). These detritic deposits were named Sambu Beds (Isaac 1965) and their facies are similar to those of the Peninj Group. The maximum thickness is found in the valley of the proto-Peninj river (Isaac 1965). Although in these deposits some vertebrate remains have been found (Isaac 1965, 1967; Kapitsa 1968), our survey in the same deposits did not find any.

The Sambu lavas are the last lavas of any great extent produced in the basin. After them, a period of tectonic and volcanic stability followed, with minor volcanic activity. The activity of the Sambu fault created a 10-30 m high step which was later filled with sediments. These sediments outcrop marginally within the semi-graben and were named the Pre-Peninj beds (Isaac 1965) and then Hajaro beds (Isaac 1967) and could be considered the beginning of the Peninj Group. However, when sedimentation stopped in that depression, a fissural volcanic episode began to infill that part of the basin. These lavas are vesicular basalt known as Hajaro lavas, which span 10 m in thickness and have reversed paleomagnetic polarity; they give a K/Ar age of 2.1-2.0 Ma (Isaac 1965, 1967; Curtis 1967; MacIntyre et al. 1974). The relative stability after the deposition of the Hajaro lavas allowed the formation of both alluvial-deltaic and lacustrine sedimentation systems. These deposits constitute the Peninj Group as initially defined by Isaac (1965, 1967). At that time, the morphology of the basin was flatter than now, although the Sambu escarpment was at least 30 m high. The progressive filling of the basin eventually overflowed the escarpment, depositing sediments on top of it.

Tectonic, volcanic, climatic and sedimentary changes are reflected in the lacustrine changes of this endorrheic basin. These changes allow the division of the Peninj Group into several sedimentary units (Isaac 1965, 1967). The size of the basin during this period is unknown, since there are no vestiges of this period either to the east, north or south. The infilling began about 2 Ma and continued until 1.1 Ma, when the Sambu fault was re-activated and the modern morphology of the basin appeared (Isaac 1965, 1967; MacIntyre et al. 1974; Dawson and Powell 1989). During the Middle and Late Pleistocene this activity continued, making the formation of new deposits difficult. One of these deposits can be found in the Oloronga Beds in Kenya. It was formed during a high lacustrine level which occurred >0.78 Ma, when the Magadi and the Natron paleolakes joined (Baker 1958, 1963, 1986; Crossley 1979; Eugster 1986). No remains of these beds have been recorded in the Natron basin. With the tectonic movements at the end of the Early Pleistocene, the erosion of the Peninj Group began.

The eruption of the Younger Extrusives began in the Middle Pleistocene: first Kerimasi (0.6-0.4 Ma), followed by Oldoinyo Lengai (0.37-present) and several smaller cones (Dawson 1962, 1964, 1989; Hay 1986; Baker 1986; Lubala and Rafoni 1987; Dawson et al. 1994). In other areas of the rift, large cones like Meru, Hanang or Ufiome were formed. After the formation of Oldoinyo Lengai, an intense fracturing of the rift axis occurred, creating a network of horsts and grabens oriented N-S with an average length of 1.5 km. These fractures constituted a typical grid faulting along many parts of the rift axis (Baker and Mitchell 1976; Baker et al. 1972; MacIntyre et al. 1974; Crossley 1979). The lake level fluctuated greatly during the Middle and Upper Pleistocene as a result of climatic changes, as is also observed in most East African lakes (e.g., Butzer et al. 1972; Roberts 1990; Casanova and Hillaire-Marcell 1987; Roberts et al. 1993; Williamson et al. 1993). Tectonics also played a role in these lake-level changes.

The Peninj Group: Stratigraphic units

The Peninj Group consists of sand and clay deposits with interbedded volcanic layers, spanning almost 80 m of sediments outcropping to the west of Lake Natron (Isaac 1965, 1967). Isaac described these deposits as lacustrine and paralacustrine deposits in a broad, shallow basin. These sediments had also been briefly described by Uhlig and Jaeger (1942) as sandy and tuffaceous deposits of fluvial origin. Guest (1953) described them as diatomites deposited during high lake levels. The maximum thickness of the sequence is found in the few outcrops at the foot of Sambu Fault, especially in an area called Type Section Maritanane, where Isaac defined the Type Section of the sedimentary sequence and where fossiliferous localities are most abundant. Isaac (1967) divided the Peninj Group into several units and sub-units overlying the Hajaro lavas (Figure 2.12): broadly, these are the Humbu and Moinik Formations. The members of each of these formations are described in detail below.

First, the Humbu Formation (formerly called "Lower Series" in Isaac 1965) is composed of 40 m of sandy deposits in paraconformity with the underlying Hajaro lavas. This formation is composed of alluvial microconglomerates, sands and clays derived from the erosion of the Precambrian relief and interbedded with ash layers of basaltic origin, containing very zeolitic (analcime) orange ashes (Isaac 1965, 1967). In the mid-section, frequent ash layers occur

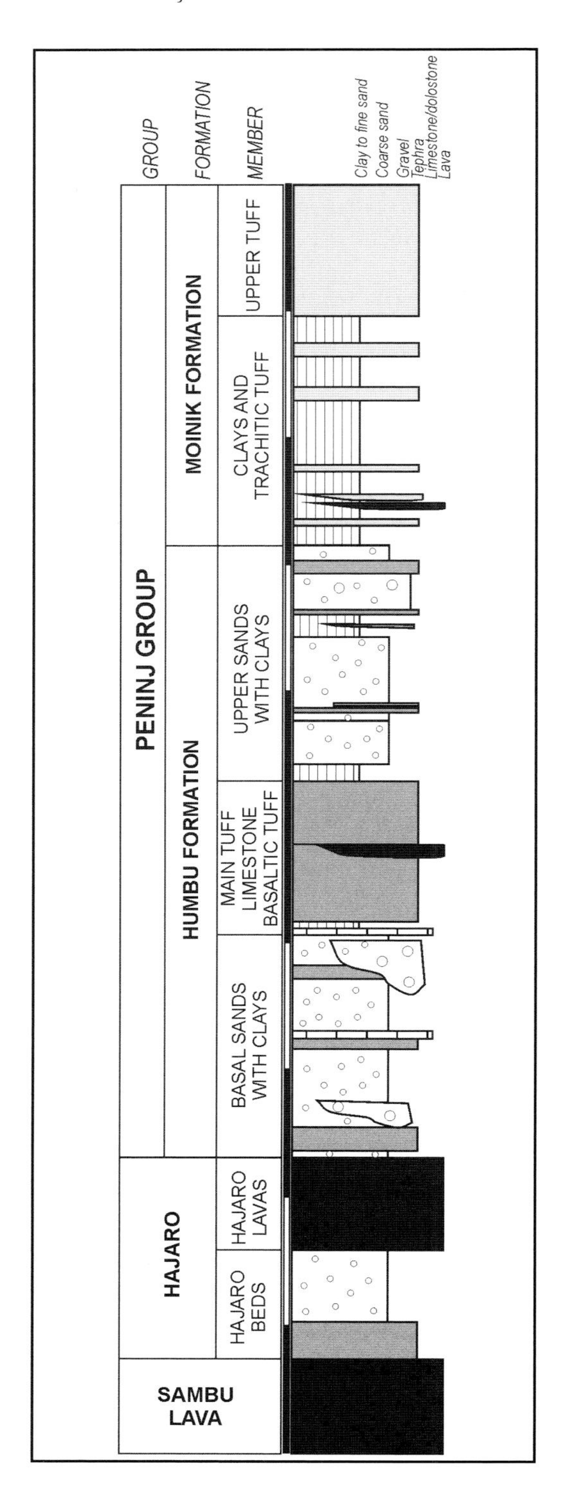

interbedded within a flow of olivinic basalt. The sandy deposits are usually composed of quartz grains, pyroxenes, amphiboles, and a low presence of feldspars in a clayey matrix (illite, smectite and analcime) with some interbedded dolostone and limestone layers (Icole et al. 1987). This facies and its authigenic minerals are indicative of a lacustrine freshwater to alkaline environment (Isaac 1965; Hay 1966; Icole et al. 1987). The Humbu Formation is very rich in vertebrate fossils and lithic remains. A Paranthropus mandible was found in the lower part of this unit (Leakey and Leakey 1965).

Second, the Moinik Formation (Formerly Upper Series, Isaac 1965) is an approximately 30 m thick sequence, with clays in the south and sands in the north with intercalated layers of altered trachytic tephra layers, which appear very zeolitized (erionite) and whose color ranges from white to yellow. Some limestones and dolomites, as well as abundant chert nodules (magadite), gaylussite, pyrite and trona casts appear interbedded, which together with the erionite suggest a stratified lacustrine saline-alkaline environment (Isaac 1965, 1967; Eugster 1967, 1986; Hay 1968; Icole et al. 1987; Schubel and Simonson 1990). This formation represents a high lake level and a receding phase of alluvial facies in the west and north. At the bottom, there is a nephelinite layer in southern exposures, and a thick tephra layer caps the sequence. Isaac (1965) noted that considerable lateral variation exists, with clays, shales and laminated tuffs in the south and east but deltaics and alluvials in the north and west.

Figure 2.12 (left) Stratigraphic section of the Peninj Group deposits, based on Isaac (1965, 1967). Maximum thickness is 70 m.

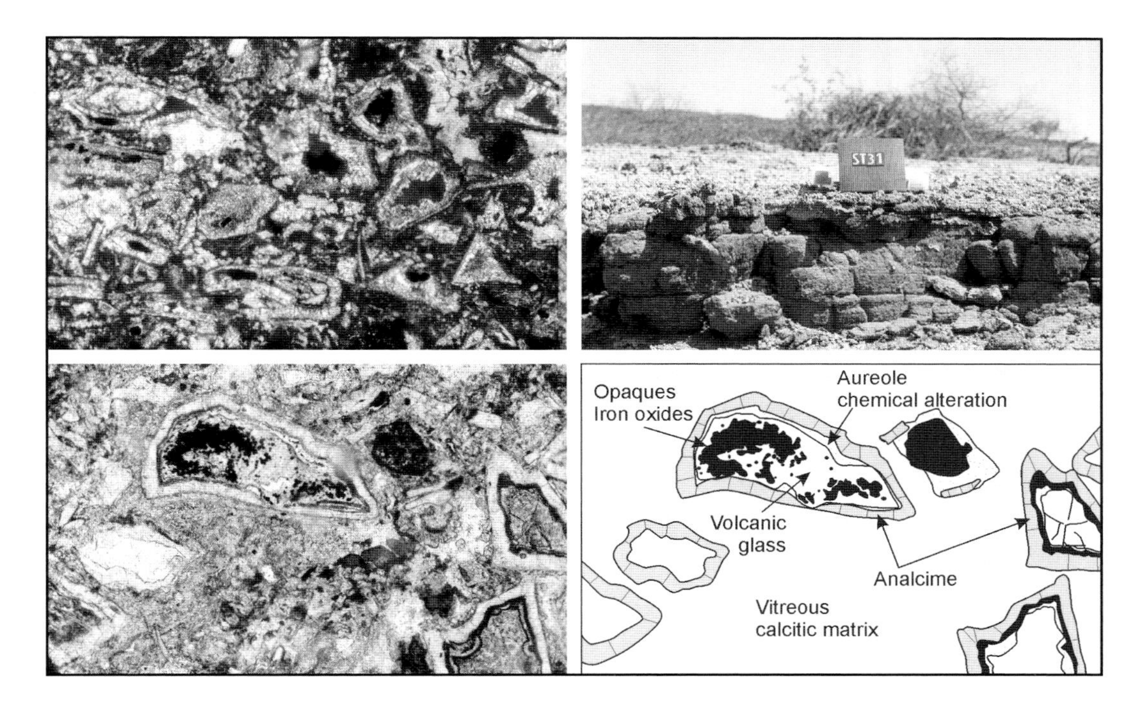

Figure 2.13 Tephra deposits in the Humbu Formation. T-1 basaltic tephra (top right); thin sections of ash-fall deposits showing altered volcanic glasses in a microcrystalline matrix (bottom left); interpretation of the petrographic features observed in the lower thin section (bottom right).

Humbu Formation

The Humbu Formation is undoubtedly the sedimentary unit richest in fossils and archaeological sites. It was divided into three members (Isaac 1967).

Basal sand with clay member (BSC)

This consists of approximately 10–15 m of sands deposited only in the most internal part of the graben, to the east of the Sambu escarpment. In Type Section, this sub-unit is composed of sands and micro-conglomerates showing several phases of deposition intercalated with three basaltic tephra layers (TBS-1 to -3). In the outcrops to the south of the lake, this sub-unit is clayey and lacustrine, and in the northernmost exposures it is more sandy but clayey, probably representing a swampy environment. It has never been dated and has reverse polarity in the uppermost section (Thouveny and Taieb 1986, 1987). It overflows the Sambu escarpment only in the lowest sections and valleys, but at the foot of Moinik ramp this member has been widely deposited. It is generally poor in fossils and lithics, although some channel infills can contain abundant bone remains. The sands are mostly quartz but they also contain mica, K-feldspar, pyroxene, amphibole, volcanic glasses and zeolites (analcime). Clays are illite and smectite, and other zeolites like mordenite and erionite also exist. Limestone and dolostone are interbedded (Icole et al. 1987).

Main Tuff (Limestone and basaltic tuff member or Main Basaltic Tuff)
This is a volcanic ash stratum ranging from 1 m to 6 m thick, locally containing coarse lapilli, altered to authigenic zeolites that show traces of mechanical deformation (slumping). The tuff become thicker towards its origin in the south. In the central area, it has two layers: the lower one is fine, laminated and folded; the upper one is darker and rich in coarse pyroclasts and sometimes in volcanic bombs. The latter intercalates limestone layers rich in gastropods and ostracods. Sometimes fish remains are also abundant, suggesting a lacustrine depositional environment. This member, as with the remaining Humbu ash layers, comes from a pyroclastic cone situated in the south of the semi-graben near the outflow of the Moinik river to the lake. At the bottom of this member there is a limestone layer, 1-5 cm thick, that contains abundant gastropods (Gabbia) and ostracods and which is covered by a few centimeters of clay. The limestone layer can be traced for 7 km along the Peninj valley and 20 km along the Mugure valley. To the west, it becomes more continental and gastropods disappear (Isaac 1965). This tuff covers the basal sands and lies immediately over the Sambu lavas on the escarpment. To the south, it has an embedded lava flow called Wambugu basalt which was dated by K/Ar and 39Ar/40Ar and shows normal polarity.

Upper sand with clay member (USC)
This member contains a great variety of facies, offering rich information about paleoenviromental changes. In general, sediments are very sandy, although they also contain clays. Diagenetic and pedogenetic processes are very abundant, such as the formation of calcretes, limestone nodules, iron crusts, mudcracks, rootcasts, burrowing and reddened soils. This member is 20 m to 30 m thick. It presents five levels of zeolitic tephra (T1-5) and an hematitic level which has been correlated with Olduvai Bed III (1.15-0.8 Ma; Isaac 1967). Archaeological sites are abundant. The thickness of USC varies greatly depending on the outcrop. In the south, it is very thick and clayey. Near the Type Section, it is thick, sandy, and shows at least four different depositional events ranging from fully lacustrine to alluvial delta plain and channels. A hiatus has also been recorded in the outcrops of the uplifted horst of the Sambu Fault. The northern exposures show erosion at the top of the sequence and the member shows less thickness and intense pedogenic processes.

Moinik Formation
The Moinik Formation is deposited over the Humbu Formation conformably in the south and nonconformably in the north. It represents a clear change in sedimentation dynamics in the basin. Lacustrine facies become predominant in the south and deltaic facies are predominant in the northernmost exposures. It was divided into two units, described below.

Clay with trachytic tuff member
This consists of zeolitic clays (mainly erionite but also analcime, chabazite, mordenite and phillipsite) with six interbedded volcanic ash layers (TM-1 to -6) of highly erionitic composition, as well as limestone levels with fewer dolostones (Icole et al. 1987). Towards the north, sandy layers become more abundant. This member lies discordantly over Humbu Formation, showing a clear erosional surface. In the lacustrine deposits, there is no evidence of fauna or lithic tools, but in the deltaic facies in the north, there are abundant lithic remains and a few faunal remains.

Upper Tuff member. This consists of a yellow trachytic laminated tuff, varying in thickness from

15 m in the Type Section to 1.5 m in the northern outcrops, where it is very sandy. In Maritanane, it reflects periods of a lack of ash-fall with the emergence of an erosive paleorelief, detritic infill and two stromatolite crusts. Sometimes, the tephra appears extensively reworked and it contains very few fossils but some lithic remains, none of which are found *in situ*.

Previous dating of the Peninj Group

The tephra layers of the Peninj Group are of a low quality for dating because they have been profoundly diagenetically altered, which has increased the amount of analcime and erionite, whereas minerals rich in potassium (mainly feldspars) are scarce or altered (Figure 2.13). According to Isaac (1965), using information provided by Hay, pyroclastic deposits originating from basaltic magmas are altered to authigenic analcime, whereas tephra originating from trachytic magmas are altered to erionite and to a lesser extent, analcime and K-feldspar. Thin sections show that volcanic glasses have been partially or totally dissolved or substituted. Remineralization produces opaque minerals (iron oxides), calcite and zeolites (analcime in Humbu and erionite in Moinik), whose pores are filled with calcite. Often, analcime grows radially around altered glass (Figure 2.13) as a result of the action of alkaline waters over volcanic glasses and feldspars (Hay 1966). This alteration can be produced through a precursor of zeolite, which can be erionite rich in sodium.

Vitreous pyroclasts in the Humbu tephra layers show a porosity of nearly 20%, their size ranging from 125 to 750 microns. The tuff matrix is 70%–80% glasses and 20%–30% pyroclasts and altered minerals like plagioclase or pyroxene. It is not unusual to find quartz grains mixed with volcanic products, which indicates a period of exposure followed by reworking and mixing with detritic sediments. The mineralogical composition is clays and diagenetic zeolites, based on data obtained through x-ray diffraction (Table 2.1). According to Manega (1993), the Peninj tuffs are mostly reworked and no primary ash-fall or ash-flow deposits have been identified. Despite this, several attempts to date the tuffs have been made, including Manega's radiometric dating and paleomagnetic studies (Curtis 1967; Isaac and Curtis 1974; Thouveny and Taieb 1986, 1987).

In the Humbu BSC member tuffs, the most abundant mineral is illite, followed by analcime. As primary minerals, pyroxene and acmite-augite, and a plagioclase such as albite occur in the same tuffs. Secondary minerals are hematite, calcite, and kaolinite. The Main Tuff presents a similar composition, although it shows lateral variations. Illite and analcime are dominant minerals, followed by pyroxene and plagioclase and to a lesser extent, hematite. The USC member contains pyroxenes and albite, with abundant authigenic analcime, illite, smectite and hematite, but in contrast with the previous tuffs, there is a higher calcite cement component (18–57%), especially in tuffs T-3 and T-4. In T-4, there is a compositional change with an increase of smectite and pyroxenes (acmite-augite).

The first tephra from the Moinik Formation is TM1 in Type Section. This layer was also called Bird Print Tuff (Manega 1993). It is a white-grey tuff with bioturbation which has been interpreted as bird footprints. Its composition is similar to that of the upper Humbu Formation ash layers. Between that tuff and the overlying ones, a mineralogical change can be observed. Erionite increases dramatically, and different zeolites, like chabazite, gismondine, phillipsite or gmelinite, are documented. Erionite appears as acicular crystals or tangled fibrous masses. The amount of albite is highly variable and pyroxene disappears. Towards the upper tuff, the amount of erionite

TEPHRA	VOLCANIC MINERAL		AUTHIGENIC - SECONDARY MINERALIZATION												
				Detr.	PHYLLOSILICATES				ZEOLITES					OTHERS	
	Plagiocl.	Pyroxen.	K Felds.	Muscov.	Illite	Palygors.	Sepiolite	Caolinite	Analcim.	Erionite	Gismond.	Natrolite	Morden.	Haemat.	Calcite
UPPER TUFF															
MT-4															
MT-3															
MT-2															
MT-1															
T-5															
T-4															
T-3															
T-2															
T-1															
MAIN TUFF															
TBS-3															
TBS-2															
TBS-1															

Table 2.1 Mineralogical composition of the tephra layers from Peninj Group deposits based on XRD analysis. Note the sharp compositional change between Humbu and Moinik formations.

becomes so dominant that in some places it comprises 100% of the sample. Calcite, by contrast, is not always present. There are strong lateral variations in mineral composition because of local diagenetic factors.

Previous dating efforts have provided contradictory results. In some cases volcanics present dates that are more or less coherent, which has been crucial for the interpretation of the Peninj Group sequence so far (Figure 2.14). The following are dates obtained for the units underlying the Peninj Group: for the Sambu Lavas, dates obtained are 2.02 Ma (Isaac 1965, 1967; Curtis 1967), 2.0-2.5 Ma (Kapitsa et al. 1968), 1.9-1.77 Ma and 3.5 Ma (Isaac and Curtis 1974), and 2.99 Ma (Manega 1993). The dates reported by Foster et al. (1997) of 1.26-1.45 Ma appear to lie outside the previously obtained date ranges. For the Hajaro Lavas, dates obtained are 1.9-2.0 (Isaac 1965, 1967; Curtis 1967), 1.7 (Kapitsa 1968) and 3.5 Ma (Foster et al. 1997).

In the Peninj Group, previously obtained dates for each of its members are as follows (methodological problems are treated in each reference):

1) In the Basal Sands, only the first tephra (TBS-1) has been dated, providing an age of 1.7 Ma (Manega 1993). This tuff shows two eruptive episodes.

2) In the Main Tuff, dates cluster into three distinct groups, one around 1 Ma, another around 1.9 Ma, and an intermediate group represented by the Wambugu basalt at 1.4-1.6 Ma (Isaac 1965, 1967; Curtis 1967). The dates obtained are: 0.96-1.21 Ma, 1.0-1.5 Ma and 1.91 Ma (Isaac and Curtis 1974); 1.52 Ma (Howell 1972); 0.91-1.21 Ma (MacIntyre et al. 1974); and 1.56 Ma (Manega 1993, based on averaging two results: 2.28±0.06 Ma and 0.97±0.1 Ma). Magnetic polarity is normal and is associated with the Olduvai event (1.95-1.78 Ma, Berggren et al. 1995).

3) The USC have not been dated due to the poor quality of their tuffs. They show reverse paleomagnetic polarity (Thouveny and Taieb 1986, 1987).

4) At the bottom of the Moinik clays, the Intra-Moinik basalt shows a very coherent date of 1.35 Ma (Isaac and Curtis 1974) and 1.33 (1.13-1.41) Ma (Manega 1993). A little above this tuff, the Bird Print Tuff offers an age of 1.26 Ma (Manega 1993) and a basalt level which could correspond to Intra-Moinik provides an age of 1.28 Ma and 1.31 Ma (Manega 1993). Two normal paleomagnetic episodes seem to alternate with reverse ones.

5) The Upper Tuff has not been dated. It presents another alternation of normal and reverse paleomagnetic events (Thouveny and Taieb 1987).

6) The Peninj Group sedimentation finishes when the tectonic activity shaping the modern basin occurs. This is estimated to be between 1.15 and 1.2 Ma (MacIntyre et al. 1974).

The faunal data show chronological contradictions. Micromammals correlate with Olduvai upper Bed I (Denys 1987), whereas macrofauna suggest a correlation with upper Bed II to lower Bed III (Geraads 1987). This contradiction can be accounted for by the difference provenience of fossils. Microfauna were obtained in the BSC member (Denys personal communication), whereas most macrofauna comes from the USC member (Geraads et al. 1987).

Recently, A. Deino (in prep.) has sampled the complete sequence of the Peninj Group and has run some of Isaac and Curtis's (1974) samples, producing different results. The sequence including the Main Tuff and USC could represent the Cobb Mountain paleomagnetic event (1.25-1.10 Ma) and the dates of this 20 m deep deposit could be between 1.3 Ma and 1.2 Ma.

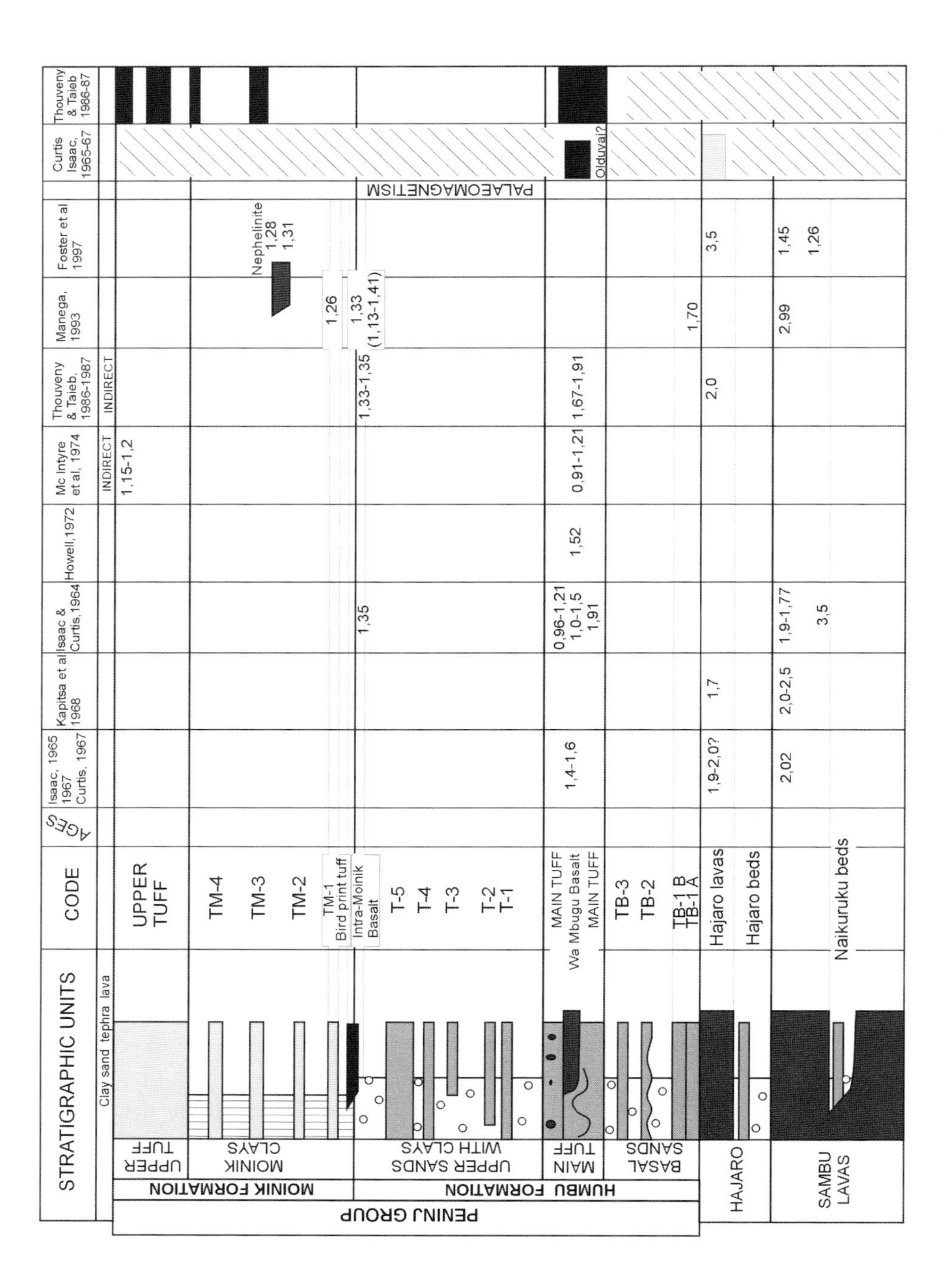

Figure 2.14 Preliminary data on Peninj Group geochronology based on various authors (see references in text and this figure).

The Peninj Group outcrops

As was mentioned above, the Peninj Group only outcrops in the western margin of Lake Natron. The deposits cover the Plio-Pleistocene basaltic lava strata formed by the Crater Highlands and by the Mozonik and Oldoinyo Sambu volcanoes. Most of the surface overlies the Moinik ramp, above the Sambu escarpment in an area of about 15 m x 4 km. The sediments show an erosive regression towards the west, creating a structural surface over the lavas. Thus, several Peninj Group sediments outcrop in the form of a thin line, a few meters wide and 20 m high, produced by the erosion of the upper part of the Sambu and Sanjan escarpments and related valleys. Stronger resistance to erosion by the well-cemented upper strata of the Peninj Group has produced the plateau exposures seen over the escarpment (Figure 2.9). The relief morphology is very abrupt, with gullies and wide valleys throughout the area.

Another area of exposure lies at the foot of the Sambu volcano. There, sediments outcrop from east to west and have a smooth relief. The area with the greatest exposure is found near the modern Peninj river delta. Over a surface of 3 km^2, sediments from the entire sequence are exposed. As has been previously suggested, based on the tilting of layers, this block probably is the upper section of a ramp parallel to the Moinik ramp, covered by alluvial and lacustrine sediments. The outcrops only show the sedimentary record of a small portion of the paleolake. There is no record of the limits of the basin, due to either erosion or continuous sedimentation.

Although we have primarily worked in the same areas as Isaac did, we have also surveyed other previously unstudied areas to better understand the paleoenvironment of the former basin. We have used our own nomenclature for each of these three areas: Type Section (Maritanane) remains the same, but Isaac's MHS is called South Escarpment and Isaac's RHS is called North Escarpment (Figure 2.15). Isaac (1965:Fig. 3) summed up the stratigraphic relationships among different fossiliferous outcrops in his early research, and our work has added detail in some areas (see Figure 2.16). Our analysis, which correlates all units by using all volcanic tuffs, has enabled us to reconstruct paleoenvironments. Ultimately this helps us understand the evolution of landscape and its influence on hominid behavior and site formation. The following are descriptions of each major outcrop area.

Type Section (Maritanane)

The core area of Type Section is called Maritanane, with Kamare and Kipalagu lying north and south of it (Isaac, unpublished notes; Mturi 1991). As noted by Isaac, Type Section is the thickest exposure of the Peninj Group sediments. This is because the lower units were deposited inside the semi-graben, following a tectonic structure similar to that which can be observed today in the Sambu escarpment. Type Section has also undergone less erosion than other areas due to its subsident position. This indicates that this fault has been present since the end of the Pliocene (2.0–1.7 Ma). Additionally, most of the volcanic layers in the Peninj Group can be found in Type Section: the Main Tuff and the trachytic Moinik tuffs are well-represented as well as the tephra in BSC (from TBS-1 to TBS-3) and in USC (from T-1 to T-5).

South Escarpment

This is a small area of outcrops. The BSC are missing in most of the area, only visible as remnants about 2 m thick near the Sambu escarpment. The Main Tuff is well-represented and lies nonconformably over the Sambu lavas, which appear very weathered. For the USC, only the upper part of the sequence can be found, in the form of

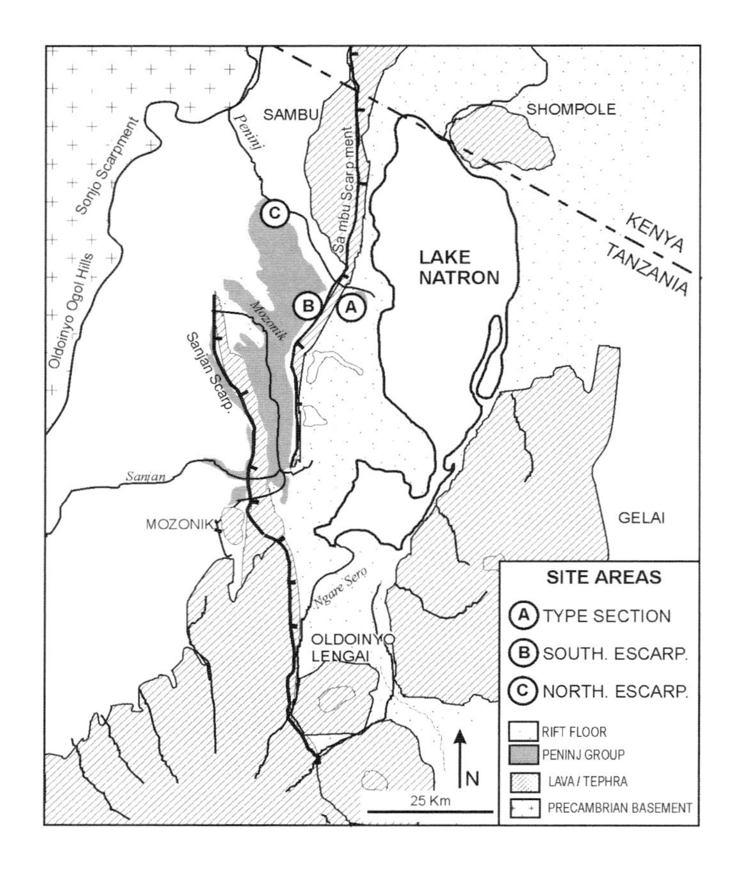

Figure 2.15 Working areas of the Peninj Research Project: Type Section; South Escarpment; North Escarpment.

sandy muds, clayish sediments and coarse sands which include the T-4 tuff, and the top of the Humbu Formation with part of the Moinik Formation. The latter is well-represented with lacustrine facies similar to those found in Type Section and southern exposures, including laminated clay, limestone and dolomite levels as well as trachytic tuffs and chert nodules (Figure 2.16).

North Escarpment. This area also presents a fairly complete sequence of the Peninj Group, but with much thinner strata from the Humbu Formation, which appears strongly eroded by the overlying Moinik Formation (Figures 2.15–2.16). The BSC are muddy and contain several edaphic levels. The Main Tuff contains several volcanic bombs and is very thin in comparison to other areas, but it also contains the carbonate with gastropods and ostracods. The USC contain several of the tuffs documented in Type Section and they show a very intense alteration. The sedimentary rate is probably much lower than at Type Section. The lowermost Moinik Formation channel-fill deposits contain lithic tools and faunal remains from Humbu Formation deposits due to erosion and reworking of these deposits in a later fluvial/deltaic

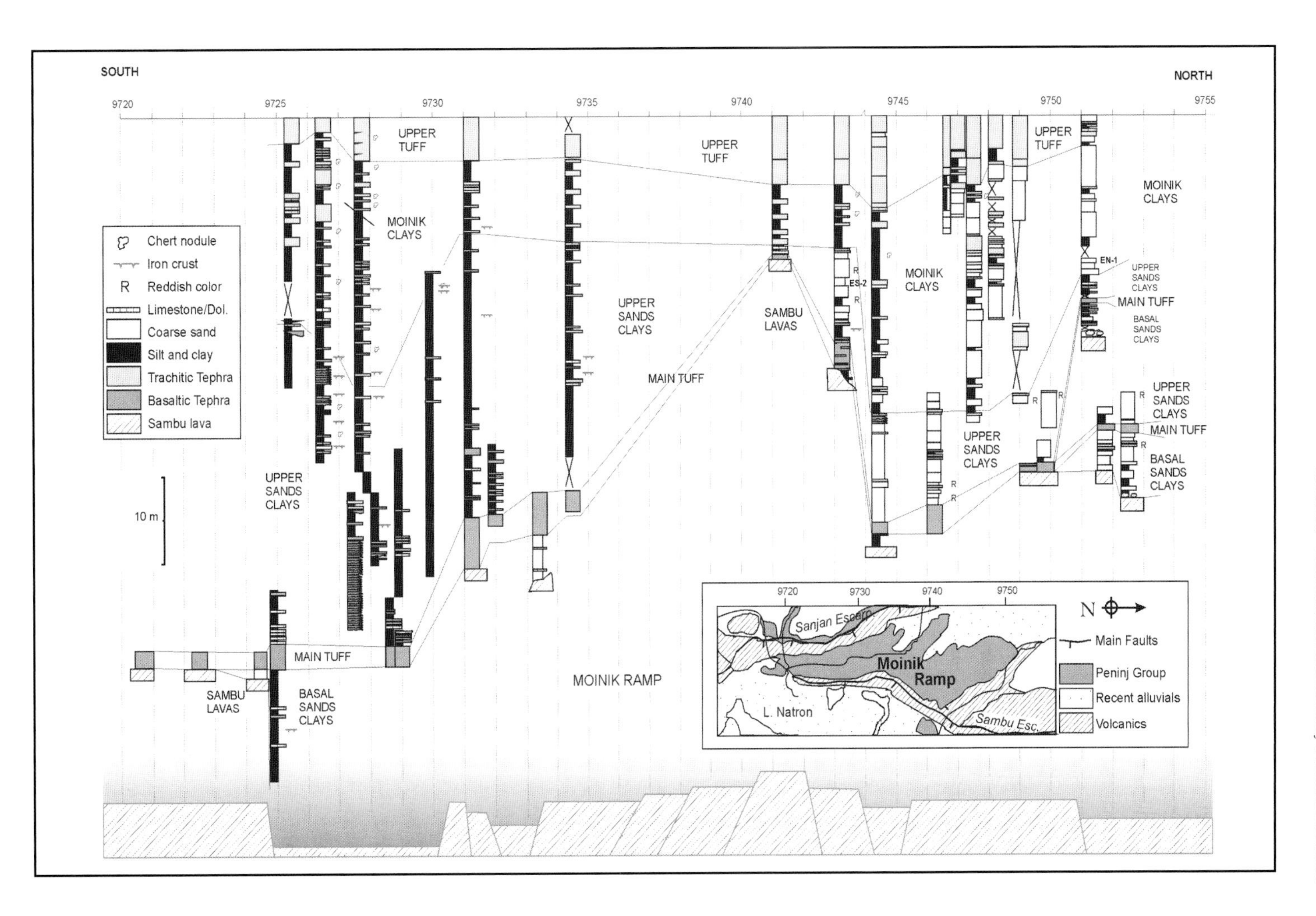

Figure 2.16 Stratigraphic correlation of some sections obtained along the Moinik Ramp from southern to northern exposures of the Peninj Group. Note the fine lacustrine deposits in the south compared with coarser alluvial-deltaic deposits in the north and east. Significant differences in sediment thickness suggest major synsedimentary tectonic movements.

environment. Only a few remnants of the upper tuff can be documented, as evidence of the typical trachytic volcanism of Moinik.

The largest outcrops in the Moinik ramp provide important paleoenvironmental information (Figure 2.16). However, from an archaeological point of view, they are of little interest, since sediments from both the Humbu and Moinik Formations are lacustrine and lack fossil vertebrate remains or lithics. However, the sediments show interesting reliefs. These paleoreliefs seem to follow structures of previous elongated horsts and grabens running N-S, as well as W-E faults with normal movements sinking southwards. The network of fractures that affected the Moinik ramp can also be observed between the North and South Escarpments. In places, the Sambu lavas can be overlaid by BSC, Main Tuff or USC, frequently acting as a marked paleorelief. Sedimentological differences and the absence of the lowest units of the sequence in neighboring areas indicate that the North Escarpment could have been isolated from the South Escarpment by an elevated tectonic block, as Isaac (1965) suggested. The most evident facies change between the South and North Escarpments can be observed in the Moinik Formation, since it switches from a strictly lacustrine environment to a deltaic environment as one moves northward. The outcrops on the margin of the escarpment show this transition (Figure 2.17). The preservation of the volcanic strata is distinct in each area, since lacustrine environments preserve strata whereas deltaic environments rework them.

Discussion

Endorrheic basins are highly sensitive to climatic changes, sedimentation rates, volcanism and tectonic movements. The basins of East Africa are textbook examples of this, preserving an excellent record of past changes to the landscape. The occurrence of a large number of basins inside the tropical belt, especially within the Gregory Rift, favors the development of ecosystems rich in fauna, which often contribute to the sedimentary record as fossils. The great diversity of sedimentary environments in relatively small basins means that taphonomic processes are quite diverse. The permanent tectonic and sedimentary activity and its temporal changes (velocity, rate, polarity, trend) since the end of Miocene have led to fossil burial and exposure. Consequently, a large number of paleontological and archaeological sites have been preserved in this area.

Today the Natron basin is filled with sandy sediments transported by rivers, which bring detritic sediments from the nearby Precambrian reliefs, volcanic escarpment shoulders and reworked quaternary sediments, delivered via alluvial cones and fans. Clays and trona-like evaporites accumulate in the middle of the basin, due to the lacustrine physical-chemical characteristics discussed earlier in this chapter. The study of modern deposits shows the sensitivity of these basins to climatic changes, since there are remains belonging to several Holocene deltas 20 m higher than the modern lake. Other geomorphological features such as hanging valleys, fluvial captures, inactive alluvial fans and stromatolite belts are also a clear sign of rapid tectonic activity and climatic changes. The trend towards aridity recorded in Africa over the past 6000–4000 years (e.g., Kutzbach and Street-Perrot 1985; Casanova and Hillaire-Marcell 1992; Damnati 1993) has resulted in the lowering of the lake level and has led to the dismantling of older deposits.

Isaac's stratigraphy of the Peninj Group units has been confirmed by our work. In fact, the sites we have selected for study, despite extensive and intensive survey of other parts of the basin, are those discovered by Isaac and his team. Our detailed stratigraphic work has enabled us to

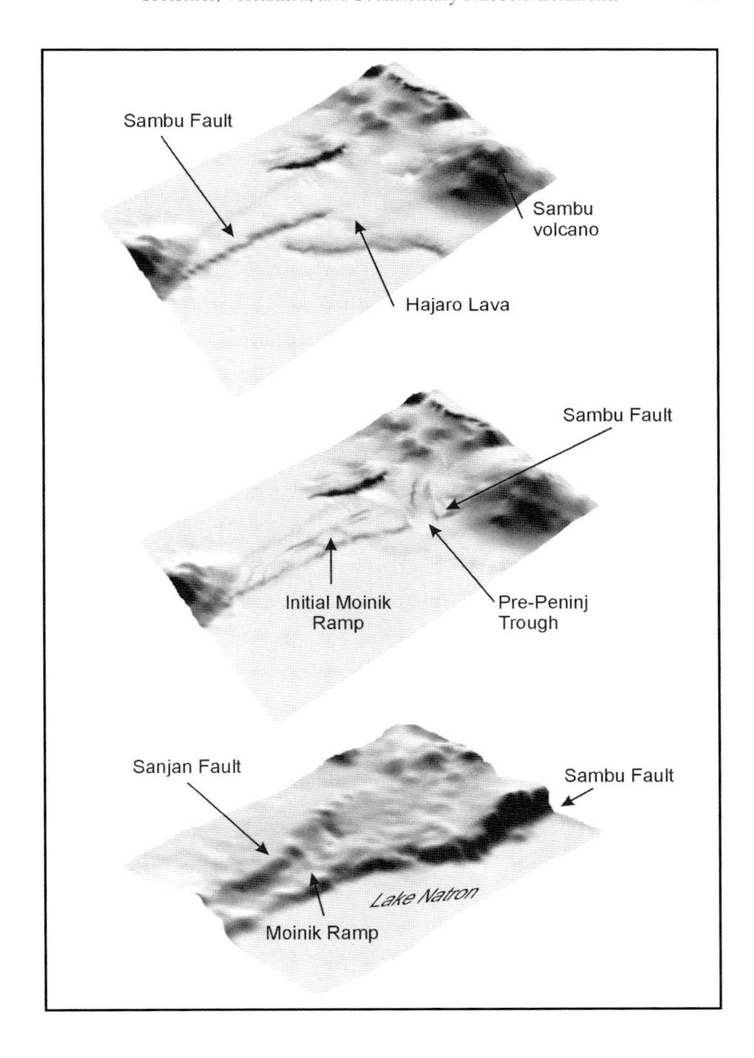

Figure 2.17 Evolution of Sambu escarpment and Moinik Ramp during the deposition of the Peninj Group and present. Southward subsidence along a steep ramp, as well as proto-grid faulting that gave rise to small horst-graben troughs and sub-basins, are inferred.

determine different tephra levels with which we have used to correlate different areas: the BSC contain three levels (TBS-1 to TBS-3) and the USC five (from T-1 to T-5). We have therefore confirmed that the area with the best representation of Peninj Group sediments is Type Section, as described by Isaac (Figure 2.18).

In the South Escarpment, most of BSC is missing as well as a substantial amount of the lower section of the USC. This suggests two tectonic movements: one pre-Hajaro (described by Isaac) and another one post-Main Tuff and before the formation of Tuff 4. This created the erosion of materials over the Sambu Escarpment. Other minor environmental changes have been determined from the stratigraphic study, but it is more difficult to distinguish if they are the result of tectonic or climatic changes. The upper section of the USC is also missing in the North Escarpment, due to intensive erosive processes. The Humbu Formation appears there in a very thin sequence compared to its representation in Type Section. Ramp tectonics, causing differential sinking towards the south, partially explain this process.

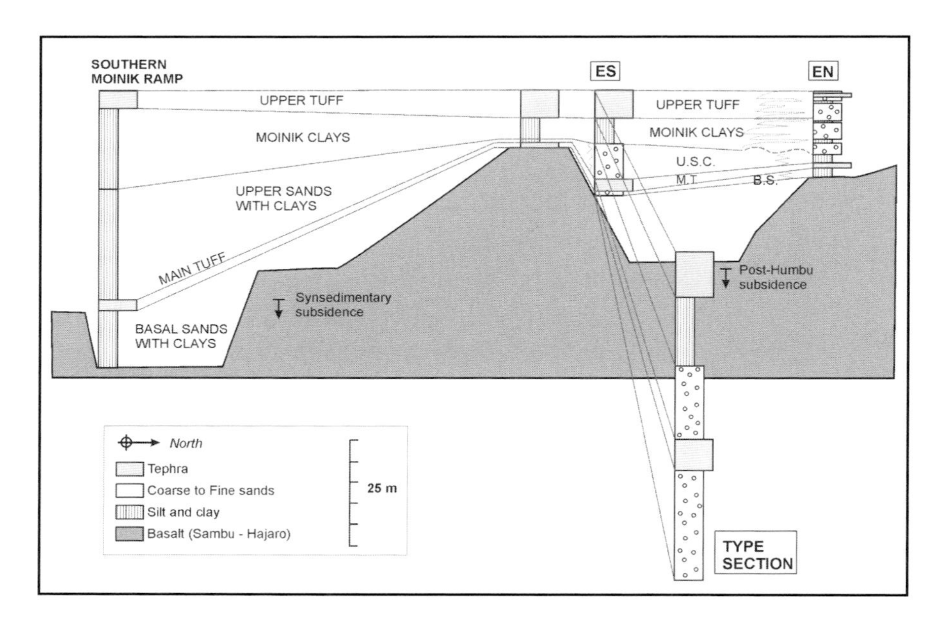

Figure 2.18 Schematic distribution of facies and stratigraphic units along the main Peninj Group outcrops. Note the greater sediment thickness in most subsiding areas (south Moinik Ramp and Type Section) and erosive processes between the Humbu and Moinik Formations in the higher northern areas. ES = South Escarpment; EN = North Escarpment.

The facies distribution suggests the sedimentary independence of North Escarpment from the other areas with outcrops, as Isaac (1965) already suggested (Figure 2.17).

Chronological context is essential to understanding the stratigraphic context of archaeological sites. As noted earlier, previous geo-chronological dates did not accurately determine the precise age of the most fossiliferous layers, but rather gave only a general idea of their age; recent dating efforts have improved on this. Biostratigraphical data are scarce and sometimes contradictory. Following both radiometric and paleomagnetic results, the beginning of the Peninj Group is suggested to be around 2.0 Ma with the deposition of the BSC member in Humbu Formation; this was deposited up until the beginning of the Olduvai paleomagnetic event (1.95 Ma following Breggren et al. 1995). An intense period of effusive volcanic activity that gave rise to the Main Tuff member could last from 1.9 to 1.75 Ma, according to the Olduvai event chronology (which ends at 1.78 Ma) and correlating with a lake retreat period detected at Olduvai and Lake Turkana around 1.75 Ma (Hay 1976; Bonefille 1995). Alternatively, as recent data suggest (Deino, in prep.), the Main Tuff could be dated to 1.2 Ma and could represent the Cobb Mountain paleomagnetic event.

The USC deposition period lasted up to 1.2 Ma, spanning ca. 400,000 years of volcano-sedimentary deposition, distributed in at least four pulses, separated by erosive processes or lying in paraconformity with one another after lake flooding. The end of the Humbu Formation (at the end of the USC deposition) is probably due to a sharp southward sinking of the Moinik tectonic ramp and/or climatic changes bringing an increase in precipitation. The overlying Moinik Formation shows paraconformable contact with Humbu Formation in the south and lies unconformably in the north. Moinik Formation sedimentation may have lasted from 1.2 to 1.15 Ma. The lower part of the formation has shown coherent radiometric results, but both radiometric and paleomagnetic data fail to show clear, definitive results in the middle and upper parts of the sequence (see references cited in the discussion of chronology above).

Previous paleoenvironmental interpretations described a lacustrine and para-lacustrine environment in a shallow basin during the deposition of the Peninj Group (Isaac 1965). Following Isaac (1965, 1967) and Icole et al. (1987), depositional facies in the Humbu Formation are fluvio-deltaic in a freshwater lake. The lacustrine environment where ash-fall deposits of the Main Tuff member were formed could signify a shift towards more humid conditions and a lake-level rise of around 33 m (Isaac 1965). On the other hand, the Moinik Formation must have been deposited in a stratified, highly alkaline lake in a dry climate similar to the conditions of the current Lake Natron. Calcareous deposits may have been precipitated in lakeshore or deltaic areas as occurs today (Manega and Bieda 1987).

Taking all of these data into account, we can summarize that sites are formed or preserved in environments with a high rate of sedimentation, especially during depositional moments in which sediment supply is increased due to tectonic or climatic events. Channel-fill or sheet-flow deposits, in prograding sedimentary environments following lake retreat, are the most common fossiliferous facies. On the other hand, periods of low sedimentation would create climate and landscape conditions unfavorable to bone preservation, since lower sedimentation rates would mean a long period of exposure on dry ground. Other more lacustrine (mudflat) environments rich in fossils that have been described in different rift basins (such as Olduvai) have no equivalent at Peninj.

A tentative comparison in climatic data has been carried out using our data and the oxygen isotope stage curve developed by Shackleton et al. (1995), assuming at least partial equivalence between warm periods at high latitudes and increased precipitation in tropical and equatorial areas (Figure 2.19). The correlation does not fit well, and previous chronological interpretations show that a great number of climatic changes have not been recorded or registered in the Peninj sequence. This approach shows that during the formation of the BSC member, a high lake level followed by a lake retreat and erosion could be related to warm and cold periods respectively, but we suggest a tectonic origin for this change. During the deposition of the USC, a great number of sedimentary changes have been recorded which can be attributed to factors other than climate change.

Following the previous interpretation, the beginning of the Moinik Formation (Manega 1993) coincides with the beginning of warmer conditions at high latitudes, which could signify increasing humidity in the area. New dating and stratigraphic research make us attribute this change to tectonic movements, but the coincidence of both factors could account for the major environmental shift recorded between the Humbu and Moinik Formations. After that, five minor climatic oscillations could be related to a similar number of graded sequences found along the Moinik Formation between North Escarpment and South Escarpment (Figures 2.16–2.17). Within the Upper Tuff member, a change to drier conditions is recorded, showing two smaller pulses of flooding.

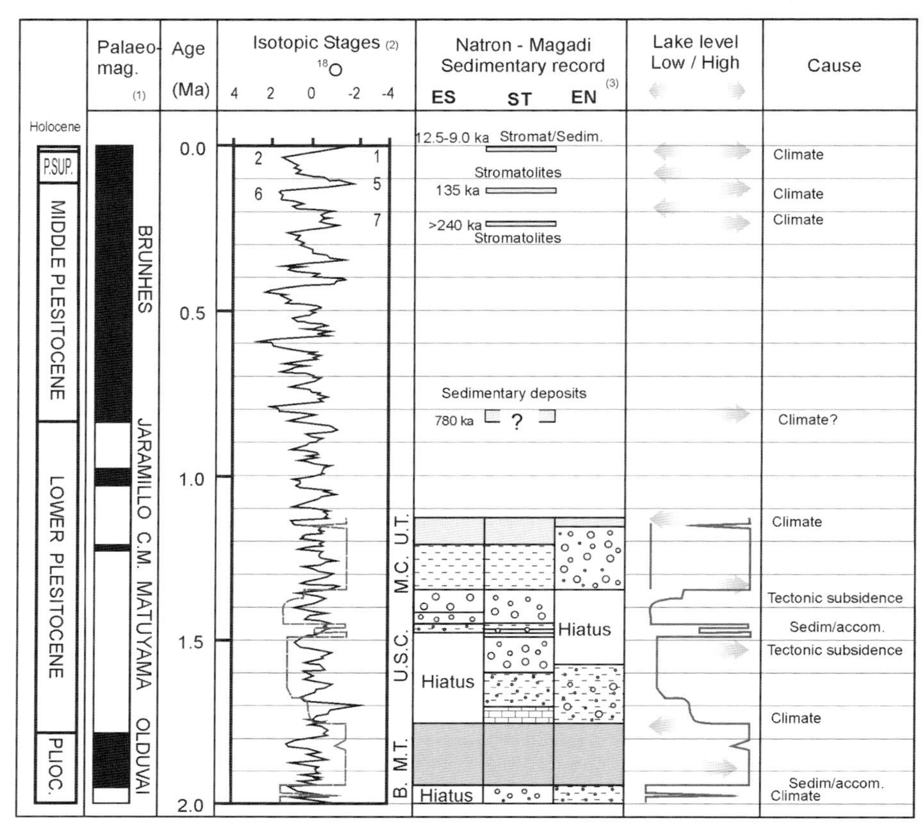

Figure 2.19 Tentative correlation between isotopic warm stages based on Shackleton et al. (1995) and lake-level changes deduced from the sedimentary record, using the available geochronological data from the Peninj Group deposits. An interpretation for lake-level changes is also included.

In summary, the Peninj Group preserves a record of the evolution of an East African endorrheic basin over a period of more than one million years, and its sediments reflect major changes in the landscape due to tectonics, volcanism, sedimentation and climate changes. These changes can be linked, particularly using the Peninj Group deposits in Type Section (Maritanane), to the environment in which the archaeological remains at Peninj were deposited. In the next chapter, we explore the geology of this specific section of Peninj, in an attempt to reconstruct the environment during the hominid occupation of the area.

Endnote

1 Also referred to as Maritanane by Isaac (unpublished), Maritanane is the local Maasai name for most of the area where there are horizontal exposures of Humbu Formation sediments in Type Section.

3

The Peninj Group in Type Section (Maritanane): An Analysis of Landscape Evolution

Luis Luque, Luis Alcalá, and Manuel Domínguez-Rodrigo

Introduction

Isaac (1965, 1967) defined the Peninj Group in some exposures outcropping along the Peninj river delta at the foot of the Sambu fault escarpment in what he called Maritanane (also known as Type Section; Figure 3.1). This 1.5 km^2 area consists of a tectonic block that seems to form part of a partially buried tectonic ramp subsiding southward and parallel to the main Moinik ramp. This block is deeply fractured due to a dense network of faults and is crossed by a number of gullies with seasonal streams. The relatively loose sediments and the tectonic movements gave rise to a large surface of vertical walls and plateaus rich in Peninj Group sediments where sites were exposed (Figures 3.2-3.4).

These outcrops enable the study of both the underlying Sambu lavas and Hajaro beds as well as all the members of the Peninj Group deposit. Due to intense faulting, it is not possible to observe a complete continuous sequence, but correlations have been made based on neighboring exposures. The maximum observed thickness in one section is about 25 m. Type Section not only shows the greatest thickness of the fossiliferous Peninj Group, but also shows later deposits in this area (Figures 3.3, 3.5). At least two different phases of Middle to Upper Pleistocene deposits have been recorded plus some deltaic-lacustrine Holocene clay beds that are widely exposed in several places. All of them lie unconformably over the Peninj Group

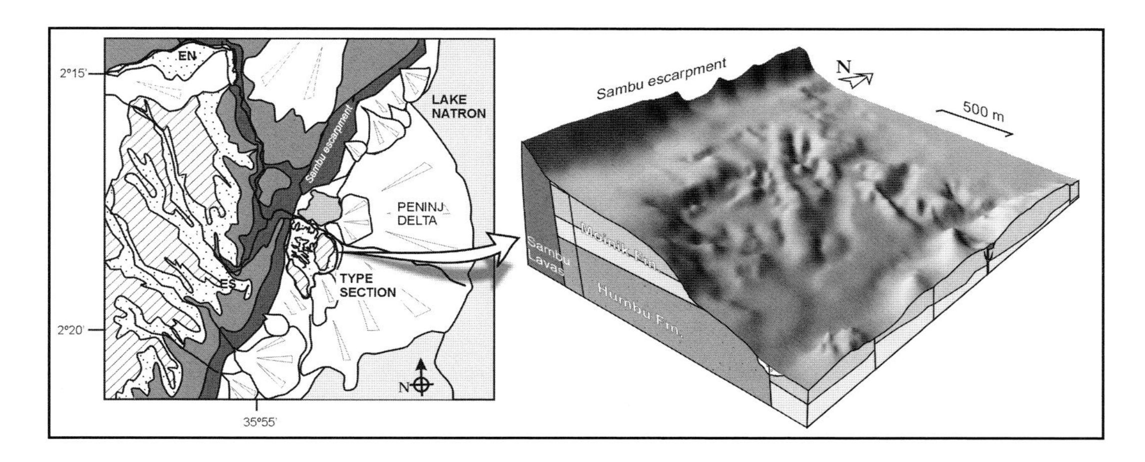

Figure 3.1 Type Section outcrops at the foothill of the Sambu Escarpment, surrounded by sands and gravels of the modern Peninj river delta.

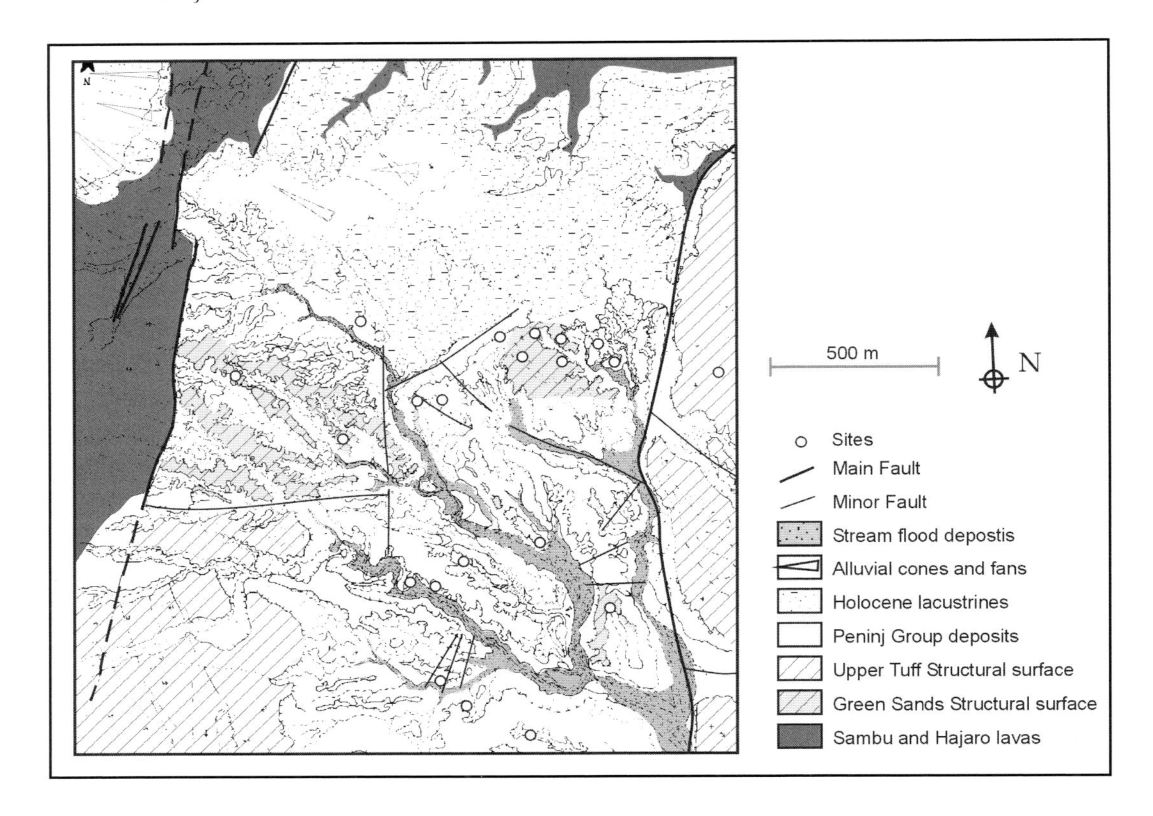

Figure 3.2 Map of Type Section, a 1.5 km² area. Deep gullies are incised by the action of seasonal streams. This map shows the distribution of some of the most important sites recorded in the area.

deposits. Previous shorelines of higher lake levels are recorded as a thin rim of stromatolitic crusts and bioherms (Casanova and Hillaire-Marcel 1986, 1987).

Isaac described the important role of the Sambu fault during the formation of the Peninj Group in Type Section. This fault sank the half-graben by about 30-40 m before the deposition of the Peninj Group began. In fact, the Hajaro lavas and the Hajaro beds are restricted to the inner part of the half-graben. The BSC member did not reach beyond the escarpment shoulders and did not infill the lower part of the tectonic structure. The activity of the Sambu fault lasted at least until 2.0 Ma. For this reason, the only outcrops where the complete sequence can be observed are in the downthrow block as well as in the southern part of Moinik Ramp. Fossiliferous sediments are restricted to the northern areas, where Type Section is located. The displacement of the fault is the same as in the BSC member (in thickness). No continuous record of this member can be observed, but it is about 30 m thick.

Peninj Group Units in Type Section

We have expanded Isaac's (1965, 1967) original stratigraphic description with additional detail and interpretation, taking into account the spatial and stratigraphic distributions of paleontological and archaeological sites (Figures 3.3-3.4).

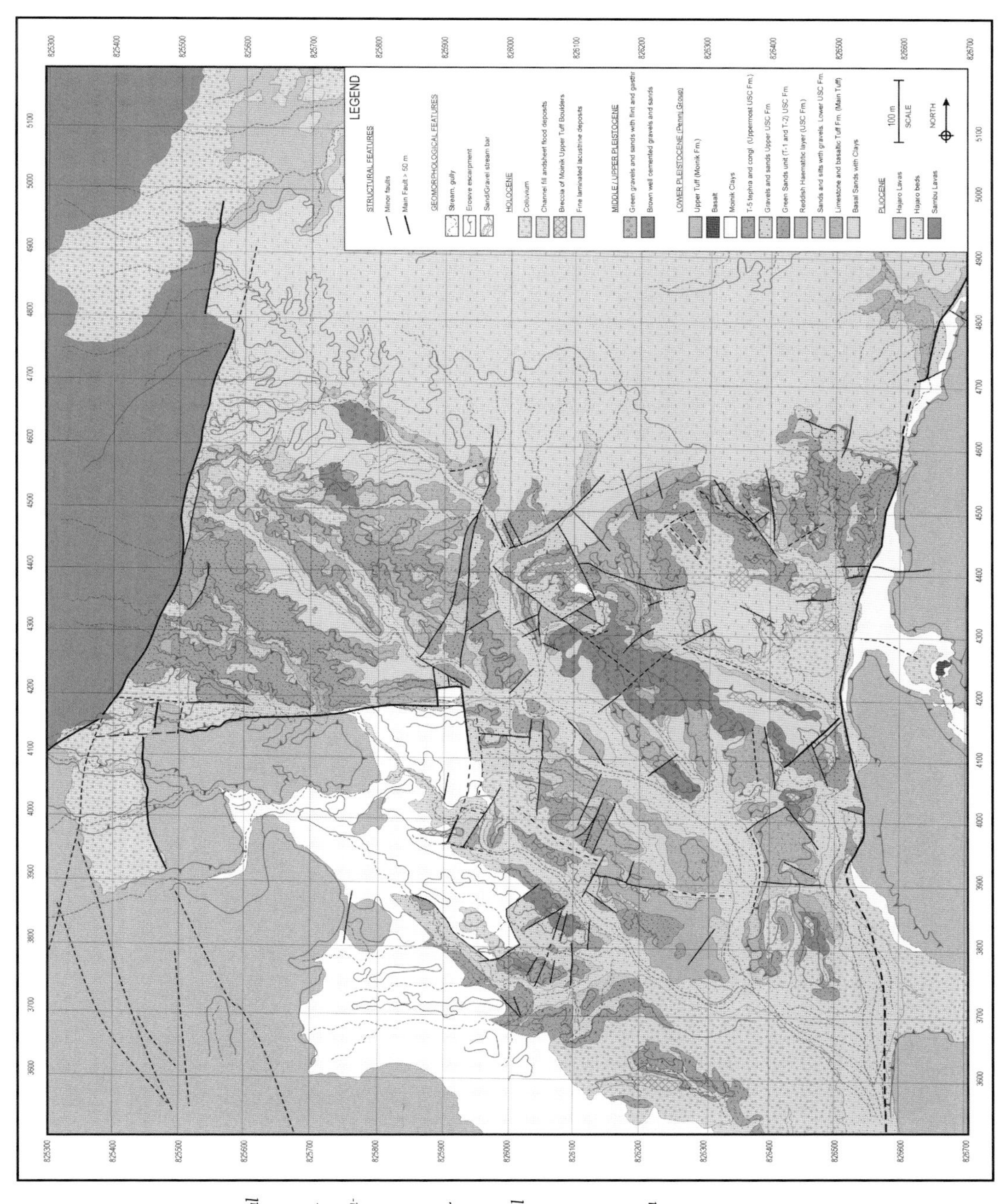

Figure 3.3 Detailed geological map showing the main sedimentary units in each member of the Humbu and Moinik Formations. Other younger units are also shown, as well as the main tectonic features. See Figure A.1 in the Appendix for a color version.

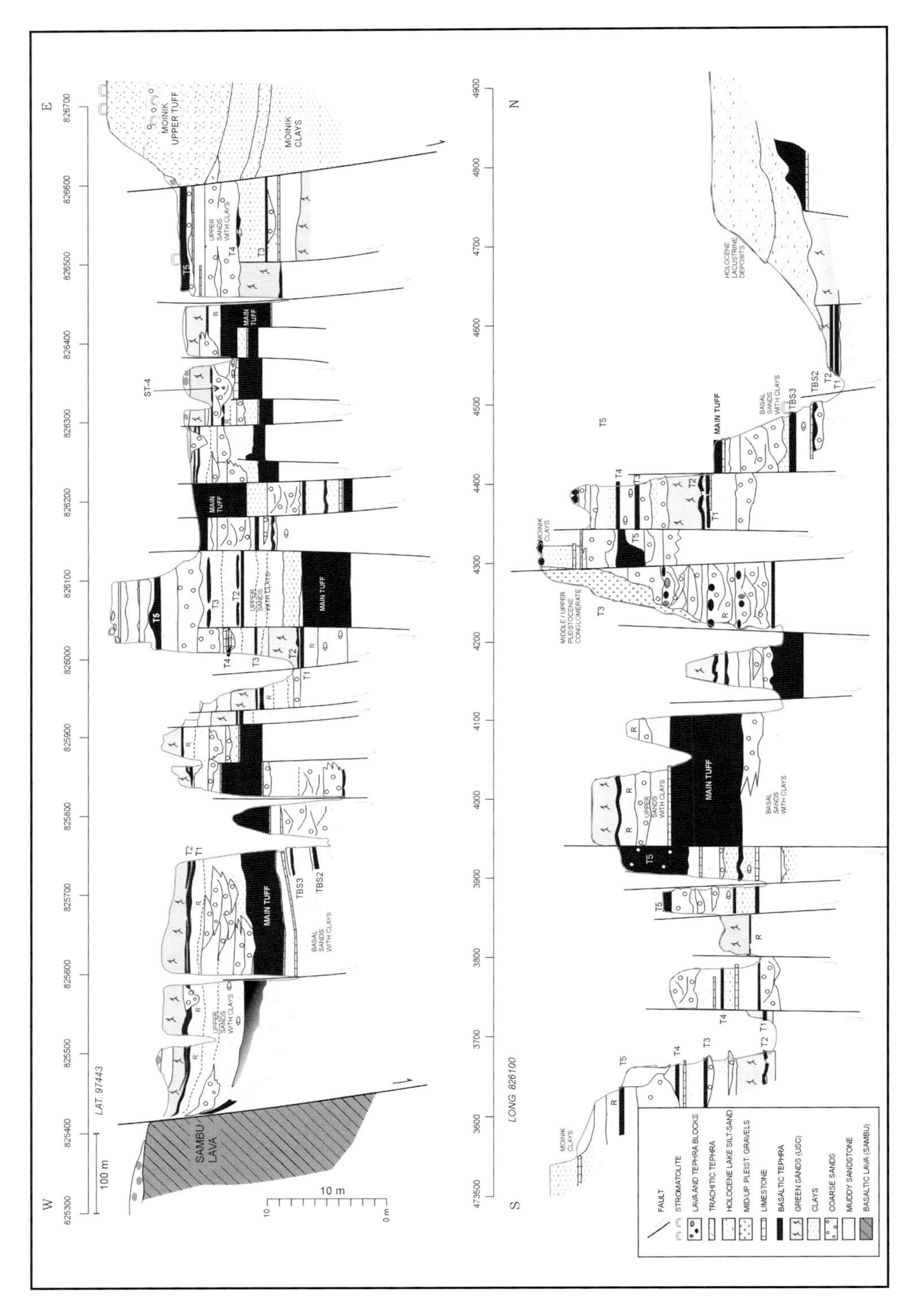

Figure 3.4 North-South and West-East sections through the Type Section exposures. Vertical scale is 10 times greater than the horizontal scale. Note the importance of both post- and synsedimentary tectonics.

Mineralogical analyses are widely concordant with those previously made by Isaac and by Icole et al. (1987). Tephra unit composition has already been shown in Table 2.1 in Chapter 2. The following is a stratigraphic description of the units (Figures 3.6–3.7).

A. The Humbu Formation

1. Basal Sands with Clays member (BSC). As Isaac (1965, 1967) previously suggested, the deposition of BSC occurs soon after the Hajaro lava flooding at 2.0 Ma. A fresh surface outcropping in the Type Section (Figure 3.5d–e) seems to confirm that idea. The first deposits filled in the irregularities of the paleorelief, which were up to 1 m deep. These are coarse and brownish sands, poorly sorted, which eventually incorporated some lava boulders. They were covered by two sandy tephra layers jointly called TBS-1, 0.15 and 1.1–1.3 m thick respectively, which were probably reworked and are grading upward. These volcanic layers contain some blocks of lava and reworked cemented tephra boulders, and are separated by a thin layer of sandy silts. The tephra layers sometimes appear reddened because of their hematite content. In southern areas, Manega (1993) dated these tuffs contaminated by argon gas to an age of 1.70 ± 0.02 Ma. He noted two eruptive stages and high reworking of the volcaniclastic deposit.

Sediments overlying TBS-1 consist of 5 m of poorly-sorted, reddish, muddy coarse sands. Some lenses of coarser well-cemented sands and gravels are interbedded as channel-fill deposits. Isolated large, sand-filled rootcasts, as well as limestone nodules, are common. In some places, large accumulations (about 1 m thick) of blocks of lava and tephra in a sandy matrix are found, due to debris produced near the Sambu escarpment. These deposits are sometimes highly edaphized and reddened.

Tuff TBS-2 overlies these reddish sands. This is a 0.15–0.3 m thick, orange, wavy well-laminated tephra layer. In some places, this layer is strongly altered or weathered and it is usually overlaid by coarse, well-cemented sands. Not far away, these sands turn into sandy limestone. The wavy structure may be a seismic feature, perhaps of volcanic origin in a swampy or wet environment. There are few outcrops showing the sediments overlying TBS-2. Where these are found, they mostly consist of muddy sands rich in limestone nodules at the base and showing interbedded lenses of coarse sands and gravels, as well as burrowed clays which are sometimes folded. Their thickness rarely exceeds 1.5 m.

Tephra layer TBS-3 overlies these sediments. It is usually thick (0.45 m) but thins laterally to a few centimeters in thickness. It is orange and finely laminated. This ash-fall deposit is overlaid by stromatolite-like algae, limestone, and green clays. Ostracods are very common. Above the TBS-3 tephra, there are 4 m of muddy coarse sands interbedding with some white coarse sands and gravels (channel-fill lenses), reaching up to the next member of the Humbu Formation. These muddy sands are sometimes topped by more clayey sediments.

In the main part of Type Section, the upper part of the BSC member has been deeply eroded and overlaid by more homogeneous, coarse channel-fill deposits. This represents a long sequence of upgrading sandstones and muddy sands showing crossed lamination and southerly flow features that resemble a deltaic depositional environment. These channel-fill deposits show scattered fossils distributed as clasts in the channels. Among these, a Paranthropus mandible was found during Isaac's first field season (Leakey and Leakey 1965). This later deposit seems to correspond to a second period of sedimentation after an erosional moment during the formation of the BSC member. At the

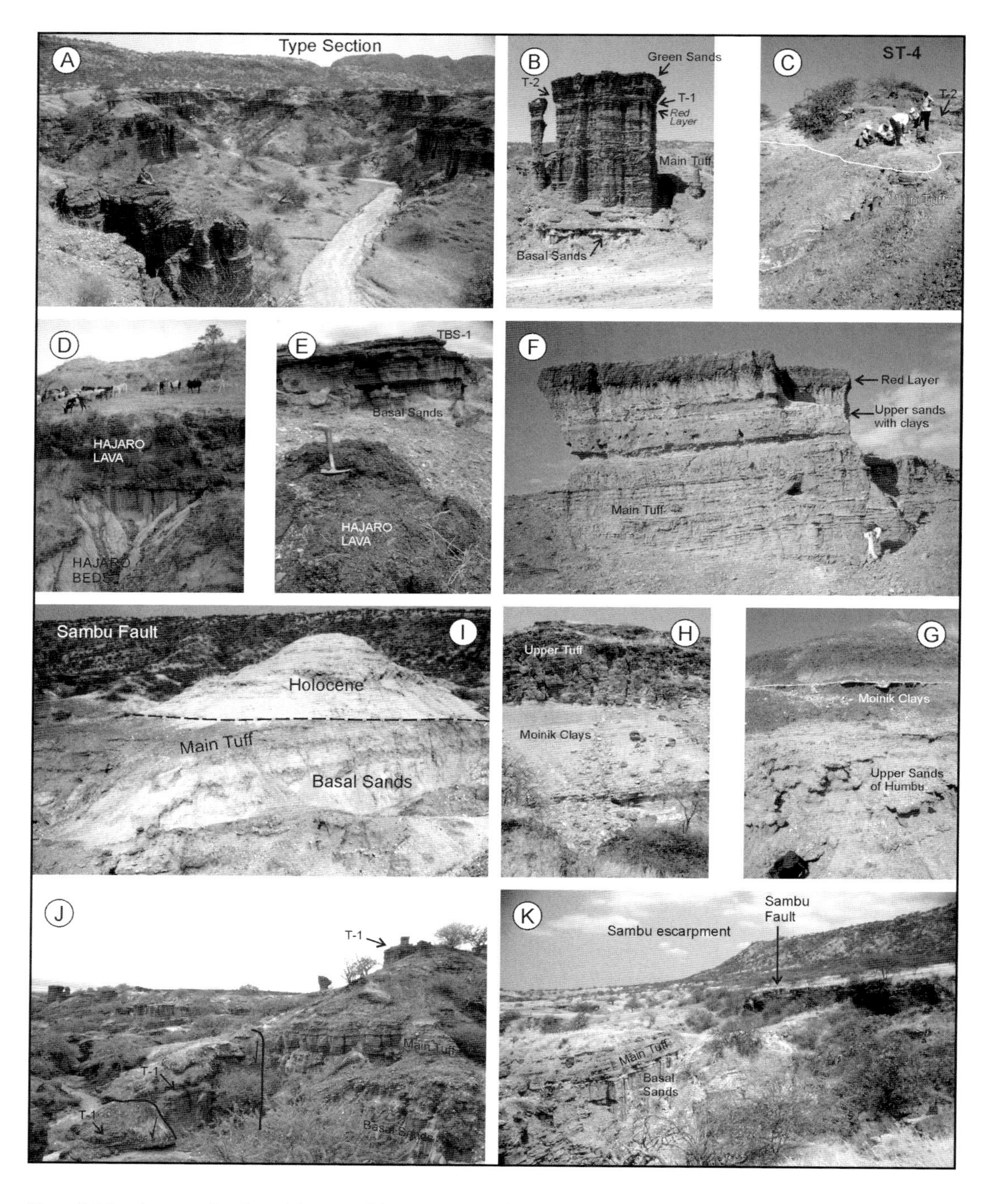

Figure 3.5 Landscape and geological features of the Type Section: A) gullies incised in Pleistocene deposits (mainly USC member); B) vertical exposure of the Humbu Formation; C) ST-4 site at the beginning of the works (1995); D) Hajaro beds (sands and lava) below Holocene deltaic-lacustrine deposits; E) fresh surface of Hajaro lava below BSC member, including TBS-1 Tephra; F) USC

(Figure 3.5, left, continued) exposures showing a structural surface on the well-cemented Red Layer of Humbu Formation; G) paraconformable contact between Humbu Formation sandstones and Moinik Formation clays; H) Moinik Clays and Upper Tuff contact at the top of the Peninj Group; I) nonconformable contact between Holocene fine lacustrine-deltaic deposits overlying tilted Humbu Formation sedimentary units; J) intense faulting of the Type Section (note position of the T-1 tephra layer in the landscape); K) main faulting at the foothill of the Sambu escarpment, and contact between Peninj Group deposits and Sambu lavas.

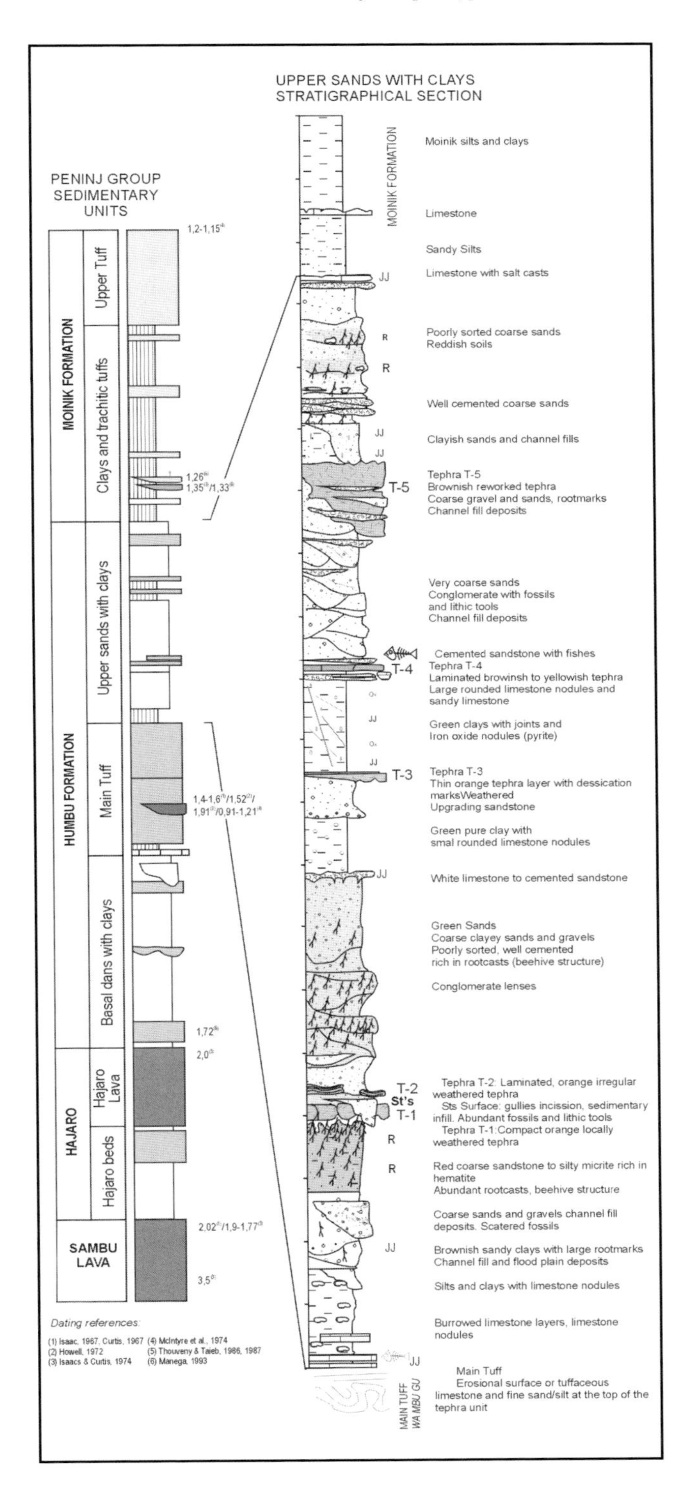

Figure 3.6 (right) Stratigraphic sections of the Peninj Group and Upper Sands with Clays (USC) member of the Humbu Formation. Unit thickness is approximate.

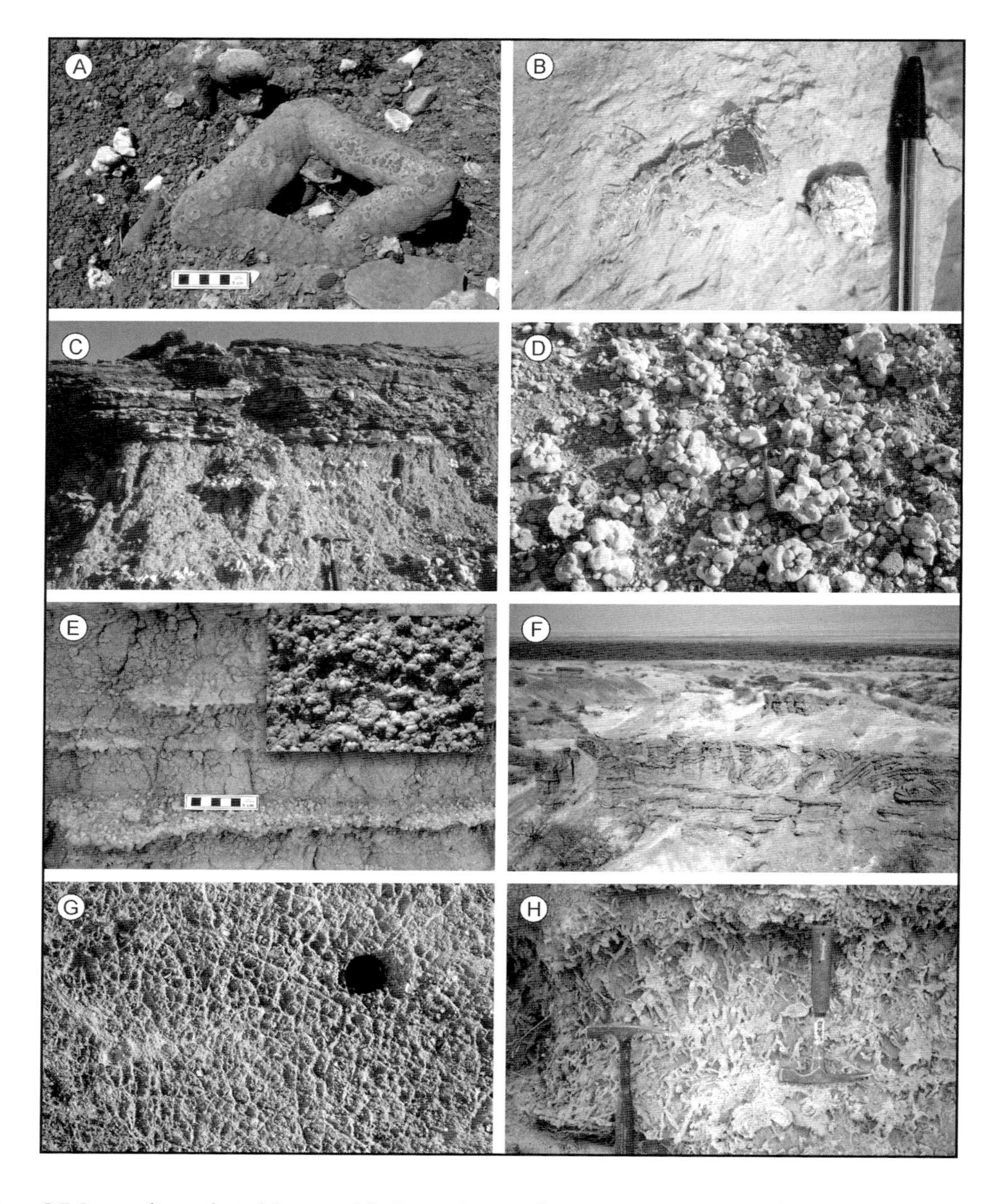

Figure 3.7 Some sedimentological features of the Peninj Group in the Type Section exposures: A) stromatolite growth around branches and boulders in a paleo-shoreline; B) fish remains in fine deltaic-lacustrine Holocene deposits overlying the Peninj Group; C) magadite chert nodules interbedded in erionitic tephra (above) and laminated clays (below) of Moinik Formation; D) large accumulation of magadite chert nodules; E) thin gastropod- and ostracod-rich limestone layers, intercalated between the muddy tuffaceous MT member of Humbu Formation; F) folded layers of tephra, resulting from slumping after the deposition of the Main Tuff member; G) bee-hive structure in a horizontal section of the Green Sands Unit of the USC member; H) lateral view of the intensely root-marked structure of the Green Sands Unit.

top of this sequence, Denys (1987, and personal communication) sampled the sediments where micromammals were obtained.

2. Limestone and Basaltic Tuff member or Main Tuff member (MT). The Main Tuff (MT) volcaniclastic layer is the main correlation level of the Peninj Group deposit. Previous researchers made great efforts to date this member because it is located in the middle of the Peninj Group (see Chapter 2). MT is an orange tephra layer found in many outcrops that varies between 1 m and 4 m in thickness. Multiple tephra layers overlie a thin (2-5 cm) and continuous limestone level full of small gastropods (Gabbia and Isaac 1965) and ostracods. This feature characterizes the member and distinguishes this layer from other tephra layers of the Peninj Group. Overlying the limestone, there is usually a 0.1-0.2 m brown clay layer. The tephra can be subdivided into two layers: the lower one is homogeneous, finely laminated, with light orange lapilli; the upper one is darker and includes coarser pyroclasts. In places, thin layers of limestone rich in gastropods, ostracods and fish are found (Figure 3.7e). In many areas of Type Section, the MT member shows folding and deformation attributed to slumps (Isaac 1967) that could be seismic in origin (Figure 3.7f).

The origin of the pyroclasts is outside of Type Section (as Isaac originally suggested). The old vent is located in the southernmost exposures of the Peninj Group, where a thick interbedded lava flow occurs. This lava flow, called Wambugu basalt, has been directly dated (see references in Chapter 2) as well as analyzed for paleomagnetic dating (Isaac 1967, Thouveny and Taieb 1986). The MT member was deposited in a lacustrine environment, as evidenced by limestone and rich aquatic organisms, as well as by deformation features. In some areas, the top has been eroded, giving rise to a paleorelief that was later filled in with fine to coarse sediments. Tectonic movements through synsedimentary faults are also evidenced by changes in thickness and distribution. Some sandy layer interbedding has also been found. All these data suggest that some lake-level change cycles as well as different eruptive and tectonic events occurred during the formation of the MT member.

3. Upper Sands with Clays member (USC). The USC member is the most important from an archaeological point of view due to its richness in sites and its paleoenvironmental record (Figures 3.5c, 3.6). In the Type Section, USC is the richest unit both in lithics and fossils of the whole Peninj Group sequence (Figure 3.5a–c, f). This unit consists of sands and clays divided into at least four different depositional periods conditioned by tectonic and climatic or paleogeographic pulses. These sediments show five volcaniclastic layers (tephra T-1 to T-5) that are useful to correlate different outcrops in Type Section and other areas.

After the deposition of the Main Tuff, the paleorelief was filled in by sediments originating at the rift shoulders. The more deeply eroded areas were filled with coarse sediments. When the paleorelief reached a nearly flat surface, a succession of fine sands, silts and clays as well as limestone layers covered the volcaniclastics. These limestones consist of highly burrowed white layers, 0.1-0.2 m thick, of lacustrine origin, suggesting a continuation of the lake environment. Sometimes, fine sediments are highly tuffaceous. This para-lacustrine environment is progressively filled with sediments and alluvial progradation is clear, with coarser sediments increasing along the sequence. Carbonates are common, but usually appear as pedogenic nodules that can form calcretes, due to changes in the water table. Well-cemented rootcasts are also common. As alluvial progradation continues, sediments become coarser and carbonates scarcer. Lenses of coarse sediments like sands or gravels are interbedded as channel-fill deposits. This coarse grading unit is about 1.4 m thick from the top of the MT member, but its thickness is highly variable. In fact, the uppermost contact is unconformable and erosional, suggesting a sharp change in depositional environment. This unit can be reduced to 0.5 m in thickness in some areas.

Overlying this erosional surface are 1–2 m of poorly-sorted, coarse muddy sands, rich in limestone-filled rootcasts. This layer consists of a sequence of channel-fill deposits with a highly erosive base and crossed lamination. Sands appear homogeneous and white. Fossils are generally scattered, but are occasionally concentrated in some channels. These fossils are distinguished by their white color and good cortical preservation. Rootcasts and burrowing are common, as are some concentric spherical sandy structures. These sandy structures are similar to others attributed to fish nests in sedimentary basins of East Africa (T. Cerling, personal communication) but could also be similar to mud mounds of seismic origin.

One of the most characteristic levels of USC is found overlying these coarse channel-fill deposits. This is a 0.3–1 m thick sandy deposit, red and rich in hematite (Figure 3.5e). It is widespread and clearly differentiated in Type Section. Sands are muddy, well-cemented, and usually show intense root-marking and burrowing. It is hardened at the top, showing clear features of edaphization. When it is clayey, it is strongly reddened and rich in rootcasts, creating a beehive-like structure. Fossils are generally rare in the red layer, but one of the few articulated carcasses from the Peninj Group sites has been found here.

The first ash-fall deposit of USC, called T-1, overlies this red layer. It is generally 0.2 m thick, but in places it can reach 0.4 m. It consists of fine, orange, well-cemented ashes that can sometimes include a few coarse lapilli. It is laminated and its degree of preservation changes over a few hundred meters. T-1 is usually compact and hardened with desiccation cracks, as well as partially isolated in rounded blocks and eroded by channel incision, where it is filled with coarse sands and gravels (Figure 3.8). This suggests that T-1 acts as a paleorelief that has been either eroded or preserved in specific locations, depending on the topography. Most of the archaeological sites found in Peninj occur on the paleosurface overlying T-1. This tephra layer was covered by a few centimeters of fine to coarse sediments rich in fossils and lithic remains, as well as by more than 1 m of coarse upward-grading sands in the channels. This is the surface of the ST Site Complex, which will be discussed in Chapter 4.

T-1 and the ST surface are overlaid by an ash-fall deposit called T-2. In some places, T-2 directly overlies the T-1 tephra, but in other places it covers ST sediments. In areas previously cut by channels, T-1 is more rarely found, probably due to the continuous water flow in these topographically lower areas. T-2 is easily recognizable due to its lamination; sometimes it is diffuse, mixed with overlying sediments, and sometimes it has been mechanically deformed showing a wavy structure. T-2 is strongly weathered and contains reworked sands, but larger pyroclasts have also been found among the ashes, making it useful for dating.

Another characteristic unit of USC in Type Section is the Green Sands unit, overlying T-2. It consists of 1.5–5 m of widely exposed, coarse, greenish muddy sands. These sediments are intensely altered by rootcasts that sometimes show a beehive structure (Figure 3.7h). These sandy deposits contain a sequence of highly erosive channel-fill deposits, similar to those of the underlying red layer and the surface of ST,

Figure 3.8 A) Sedimentary record of the paleosurface of the ST Site Complex shows erosion and weathering of the T-1 tephra layer and sedimentary infill of the paleorelief, later coated by T-2; B) small-scale field example and interpretation of the erosional surface that gave rise to the paleorelief of the surface of the ST Site Complex (note intense weathering of the T1 tephra layer on the right and blocks of tephra fallen into the incised paleorelief, later refilled by sediments, limestone nodules and rhizoliths; C) T-1 tephra layer partially separated in blocks.

which deeply eroded previous deposits in specific places. Gravel deposits are also commonly interbedded between muddy sands. The clayey matrix is rich in analcime. At the top, sands are more massive and root casts are more scarce. Throughout the deposit, both fossil and lithic remains are documented but are scattered, only occasionally concentrating in one spot.

A sharp depositional change is recorded after the deposition of the Green Sands, where one finds a thin (0.1 m thick), sandy white limestone level, highly burrowed and showing desiccation cracking, lying conformably over this sandy deposit. This lacustrine deposit is overlaid by 3 m of fine sediments whose exposures are spatially restricted. The deposit starts with 0.8 m of pure green clays, rich in rounded limestone nodules (ca. 1 cm diameter). After this, a 0.4 m thick graded erosive sand bed was formed. Directly on top of the sands, a new tephra layer was deposited, called T-3. This level consists of 0.2 m of yellowish volcanic ashes that thin laterally, which are strongly weathered and overlaid by a thin layer of limestone, showing desiccation cracks filled by iron oxides. The T-3 layer is covered by about 1.5 m of massive green clays that contain abundant ferruginous (pyrite?) nodules, associated with sandy burrowing by endobenthonic organisms.

After this lacustrine depositional environment a new ash-fall deposit is formed, called T-4. This consists of 5-8 cm of pyroclastic deposits that show clear lamination, changing from dark brown to yellowish. Although it is very wide, this layer has been eroded in places and replaced by later deposits, making its outcrops scarce. A typical feature of this layer is the post-sedimentary formation of large, rounded limestone nodules that grow surrounding this lapilli deposit. Sometimes, in the absence of these nodules, a well-cemented coarse sandstone has been found. A few centimeters of fine sediments or sands cover T-4. This is the only tephra layer from the USC member that has also been found in the South Escarpment exposures.

The deposits overlying T-4 contain more than 2 m of coarse sands (ranging from very coarse sands to gravels) with abundant fish remains at the base. These sands show different upgrading sequences. This represents channel-fill deposits in an alluvial-deltaic environment. Some large fossils, such as elephant or rhinoceros, have been found within these coarse sediments. Sometimes bones from the same individual have been found together. At the top of this unit, a brown tephra layer mixed with coarse sands called T-5 is found. It shows channel-fill deposit structures (crossed bedding, ripples, graded bedding, and bioturbation at the top), indicating a contemporary reworking of the tephra. In some exposures, we found the same layer undisturbed and deeply cemented, reaching 0.5 m in thickness, with a lighter color and high homogeneity.

Overlying T-5, the last sediments from the Humbu Formation were deposited, ranging from 0.8–1 m in thickness. They consist of coarse muddy sands, but they are much finer deposits than those below. These channel-fill and sheet-flow deposits are reddened, root-disturbed and burrowed in places, suggesting more intense pedogenic processes. The top of the Humbu Formation is sharply in paraconformity with the overlying Moinik Formation (Figure 3.5g). Sometimes it is marked by a 0.1 m thick white dolostone layer that indicates the beginning of lake expansion, as well as by a thin grey layer rich in small calcite crystals.

B. The Moinik Formation

1. Clays and trachytic tuff member (Moinik Clays). Type Section shows a highly homogeneous Moinik Clay member. It consists of more than

20 m of silts and clays, finely laminated in the form of pale gray deposits (Figure 3.5h). These clays and silts seem to have been deposited in a highly alkaline lake environment. There are at least five interbedded tephra layers and several limestone and dolostone beds as well as *in situ* magadite-chert nodules (Figure 3.7c–d). Outside of Type Section, in an area south of our study zone, a basalt flow called Intra-Moinik Basalt has been described and dated to 1.2–1.35 Ma (Isaac and Curtis 1974; Manega 1993).

There are no great disturbances in the Moinik Clay member, except for the volcanic layers. They have been strongly altered to secondary minerals like zeolites, with alteration increasing upwards. The main tephra layers are the following: first, TM-1, also called Bird Print Tuff due to the marks on its surface that resemble bird footprints. It is a white tuffaceous limestone, 8 cm thick, rich in detritics like quartz and feldspar, and also rich in esmectite and analcime. It has been dated to 1.26 Ma by Manega (1993). About 5 m above this layer lies TM-2, a white-yellowish tephra layer up to 1.3 m thick. It is finely laminated, and shows current marks such as ripples and signs of reworking in the form of detritics. It has been intensely burrowed by endobenthonic organisms. At the base, there is a thin dolomitic layer and casts of crystals previously attributed to trona by Isaac (1965). It is mainly erionitic and it also includes authigenic clays. A couple of thin brownish ash-fall layers were sampled 2 m above TM-2: TM-2 B and C. Both layers grade upward and contain coarse to very fine lapilli. A few meters above, there is a discontinuous tephra layer, TM-3, that thins laterally from 1 m–0.05 m. TM-4 is a 1–0.5 m thick white to reddish brown ash-fall deposit, finely laminated and showing crossed and planar bedding. A thin dolomitic layer is found at the base. Mineralogical composition shows it is almost 100% secondary erionite.

2. Upper Tuff member. The Upper Tuff member consists of 4–20 m of yellowish trachitic tephra (Figure 3.5h). This tephra has been diagenetically altered to erionite. It shows current flow structures like planar and cross-bedded lamination as well as longitudinal ripples. In the middle of the sequence, an erosional event was recorded. This event signifies a lake retreat and gully incision. After that, sediments filled in the gullies, but two contemporaneous high lake-level episodes enabled the formation of two stromatolitic crusts, clearly seen in the Type Section area as well as on both the Sambu and Sanjan escarpments. Stratigraphic contact with the underlying Moinik clays member is locally erosive, probably due to both later slumping during the Middle Pleistocene tectonic movements and depositional reworking.

C. Post-Peninj Group sedimentary record

At least two sedimentary units from the Middle to Upper Pleistocene have been recorded in Type Section. The lower one consists of homogeneous well-sorted coarse gravels and conglomerates overlying a deep erosional surface, and sometimes including large reworked fossils. The upper one consists of greenish muddy coarse sandstones, poorly sorted, and includes blocks of magadite, volcanic tephra, and limestone reworked from the Peninj Group deposits. Some bones, large gastropods, and ostrich eggshells have been found inside. The age and paleoenvironmental significance of both deposits are still under study.

Fine and well-laminated deltaic to lacustrine sands and silts were deposited nonconformably over the Peninj Group and later sediments (Figure 3.5i). These loamy sediments are coarser near the Peninj river valley and finer and more laminated towards the lake, in a paleodeltaic sequence more than 20 m thick, which include occasional complete and disarticulated fish remains (Figure 3.7b).

Lake Natron paleoshorelines from the uppermost Middle Pleistocene to Holocene are recorded in at least four almost continuous lines of stromatolites around the western part of the lake (Figure 3.7a). Some of these stromatolitic growths overlap at approximately the same height. In the same area, partial accumulations of boulders of Sambu lava and Precambrian quartzite, as well as chert nodules, are found as old terraces.

The Type Section Paleoenvironmental Sequence

The Natron basin has exceptional outcrops which enable us to make paleoenvironmental interpretations of the west basin. The continental deposits that surround many of the northern reliefs to the west of the lake inform us about the more terrestrial environments where hominids and macrofauna were active. As we previously noted, most of the sedimentation occurred to the east of the Sambu fault, which was active during this time; this gave rise to the Natron half-graben being progressively filled (Figure 3.9a). Although it is a small part of the basin, Type Section provides very comprehensive information about the main landscape changes that occurred during the deposition of the Peninj Group. In the previous chapter, we explored the modern geomorphology of the basin and the main volcanic and tectonic features which conditioned its geological evolution; in this chapter, we now use this information to understand and interpret more specific areas. In the following section, the sedimentary environments and their changes during the deposition of each stratigraphic unit of the Type Section will be discussed, with special emphasis on the Humbu Formation (Figure 3.9b).

Humbu Formation

Basal Sands with Clay (BSC)

As we previously noted, the basal sands occupied the inner part of the Natron half-graben and the southern part of the Moinik valley. The complete extent of the sedimentary infill is difficult to know because outcrops thicker than 8–10 m are scarce. Isaac (1967) remarked that the Hajaro lavas, which had already filled up part of the half-graben, are very fresh under the basal sands. This means that much time did not pass between one deposition and another. The basal sand deposits correspond to two main phases, and are interbedded with at least three levels of volcanic tephra. The older phase implies at least three successive progradation

Figure 3.9 (right) A) Stratigraphic correlation between most subsiding Type Section and the upthrown South Escarpment outcrop units, showing the importance of the Sambu fault movements during the time of Peninj Group deposition; B) model of landscape evolution of the Type Section during Humbu Formation times: 1) subsiding area east of Sambu Fault, and the resulting accommodation space; 2) Basal Sands and Clays (BSC) alluvial deposition at the foothill of the Sambu Escarpment, with erosion in the higher areas; 3) basin infill by sediments and lake expansion, with the beginning of the deposition of the Limestone and Basaltic Tuff (Main Tuff); 4) after the deposition of the Main Tuff, ash-fall sedimentation inside the lake and lacustrine retreat and alluvial progradations occurred; 5) sambu fault activity, subsidence and red soils formation in the rift axis, erosion of previous Peninj Groupdeposits and sediment movement to lower areas; 6) new volcanic eruption leading to the deposition of the T-1 tephra, which is later partially eroded by gully incision, and filled with sandstone; 7) deposition of the Green Sands unit, with a greater supply of sediments and water; erosion is continuing in the uplifted block; 8) basin filling, lake expansion, mud flat and distal alluvial deposits across the entire western basin; 9) alluvial progradation, with large channels moving through delta and alluvial plains; 10) finer deltaic sediments at the top of Humbu Formation and prior to lake invasion during the Moinik Formation. See Figure A.2 in the Appendix.

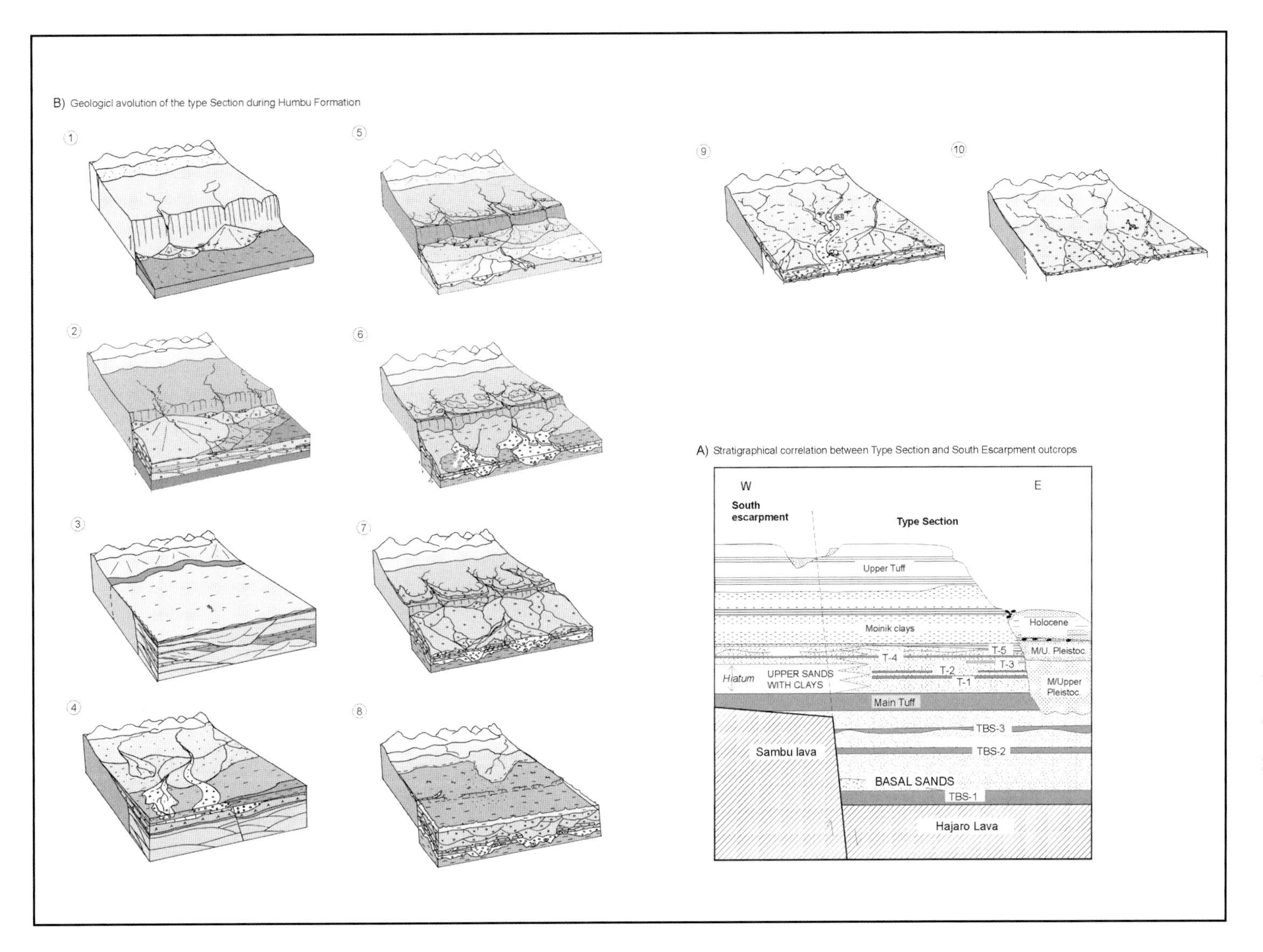
B) Geologicl avolution of the type Section during Humbu Formation
1
5
9
10
2
6
3
7
4
8
A) Stratigraphical correlation between Type Section and South Escarpment outcrops
W
E
South escarpment
Type Section
Upper Tuff
Moinik clays
Holocene
T-5
M/U. Pleistoc.
T-4
T-3
Hiatum
UPPER SANDS WITH CLAYS
T-2
T-1
M/Upper Pleistoc.
Main Tuff
TBS-3
Sambu lava
TBS-2
BASAL SANDS
TBS-1
Hajaro Lava

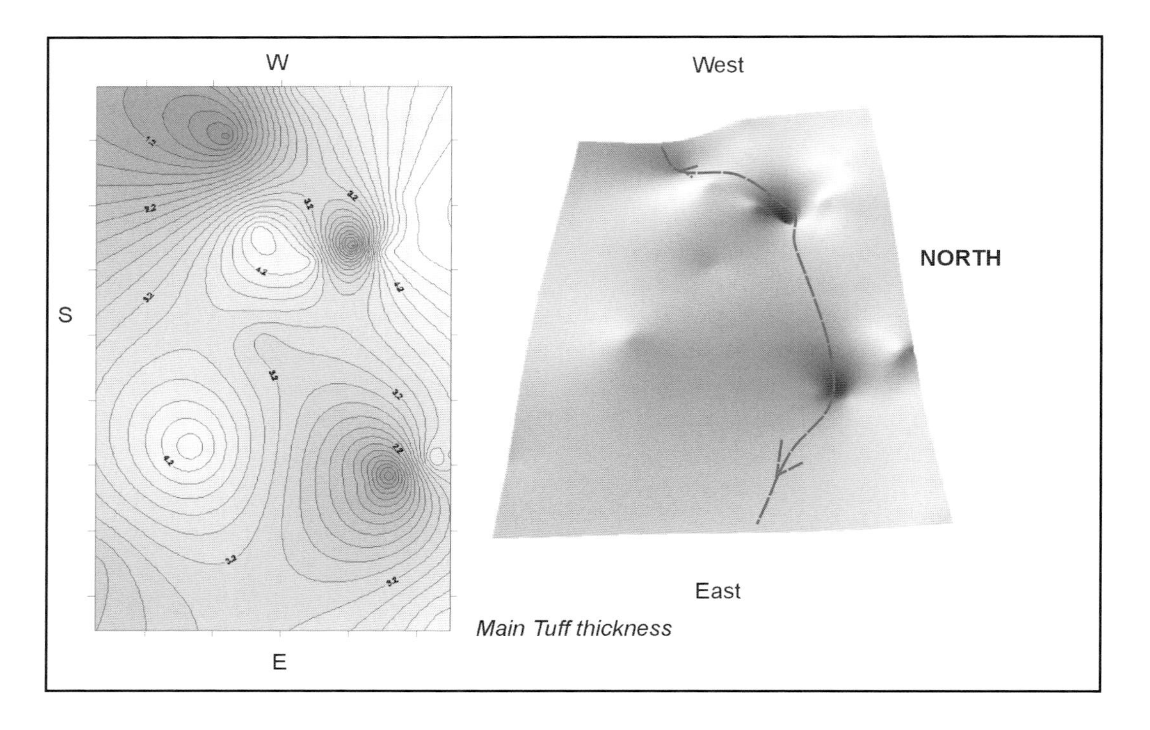

Figure 3.10 Isopachs of the Main Tuff member. The least thickness runs west-east due to erosive processes.

and retrogradation phases of the alluvial facies and expansion and retreat of the lacustrine environments in a narrower half-graben. The younger phase is at first erosive, signifying a sharp change in the depositional environment, and the later sedimentary filling was mainly due to channeling in a deltaic-alluvial landscape.

The former environment, as represented in the older phase of this unit, seems to have been quite influenced by changes in water table, with oxidation of sediments and formation of hematite, limestone calcretes, rhizoliths and isolated, irregularly-shaped nodules. The environment could have been an alluvial plain not far from the lake. The coarse-grained detritus and its abundant quartz indicate that there were probably also long rivers which came from the Precambrian reliefs. Alluvial cones near the Sambu escarpment also left lava blocks near the paleorelief. The outcrops on top of the Hajaro lavas are so scarce that no further interpretation can be provided. Coarse sand lenses, interbedded between the muddy sandstones of the alluvial plain, represent either channel-fill deposits that crossed through the floodplains or sandy sheet-flows of alluvial origin. However, in the middle and north zones of Type Section, lacustrine deposits can also be found interbedded with sand gravels and clayish sands. The most typical traits of these lacustrine facies are limestone and clay levels with ostracods. The best example is documented over the TBS-3 tephra, where stromatolitic crusts were developed. The apparent conformity of these layers indicates sharp changes in the lake extent. The former paleo-lake could have been closer to the old Sambu Escarpment and the Oldoinyo Sambu volcano, showing alluvial facies around it within a narrow mudflat.

The younger phase shows localized erosion of the underlying unit, with channel incision reaching up to 8 m in depth. This sedimentary facies shows a thick level of clayish sandstones, channels and upgrading sequences. The sedimentary flow structures seem to follow a NW–SE direction with two deep areas where previous valleys were filled by sediments. The lack of outcrops in the south prevents us from knowing if these valleys were part of a larger deltaic system, but a regression of lacustrine environments and alluvial progradation seems clear. The deep valley incision, and the folding of some layers like TBS-2 or the contact between TBS-3 and its underlying clayish sands, suggest that tectonic movements were relatively frequent during these times. The reactivation of the faults in the inner zone of the half-graben, which produced the differential subsidence inside the basin, could have displaced the lacustrine environments to the rift axis and lower areas. This would have created channel incision at the foot of the Sambu escarpment. Previous changes in lake level seem to be related to other processes such as climate change or the formation of new volcanic reliefs.

Limestone and Basaltic Tuff or Main Tuff (MT)

This member is found throughout Type Section, as well as in some areas over the Sambu escarpment. Limestone rich in gastropods and ostracods clearly suggests a sharp expansion of the lacustrine depositional environment after the sedimentary infill surpasses the height of the Sambu Escarpment. At the beginning, it must have been a rapid expansion because the detritic sediments are scarce and there is precipitated carbonate. This level is locally less carbonated and has more marl. Faunal abundance suggests environmental stability. After the lacustrine expansion, the clays started to deposit in the middle of the lake.

The lake is present throughout the deposition of the Main Tuff member, as is suggested by the interbedding of carbonate layers, with the presence of gastropods, fish remains and flow features (ripples, crossed lamination). Fluid escape structures are also frequent in these layers, because of seismic shaking of water-covered sediments. The Main Tuff varies in thickness from 4.7 m to 0.4 m, due to synsedimentary and post-sedimentary erosion. Isopach maps show that the thinnest area corresponds to a NNW–SSE band similar to the main tectonic direction (Figure 3.10). The erosion may be due to rift faulting activity that causes uplift and sinking, leading to the erosion of these high areas, or to channel incision following the rift direction. A widely exposed slump deposit supports this hypothesis, because it could be accounted for by tectonic movement and the displacement of the tephra layer into an underwater environment downslope. On top of the tephra levels, there are both detritic sediments over an erosive surface and carbonates and clays, apparently conformable regardless of layer thickness. Limestones are distributed in a southern band, which means that detritus came from the north. The filling of the depressions is generally alluvial but can sometimes be lacustrine. Clays with many limestone nodules show possible changes of the water level during the time of sediment deposition.

The lacustrine expansion reached the same height as the top of the Sambu paleo-escarpment. This filling of the inner basin may have created the plain landscape. Shortly after the deposition of the Main Tuff, the lake environment continued and there is a band of limestone that shows the older shore and alluvials, indicating an inland mudflat. After this moment, there was an erosional event, probably due to tectonic movement, that gave rise to a N–S or NNW–SSE valley incision parallel to the escarpment fault.

Laterally, the Main Tuff tephra correlates with a volcano located in the southern Moinik Ramp, in an old vent that Isaac (1965) described as the origin of the tephra. This vent also includes interbedded lava flows. Ashes and lava flows overlie lacustrine deposits, showing that eruptions mainly took place in a vent inside the lake. The presence of active minor horsts and grabens inside the basin could have given rise to fissural floodings and intense eruptions in the area. Near the Sambu escarpment, pyroclastic deposits and lava flows seem to overlie high tectonic blocks.

Upper Sands with Clays (USC)

The USC member lies conformably over the Main Tuff member in Type Section. It shows several interesting environmental changes and a number of archaeological and paleontological sites which provide useful information about the sedimentary environments in which they were deposited.

The scarcity of clastic sediments and the abundance of lacustrine carbonates overlying the MT tephra show that once the eruptions stopped, the environment of the area continued to be lacustrine. Later, the continuous progradation of alluvials towards the middle of the basin in this gently sloping environment produced a progressive accumulation of clays followed by clayish sands. These detritic deposits include many calcretes and limestone nodules which indicate that the water table was varying. The thickness of this layer reaches a maximum of 1.5 m and shows the transition from a lacustrine mudflat to a distal alluvial plain. Where the record is more complete, there is another erosive unit above the previous one. This layer has very coarse-grained detritus, which implies an incision of wider channels and the establishment of more proximal alluvial environments in the area. Some of the gravel and coarse-grained sand channels locally contain abundant well-preserved fossils. Channel-fill deposits show graded sequences that end in clayish paleosols with rootcasts, indicating plants growing over old sand bars and swamp areas. Laterally this unit consists of muddy brown sandstones with many large rootcasts which correspond to flood deposits in an alluvial plain.

The characteristic red layer is very rich in hematite. This mineral is generally associated with warm environments and poorly-drained soils with a neutral pH. But the sedimentary features indicate a rich vegetation with highly carbonated soils, sometimes almost lacking sand, which suggests wetter conditions. Areas with a more intense red color are high in hematite and clays (rich in travertine-like rootcasts). This could be associated with ponds or springs near the fault. We can also link these features with alkaline hotsprings or changes in stream distribution that gave rise to ponds inside the delta plain apart from the main river courses. Fossil remains are few in this red layer but articulated remains have been found in this level, supporting the idea of a low-energy depositional environment. The top of this layer becomes redder and richer in vegetation, suggesting a recess in the sedimentation and the action of intense edaphic processes, giving rise to iron oxide dissolution and a later hematitic soil formation. In places, this layer is very sandy, implying an increase of the sedimentary supplies. We attribute this increase and the following Green Sand unit to a new Sambu fault movement, which sank the Type Section area and raised the west area, exposing it to erosion. During this time, the west flank of the rift was the main source of sediments.

Later, the sedimentation of the Green Sand unit began and a paleolandscape was formed where the most abundant archaeological remains are recorded: the surface of ST Site Complex, which overlies the T-1 tepha layer (Figure 3.11). This implies a strong increase in alluvial sedimentation and a lake retreat. These start with a widespread volcanic event which deposited the T-1 tephra layer. After ashes fell on the surface, erosive processes started with gully incisions running W-E and NW-SE. These gorges can be more than 1.5 m deep (Domínguez-Rodrigo et al. 2002) and include tephra blocks from T-1 itself. Channels loaded in coarse-grained sands and gravels deposited in different upgrading sequences are very rich in fossil remains and lithics. This sedimentary infill defines what we call the surface of the ST Site Complex, because a great number of sites found in Type Section are situated between tephra T-1 and T-2 (Chapter 4).

At the beginning, the landscape showed gullies where water flows carried coarse-grained sediments. Concurrently, the higher areas covered by tephra were eroded near the streams, bringing volcanic blocks into the valleys. The middle areas were weathered and well-cemented. Gullies progressively filled the low areas with sands and gravels. Then, channels were saturated and water overflowed through the river banks and left fine sediments on the tephra surface. Inside the gullies, some areas continued to be flooded, leaving clay deposits on the sands. These channels were occupied by plants that left rootcasts later filled by sands and cemented by calcite. Outside the channels, water flooding left fine sediments on the volcanic surface of the plain. Some parts of T-1 were not covered by sediments.

T-2 overlaid the paleorelief, showing a wavy surface. It has been partially reworked by wind and water. Some small faults affected the sediments between T-1 and T-2 deposition. The landscape was similar and erosion seems to have cut through the same areas as it previously did during T-1 times. Coarse sediments were widely colonized by plants that Isaac (1965) attributed to the genus Typha. The great thickness of the Green Sands is the result of the new inner basin sinking and the upper escarpment erosion in the west mentioned above. Sites in this unit are more scarce than on the surface of ST, but they do occur. Authigenic zeolites suggest that the lake had alkaline waters (Hay 1966; Icole et al. 1987). Fish remains are common in places, indicating the presence of water flows which were more permanent than seasonal.

After the deposition of the Green Sands, a clear change in environmental conditions occurred (see Figure 3.9b). There are few outcrops recording this moment, but there is a visible transition from high clastic sedimentation to more lacustrine limestone and clay deposition, signifying a lake expansion. When comparing the thickness of the underlying deposits to the top of the formation, there is a coincidence between this expansion and the beginning of sediment deposition in the South Escarpment area. The sedimentary record in this western area shows limestone, clays and muddy sandstone deposited in a distal alluvial plain. It may be the result of a new basin sedimentary infill, a large plain or a flat landscape that allowed water table expansion and widening of the paleolake. At the same time, in Type Section, the lake is relatively deeper as one moves offshore. Deposits are divided into two units, separated by a moment of alluvial progradation and the volcanic eruption represented by T-3. The ashfall deposits may have influenced the type of sedimentary deposition. At the end of this large lake period, the T-4 tephra was deposited.

After the deposition of T-4, a new stage of coarse alluvial progradation began. Deltaic facies of channels with upgrading beds refilled a previously eroded landscape. There is a thick section of sediments that reworked the overlying pyroclastic deposit (T-5). On these sandstone and gravel deposits, abundant lithic remains and fossil remains from large animals (some of which are

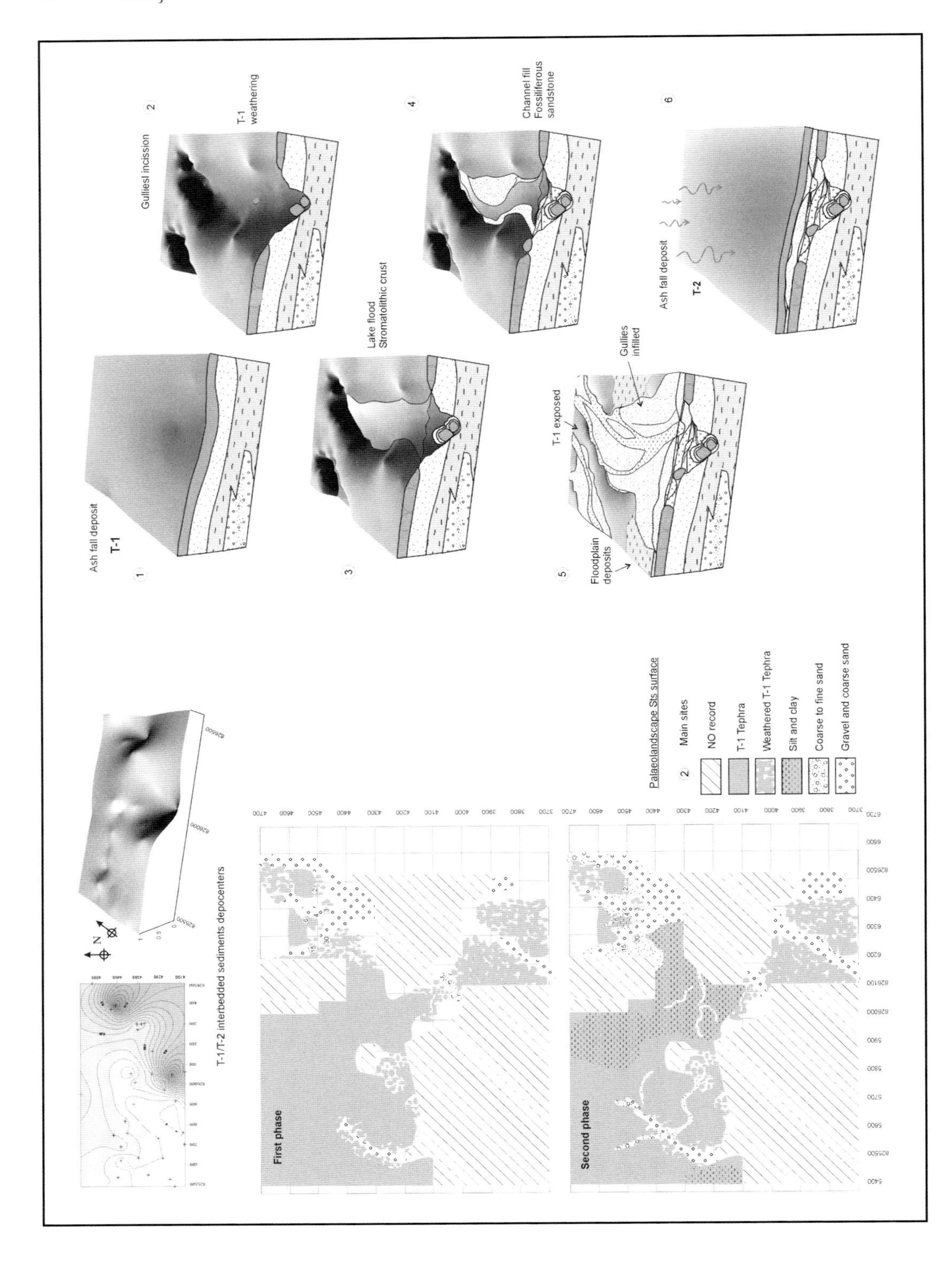
T-1/T-2 interbedded sediments depocenters
First phase
Second phase
Palaeolandscape Sts surface
Main sites
NO record
T-1 Tephra
Weathered T-1 Tephra
Silt and clay
Coarse to fine sand
Gravel and coarse sand
Ash fall deposit
T-1
Gulliesl incission
T-1
weathering
Lake flood
Stromatolithic crust
Channel fill
Fossiliferous
sandstone
Floodplain
deposits
T-1 exposed
Gullies
infilled
Ash fall deposit
T-2

from the same individual) are well-preserved. The coarse alluvial deposits of Type Section have a corresponding deposit in South Escarpment, where large fluvial channels cut through a muddy alluvial flood plain.

The last Humbu Formation deposits show a decrease in the grain size of the deposits and an increase in the action of edaphic processes. It can be interpreted as a new alluvial facies retreat, due to lake expansion or on the contrary, to the continuous progradation and alluvial flood-plain landscape in the area. The reddish color can be correlated to the top of Humbu Formation in South Escarpment by the presence of red muddy sandstone and coarse sandstones in that area.

Moinik Formation

Moinik Formation is completely lacustrine in the Type Section. The environment is interpreted as a stratified saline-alkaline lake, deeper than the modern Lake Natron (Isaac 1965; Icole et al. 1987). The Upper Tuff member geochemistry and structures indicate the same environment, except for a short moment of lake retreat and erosion, followed by channel filling by sands and the creation of a wide stromatolitic crust. The first lacustrine expansion, as described above, is probably due to a southward sinking of the Moinik Ramp. Isaac found a sharp climatic change to be a more likely explanation. Either possibility could be the cause of this abrupt change in sedimentary conditions.

Discussion

The great thickness of the Peninj Group deposits in the Type Section described by Isaac (1965) is due to the previous tectonic movements along the Sambu fault, which gave rise to a large accommodation space for sedimentation in the inner part of the young half-graben. This area, together with the southernmost zone of the Moinik Ramp and some small grabens in it, were the lowermost areas of the Natron sub-basin at the beginning of the deposition of the Peninj Group. The concentration of archaeological and paleontological sites here is explained because of the more lacustrine environments occurring southwards and the repeated tectonic movements along Sambu fault and the subsequent escarpment. Most sites were formed in depositional environments with a high rate of sedimentation, such as channel-fill deposits in alluvial and deltaic environments. All of these sediments originated in the Precambrian basement reliefs to the west and in the rift shoulders after each subsiding movement. This tectonic activity is indicated by synsedimentary features like slumps, fluid escapes, sediment and tephra deformation (wavy structures), possible sand mounds, fault displacements and lowland refill.

Figure 3.11 (left) The surface of the ST Site Complex: A) Sediment thickness between T-1 and T-2 tephra layers, showing the approximate paleorelief created after the deposition of T1. It shows at least three depressed areas that coincide with channel incision, and coarse-grained deposits; B) facies distribution of the surface of ST in Type Section, divided into two sedimentary phases; first channel infill (above), second channel infill and flood plain finer sedimentation on the margins (below); C) model of sedimentary evolution along the surface of ST: 1) T-1 ash-fall deposition; 2) gully incision and tephra alteration, with some blocks entering the small valleys; 3) lake flooding and stromatolitic crust growing around some blocks; 4) stream flooding and deposition in the gullies (reflected in reworked sediments and sand bars); 5) sedimentary filling of the gullies, with sand bars in the channels, finer silty-clayey deposits on the flat lower areas and some tephra exposures; 6) deposition of the T-2 tephra layer, which coated the sands, silts and T-1 tephra exposures.

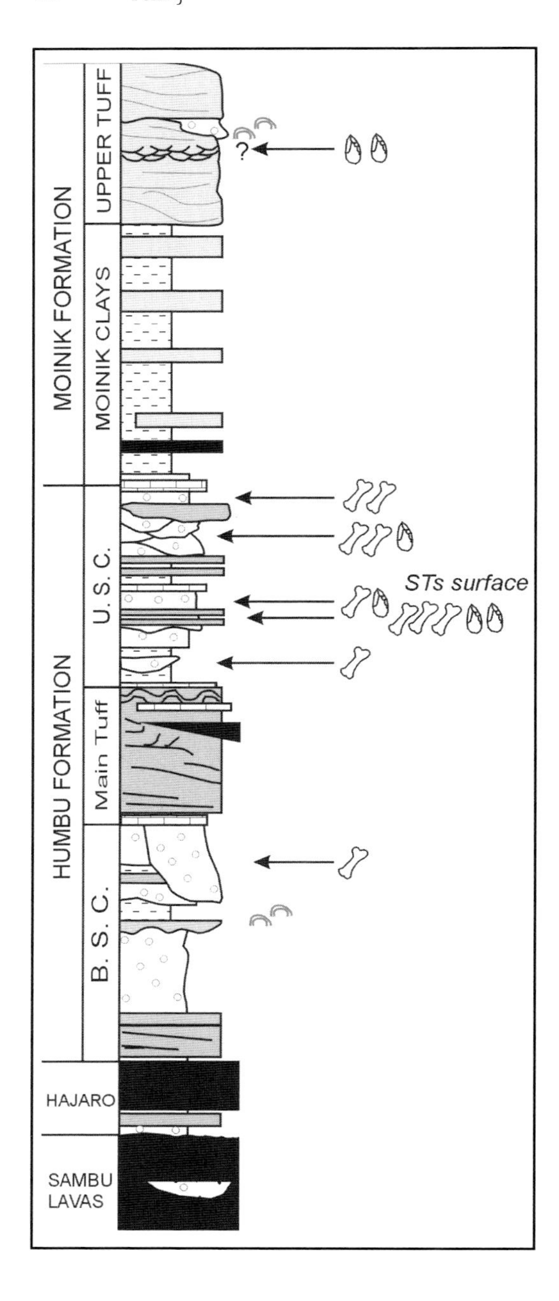

Figure 3.12 Fossils and stone tools are found in almost all sedimentary units of the Peninj Group, but the larger accumulations are restricted to some units and layers such as the surface of ST, or those areas with a greater supply of coarser sediments.

The richness in clastic sediments is also supported by the neighboring Oldoinyo Sambu volcano slopes and the relative uplift of the area with respect to the southern subsiding depositional environments.

Climatic change seems to have affected lake expansion and retreat, but had a lesser effect on sedimentation, which was more conditioned by tectonics. The main fault displacements recorded in Type Section are: Pre-Hajaro beds and the Peninj Group deposition, before the second sedimentary phase of BSC, before the Red layer and the Green Sand deposition, and in the boundary between the Humbu and Moinik Formations. The major geochemical and sedimentary changes between the two formations could signify simultaneous climatic change and volcanic activity. Minor displacements occurred throughout the deposition of the Type Section, affecting different layers such as the Main Tuff member.

The post-sedimentary faults are extraordinarily abundant in the Type Section. Some of these faults seem to have already acted during the Lower Pleistocene, moving in the same direction as later on. The distribution of these faults suggests that they ran in parallel and perpendicular directions slightly oblique to the dominant direction of the Rift. This suggests the existence of a small parallel tectonic ramp that steps from north to south, with a greater southern subsidence at the east side of the Moinik Ramp.

Although the fossil remains and the lithics are abundant along the whole sequence of the Peninj Group, the richest sites seem to be associated with specific levels (Figures 3.12-3.13). The BSC channel-fill deposits show scattered remains along the whole Member thickness. The Main Tuff presents a great amount of fish and invertebrate remains, but very few vertebrates,

Figure 3.13 Some examples of fossil remains from Humbu Formation in Type Section: A) Parmularius *skull and horn fragment in the ST-4 site; B) partially articulated skeleton of a small bovid in the Red Layer; C) molar and fragmentary maxilla of a suid* (Metridiochoerus) *in the lower USC member; D)* Parmularius *horn remain from ST-4 site; E) fragmentary maxilla with two premolars of a* Theropithecus *from the ST-4 site; F) scattered faunal remains like this metapodial are commonly found in the root-marked sediments of the Green Sands unit; G) large vertebrae from the coarse tuffaceous sandstones of the T-5 unit at the top of Humbu Formation.*

due to its volcanic character and its deposition in a predominantly lacustrine environment. The USC is undoubtedly the most fossiliferous deposit of the Type Section. In marginal lacustrine facies and distal alluvial-deltaic facies, almost no remains have been recorded. Nevertheless, along the channels crossing this plain, fossil remains can be abundant in places. The red level contains scarce fossil remains and lithics. The Green Sands were deposited in the alluvial fans and deltaic facies, with abundant water. Vegetation was very abundant, though the distribution was not homogeneous. There are isolated fossil remains in certain areas, generally associated with sandy channels. An exceptional case is the surface of the ST Site Complex, to be discussed more fully in the following chapter. This depositional environment evolved from a network of gullies incised in a flat, probably herbaceous landscape. The base of this landscape was the T-1 tephra layer, above which fauna and lithic remains were deposited, which form the basis of the analyses in several of the following chapters.

4

A Taphonomic Study of the T1 Paleosurface in Type Section (Maritanane): The ST Site Complex

Manuel Domínguez-Rodrigo, Luis Alcalá, and Luis Luque

Introduction

The paleosurface documented above Tuff 1 (T1) in the Upper Sands with Clays (USC) of Humbu Formation can be traced over an area that spans almost 1 km^2, which includes gullies and areas where the targeted surface is overlaid by younger deposits. The horizontal exposures of this paleosurface - that is, when the surface is either fully exposed or covered by less than 50 cm of younger sediments - total 0.6 km^2 approximately, and are mostly concentrated in the north of Maritanane (Type Section) and also in the southeast sector in the form of a plateau. In Maritanane, the paleosurface is widely exposed due to horizontal erosion, and is covered by a thin layer of sediments. The paleosurface sediments are highly carbonated and cemented, which has enabled the preservation of numerous fossils and stone tools, since these are firmly attached to the paleosurface and therefore resist erosion. Stone tools and fossils appear loose on the ground only when erosion and weathering have been strong enough to dissolve the carbonate and dismantle the sediment particles of the paleosurface, exposing the underlying tuff (T1).

When we first discovered the paleosurface, we could not confidently determine fossil provenience, even though several fossils were lying on flat horizontal surfaces. For this reason, we conducted thorough surveys in 1995 and 1996, collecting most surface findings to "clean" the paleosurface so that successive erosional periods would produce more surface materials that could confidently be ascribed to specific locations. Additionally, the freshly eroded materials could be indicative of landscape preferences by the diverse depositional agents involved. These surveys covered the entire exposed paleosurface and involved a minimum of eight people.

From the artifact and bone distribution observed during 1995 and 1996, it became clear that these materials, especially stone tools, appeared concentrated in patches: that is, discrete clusters of materials which contrasted with areas where materials were either absent or appeared in very low numbers and were rarely in spatial proximity to one another. The application of terms such as "patch" or "scatter" can never be defended in terms of absolute density figures, but rather in terms of the relative differences in materials distribution over any given horizontal, geologically discrete surface. Scatters on T1 were very diffuse and mostly consisted of isolated specimens which contrasted with the large amounts of materials concentrated at specific *loci* (patches). This observation was confirmed in the 2000 and 2001 field seasons. After four years, where the paleosurface had been left unsurveyed, the erosion produced by successive rainy seasons yielded freshly-exposed materials, whose distribution was very similar to that documented in 1995 and 1996.

Scatters were very uncommon. More than 90% of the archaeological materials deposited on the paleosurface overlying T1 were concentrated in the above-described patches that we consider sites. Site distribution on the T1 paleosurface and in nearby stratigraphic positions can be observed in Figure 4.1. A total of 12 sites were recorded across the total extent of the T1 paleosurface. Of these, eight appear clustered in the northeast corner of Maritanane, in what has been called the ST Site Complex (ST = Sección Tipo = Type Section; Domínguez-Rodrigo et al. 2002). Most of the sites on T1 are archaeological sites, containing stone tools associated with fossil bones. Only ST 17 and ST 18 are paleontological localities, devoid of stone tools. This further reinforces the fact that hominid stone-tool discarding activities were mostly restricted to the ST Site Complex area and more exceptionally, to the ST 35 and ST 46 sites. The latter site is in a non-secure stratigraphic position and will be excluded from the discussion of the archaeology of T1.

Surface densities of archaeological materials appear in Figure 4.2 and Table 4.1 and can be compared to the remaining fossil localities discovered in different stratigraphic positions in the Type Section. Densities documented in these sites are much higher than those reported in the remainder of archaeological sites and paleontological localities throughout the Humbu Formation sequence. All the sites with specimen densities >50 are of archaeological origin, with fossils and stone tools in association. Paleontological sites show a much lower number of specimens than most archaeological sites. This suggests special site formation processes at archaeological localities, in which hominids seem to have played a relevant role.

It could be argued that the "patches" at Peninj occur in low spots, i.e., in places where the paleosurface has been more severely eroded, if there were "empty" zones in high spots. However, the relative occurrence of patches and scatters across the uniformly eroded exposures (0.6 km^2) is not related to topographic features. More than 95% of the horizontally exposed paleosurface is devoid of archaeological materials, despite the fact that the areas where overlying sediments occur are evenly distributed and the remaining deposits are similarly eroded .

A landscape taphonomy study of the paleosurface

In order to understand carnivore dynamics in this paleolandscape, a taphonomic approach similar to the one used in modern savannas was applied (Blumenschine 1989; Tappen 1992). Bone remains in different areas of this paleosurface have been analyzed using the methodology outlined by Blumenschine (1989): total number of remains per area measured, MNE/MNI ratio, carcass size, skeletal part analysis, and carnivore damage. Four transect areas similar in size covering the paleosurface exposures have been sampled (Figure 4.3). Area A is the only one with strong hominid involvement in carcass accumulation, which explains why bone density in this area is biased compared to the others. However, given the strong carnivore involvement with the bone assemblages in all the areas (Area A included), bones from this site can still be used to estimate hyena presence in the ecosystem.

Area A has the highest density of faunal remains of the four transects due to hominids repeatedly transporting carcasses to specific spots in this area, where they butchered them. In addition, small-sized carcasses (smaller than 100 kg) show a higher MNE/MNI ratio (between three and almost nine times higher than in the other

Figure 4.1 (right) Main features of the paleosurface overlying the tuff T1 and the sites found in it. T1 is represented by grey. The grey oval stains show areas where T1 is altered. White areas indicate erosion and disappearance of T1. Closed circles show the archaeological sites on the T1 surface.

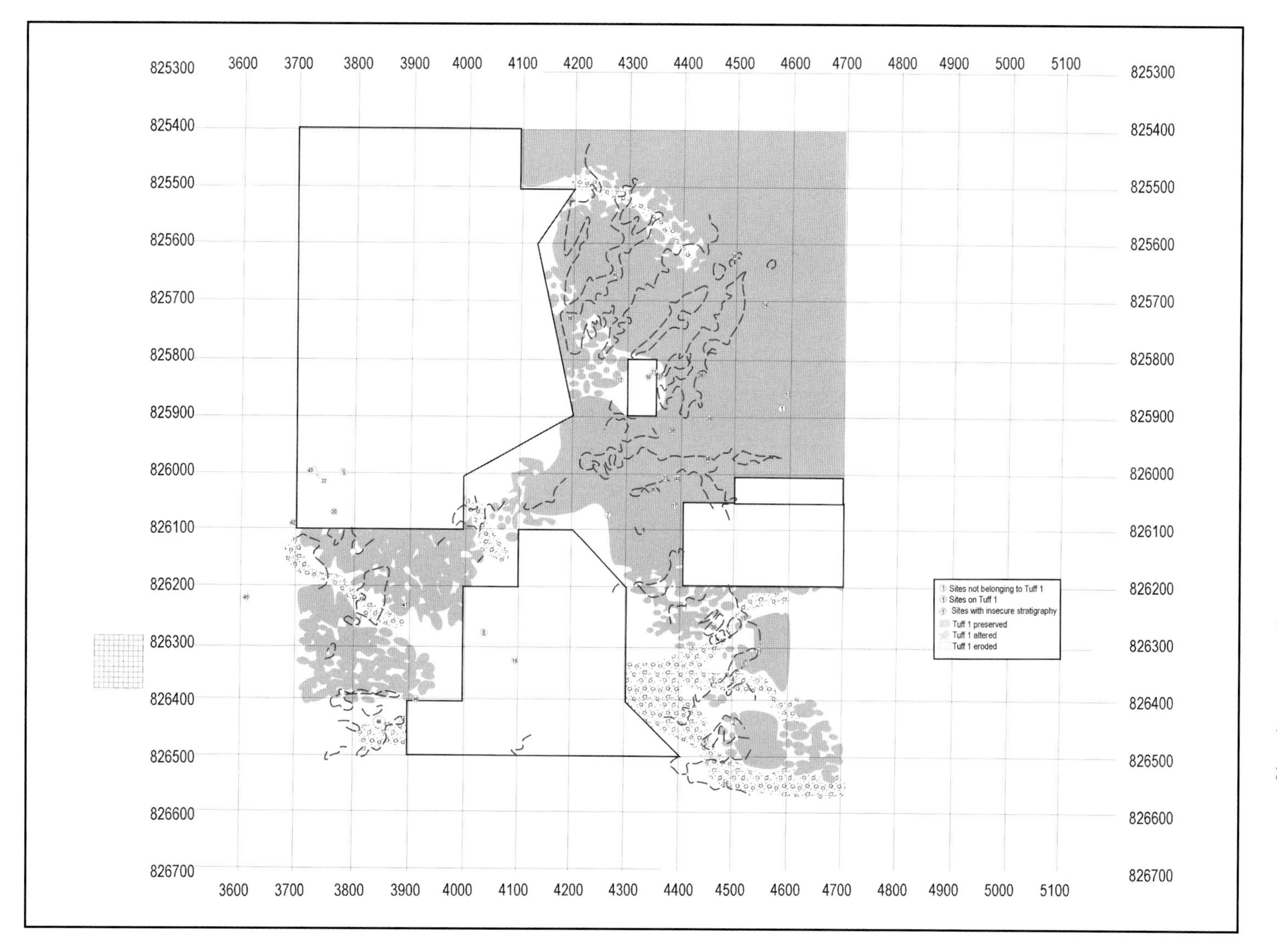
3600
3700
3800
3900
4000
4100
4200
4300
4400
4500
4600
4700
4800
4900
5000
5100
825300
825400
825500
825600
825700
825800
825900
826000
826100
826200
826300
826400
826500
826600
826700
Sites not belonging to Tuff 1
Sites on Tuff 1
Sites with insecure stratigraphy
Tuff 1 preserved
Tuff 1 altered
Tuff 1 eroded

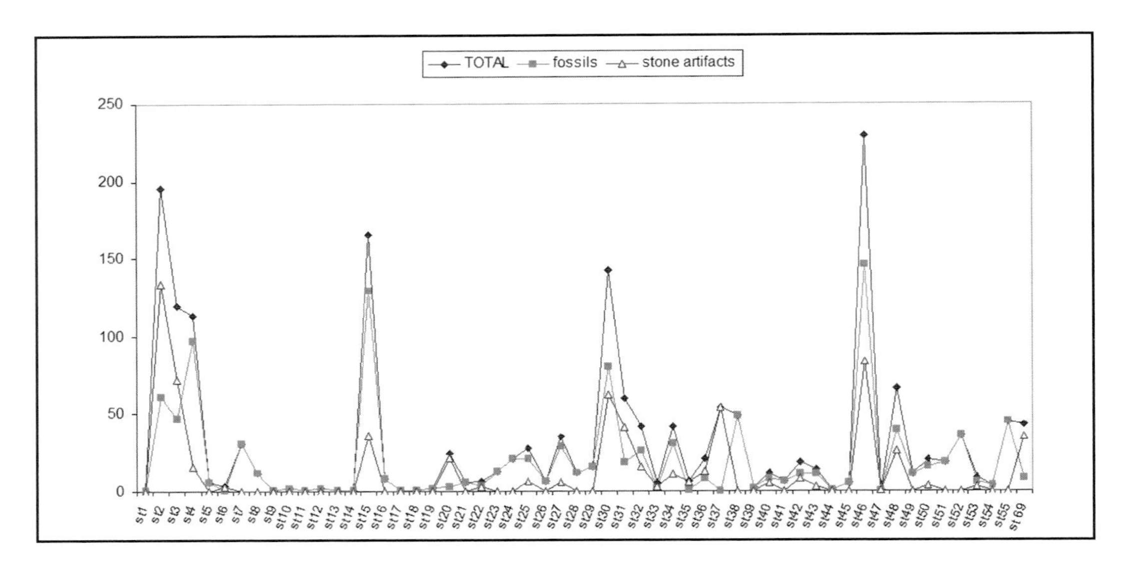

Figure 4.2 Absolute frequencies of fossils and stone tools at each site discovered in Type Section (Maritanane).

areas). Medium-sized carcasses (weighing 100–450 kg) are also represented by a higher MNE/MNI ratio (almost 50%) than anywhere else (Table 4.2, Figure 4.4). Areas B, C, and D show similar MNE/MNI ratios, regardless of carcass size. Small-sized carcasses are underrepresented (Area C has none) and most individuals are represented by just 1 or 2 MNE. Medium-sized carcasses are represented by about 5 MNE/MNI. Only large-sized carcasses (weighing more than 450 kg) maintain a similar MNE/MNI ratio across all three of the areas (Area C being the exception). Only two appendicular bones (both metapodials) were found complete among the four areas. According to Blumenschine's (1989) referential study in the Serengeti, this would suggest high competition, since carnivores tend to break most of the bones when competition is strong.

When looking at axial remains versus cranial and appendicular remains, we see that the latter are predominant and the former are only significant in large-sized carcasses. When comparing the Maritanane data to data from modern savannas, the results suggest that the setting must have been very competitive (Figure 4.5). The scarcity of axial remains and small-sized carcass remains, together with the low MNE/MNI ratio, suggests a strong carnivore presence for all areas except Area A. Areas B, C, and D do not show strong similarities with the data gathered in modern arid steppes, therefore indicating that carnivore presence in the landscape must not have been in the form of isolated individuals or small foraging groups (two to three individuals), as is the case for the Galana and Kulalu steppe ecosystems (Domínguez-Rodrigo 1996). The closest equivalent to the Maritanane data comes from Ngorongoro Crater (Blumenschine 1989), where all the taphonomic indicators (MNE/MNI ratio, axial elements, remains from small-sized carcasses) are low as a result of large groups of carnivores overlapping in the use of resources in a very open landscape. If modern trophic dynamics apply, Maritanane must have been the scene of strong

Table 4.1 Number of fossil bones and stone tools in each locality discovered at Type Section

Type Section	Total	Fossils	Stone Artifacts
st1	1	1	0
st2	195	61	134
st3	119	47	72
st4	113	97	16
st5	6	6	0
st6	4	1	3
st7	31	31	0
st8	12	12	0
st9	1	1	0
st10	2	2	0
st11	1	1	0
st12	2	2	0
st13	1	1	0
st14	1	1	0
st15	165	129	36
st16	8	8	0
st17	1	1	0
st18	1	1	0
st19	2	2	0
st20	25	3	22
st21	6	6	0
st22	7	4	3
st23	13	13	0
st24	21	21	0
st25	28	21	7
st26	7	7	0
st27	35	29	6
st28	12	12	0
st29	16	16	0
st30	142	80	62
st31	60	19	41
st32	42	26	16
st33	5	2	3
st34	42	31	11
st35	7	1	6
st36	21	8	13
st37	54	0	54
st38	49	49	0
st39	2	2	0
st40	12	8	5
st41	7	7	0
st42	19	11	8
st43	14	11	3
st44	1	1	0
st45	5	5	0
st46	229	146	83
st47	1	0	1
st48	66	40	26
st49	11	11	0
st50	20	16	4
st51	19	19	0
st52	36	36	0
st53	9	6	3
st54	4	4	0
st55	45	45	0
st69	43	8	35

Figure 4.3 Topographic map of Maritanane, showing the four areas where T1 was exposed and intensively surveyed for a landscape taphonomic study.

Table 4.2 MNE/MNI ratios, skeletal part distribution and tooth mark percentages in the faunal remains from Maritanane

Carcass Size	Small	Middle	Large	Total
AREA A	121/13 (9.3)	109/14 (7.7)	73/9 (8.1)	303/36 (8.4)
AREA B	13/5 (2.6)	29/6 B(4.8)	37/5 (7.4)	79/16 (4.9)
AREA C	0/0 (0)	11/2 (5.5)	8/3 (2.6)	19/5 (3.8)
AREA D	4/3 (1.3)	23/6 (3.8)	36/5 (7.2)	63/14 (4.5)
MNE				
AREA A				
Cran-append.	104	82	65	261
axial	17	17	8	42
AREA B				
Cran-append.	9	20	22	51
axial	4	9	15	28
AREA C				
Cran-append.	0	8	4	12
axial	0	3	4	7
AREA D				
Cran-append.	3	16	20	39
axial	1	7	16	24
NISP	Total	Damaged Cortical	Well-Preserved	Tooth-Marked
AREA A	1,166	1,014	152	47 (31%)
AREA B	211	159	52	33 (65%)
AREA C	59	19	40	29 (73%)
AREA D	261	110	151	131 (87%)
Total (ABCD)	1,697	1,302	395	240 (60.7)
Total (BCD)	531	288	243	193 (79.4)

Numbers in Numerator are for the MNE. Numbers in Denominator are for MNI

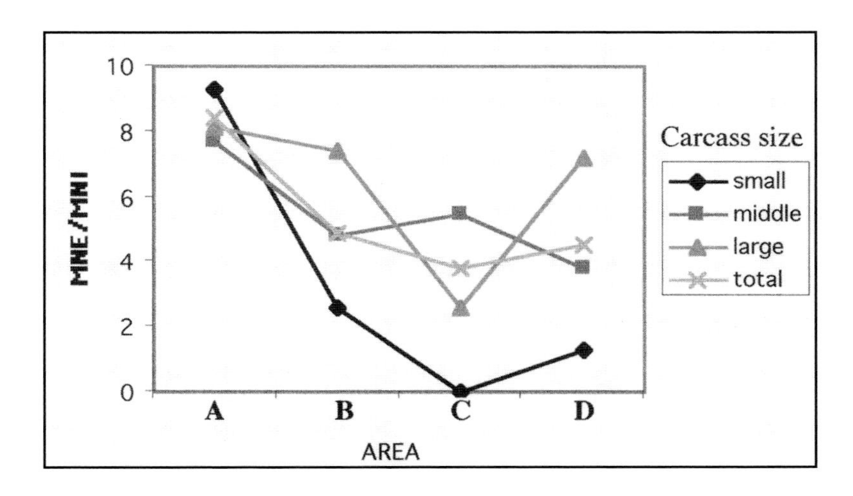

Figure 4.4 MNE/MNI ratio for each carcass size in the four areas studied.

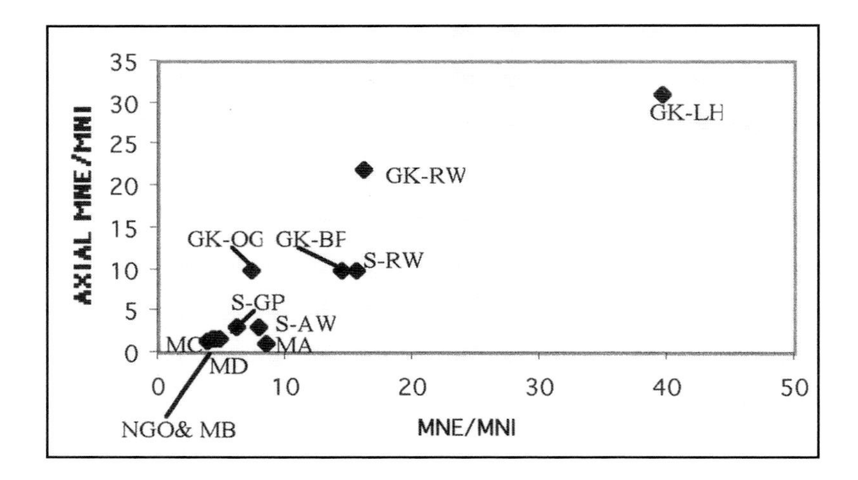

Figure 4.5 Axial MNE/MNI:Total MNE/MNI ratio for each of the four areas studied in Maritanane (MA, MB, MC, MD), compared to different habitats of modern savanna ecosystems (data from Blumenschine, 1989; Domínguez-Rodrigo, 1996): S-AW, Serengeti Acacia Woodland; S-GP, Serengeti Grass Plains; NGO, Ngorongoro; GK, Galana and Kulalu Steppe; RW, Riparian Woodland; BP, Bushy Plain; OG, Open Grassland; LH, Lali Hills.

competition among different carnivore species. Those that are gregarious by nature, such as lions and hyenas, must have been adapted in the form of extensive groups as seen in Ngorongoro Crater (Figure 4.5).

Modern observations relate group-size adaptation by carnivores to open environments (Blumenschine 1986, 1989; Domínguez-Rodrigo 2001). The greater the bushy component of the landscape, the smaller the groups of carnivores. If this observation can be used to interpret landscape ecological dynamics in Maritanane at 1.5 Ma., Area A must have been the only one with a higher degree of bush vegetation, although carnivore presence was much higher than in riparian environments in modern savannas where closed vegetation exists. Areas B, C and D should have been more open.

As further support for this interpretation, almost 69% of the bone sample in paleontological contexts (Areas B–D) is tooth-marked, which rises to 79% of bones when we exclude those specimens with significantly damaged cortical surfaces (Table 4.2). This strong carnivore involvement with bone assemblages has not been documented in any other landscape approach to Plio-Pleistocene sediments with a fossil record. This percentage is significantly higher than the percentage of tooth-marked specimens from Area A, where a strong hominid presence would have conditioned the amount of resources available for post-ravaging carnivores.

The ST Site Complex

As was previously published (Domínguez-Rodrigo et al. 2002), the ST Site Complex contains a new type of archaeological deposit not documented in other Plio-Pleistocene localities. The northeastern section of Maritanane is occupied by two gullies. Within these gullies, archaeological remains are distributed more or less regularly over an area of 3,500 m^2 of exposed paleosurface,

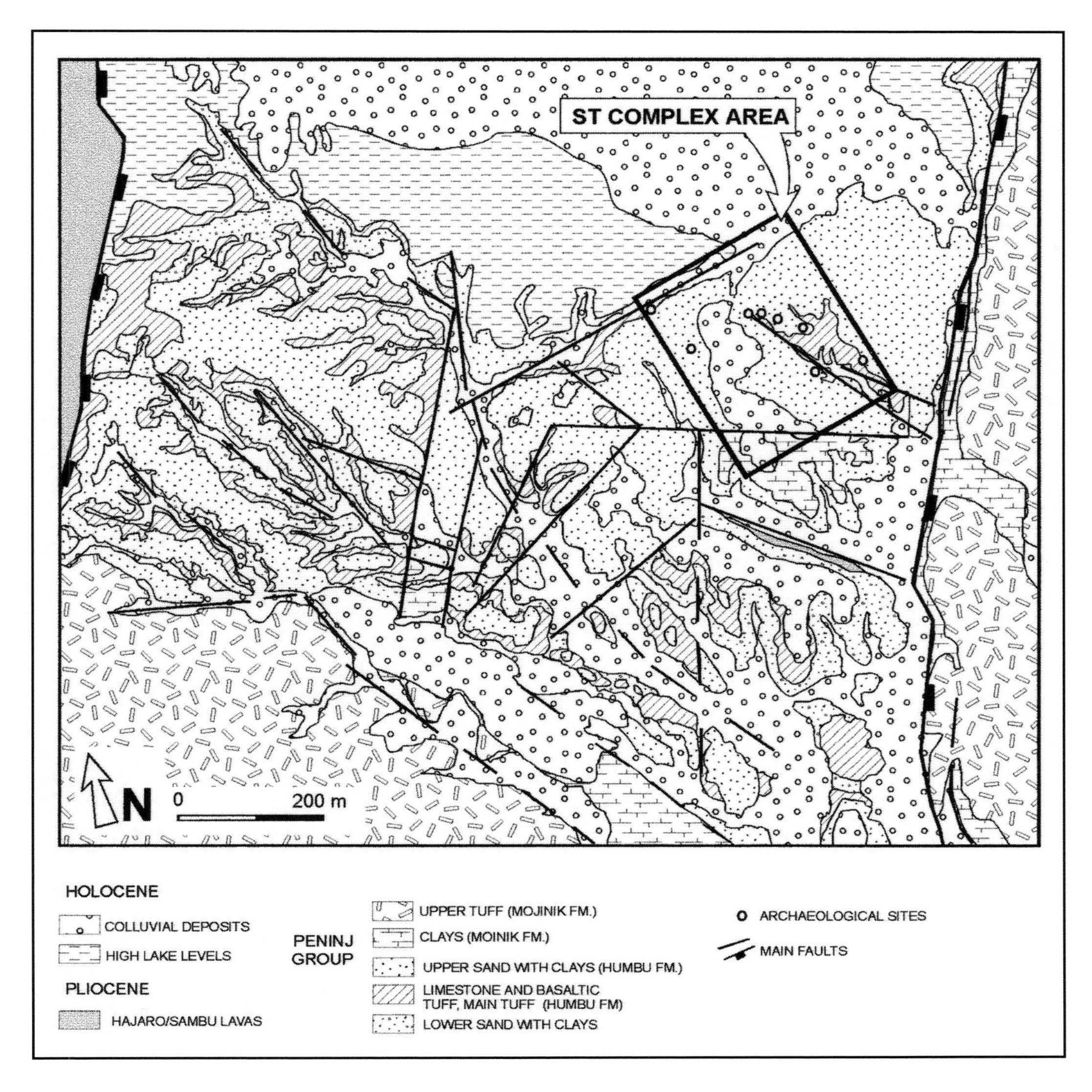

Figure 4.6 Geological map of Maritanane (Type Section) showing the location of the ST Site Complex.

with higher density spots which were formally called sites (ST2, ST3, ST4, ST6, ST15, ST30, ST31, ST 32; Figures 4.6–4.7). The sediments exposed in between these sites are also littered with archaeological remains, though in smaller amounts, making artifact and fossil distribution continuous in this constrained area.

The archaeological materials appear on a paleosol directly on the surface of the tuff (T1) in most of the area, except where the tuff is eroded or cut by river channels. In these places, artifacts and fauna appear in the sandy clays and sands of the channel-associated sediments. With the rest of the area exposed as a background, the ST Site

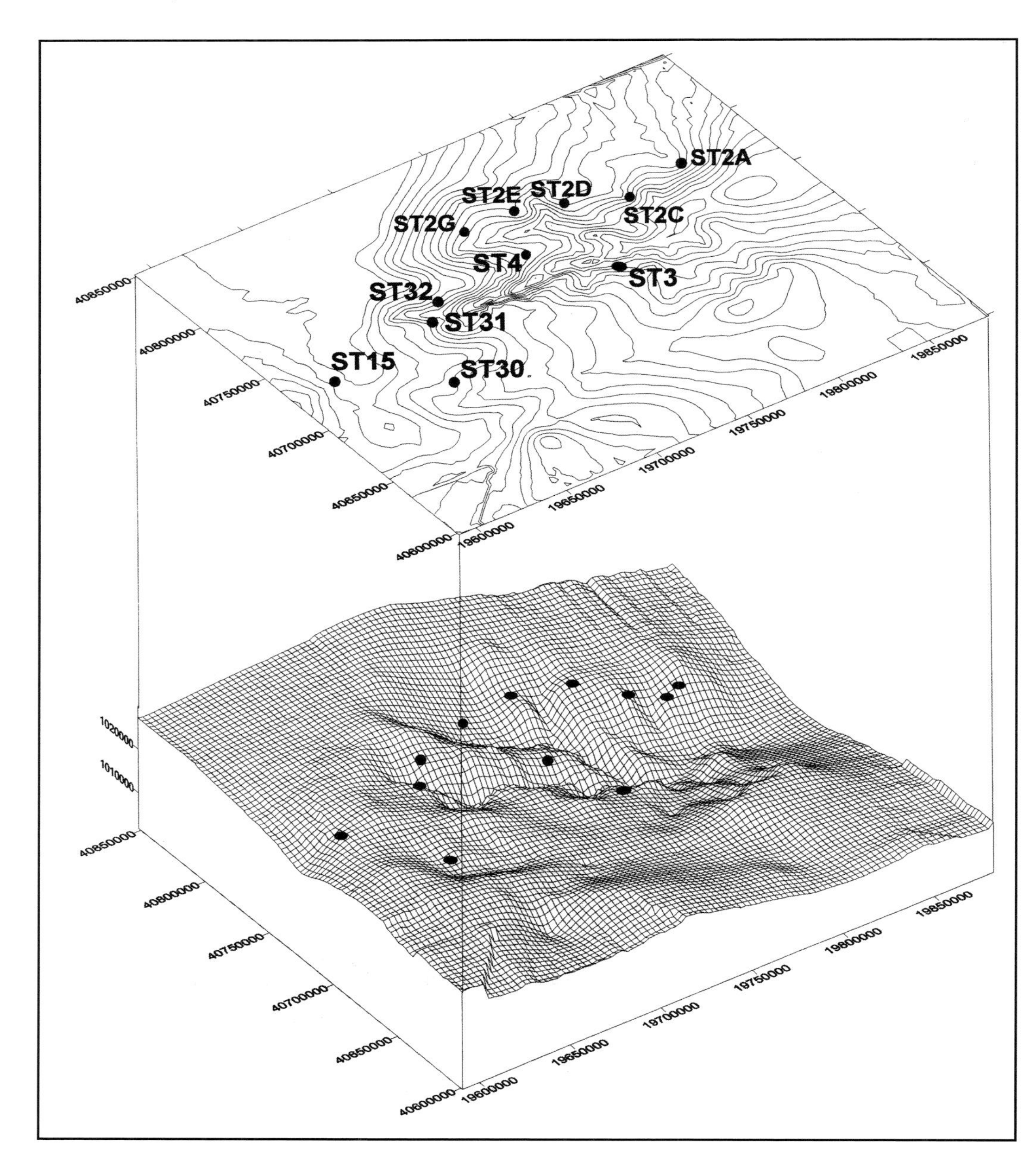

Figure 4.7 Isometric reconstruction of the modern topography of the gully where most of the sites from the ST Site Complex are located (de la Torre 2002).

Complex could be considered a sort of mega-site, with the higher density spots that we call sites as the most significant expression thereof. The rest of the surface of the exposed tuff is comparatively devoid of artifacts, which appear unevenly and in very small scatters. For some reason, hominids seem to have repeatedly chosen the small area around the above-mentioned gully to carry out activities related to carcass manipulation. Stone tool-discarding behavior seems to have been mostly restricted to that area. However, the size of this area (Figure 4.8), where materials were accumulated in higher amounts than in paleontological sites (see Figure 4.2), seems too large compared to the smaller patches documented at Koobi Fora and Olduvai. There, sites appear in much more discrete concentrations and are spatially restricted to smaller areas.

Another feature that makes the ST Site Complex unique is its occurrence directly on the surface of a tuff. Many materials appear exposed but encrusted onto the tuff by carbonates. This allows us to have better control over time-averaging processes, since the overall impression is that the whole archaeological horizon was deposited during a short time interval. This is further supported by the fact that materials are not vertically dispersed in a deposit, but rather are found in a vertically discrete position, mostly in a very thin horizontal layer on top of the tuff surface. Despite erosional processes, diagenetic carbonate adhering materials to the surface of the tuff has enabled the preservation of the original depositional configuration of a substantial number of the archaeological remains. Taphonomic analyses indicate that the concentration of archaeological remains in this area is mostly due to hominid behavior rather than post-depositional natural processes (Domínguez-Rodrigo et al. 2005). The spatial distribution of these archaeological materials therefore offers us a snapshot of a large area of the paleolandscape, enabling us to contemplate the differential distribution of artifacts and faunal remains discarded by hominids in a geologically short time period. This exceptional preservation and stratigraphic resolution is unique in its wealth of information regarding hominid behavior. In this work, taphonomic information will be used, in both in-site and off-site approaches, to explain hominid adaptation at Maritanane a little less than 1.5 million years ago.

Data collection methods

Most sites in the ST Site Complex are located on individual slopes, except for ST2 (comprising five slopes: A, B, C, D, and E). The T1 tuff is exposed along the entire area, connecting all the slopes that constitute the main gully. Most sites contain sediments overlying the archaeological horizon, varying from thick strata (ST3) or moderately thick strata (ST4 and ST30) to thin deposits (ST2A, ST15, and ST32) or a lack of overlying sediments (ST31 and ST2B-E). In the last of these cases, a significant number of remains had to be painstakingly removed from the carbonate matrix that adhered them to the paleosurface, a laborious process. This situation was also encountered in some areas of the sites where the tuff surface is partially exposed without overlying sediments (ST2, ST3, ST32, and ST4). In these cases, most of the archaeological horizon is also found covered by sediments. Test excavations have yielded the same kind of carbonated materials adhered to the tuff surface. In all these sites, archaeological materials also appeared on the surface of the slopes due to erosion. These surface materials were also considered to belong to these sites for the following reasons:

First, in Maritanane, clusters of surface archaeological materials are spatially restricted to the localities where sites have been documented to exist *in situ* through excavations. Therefore,

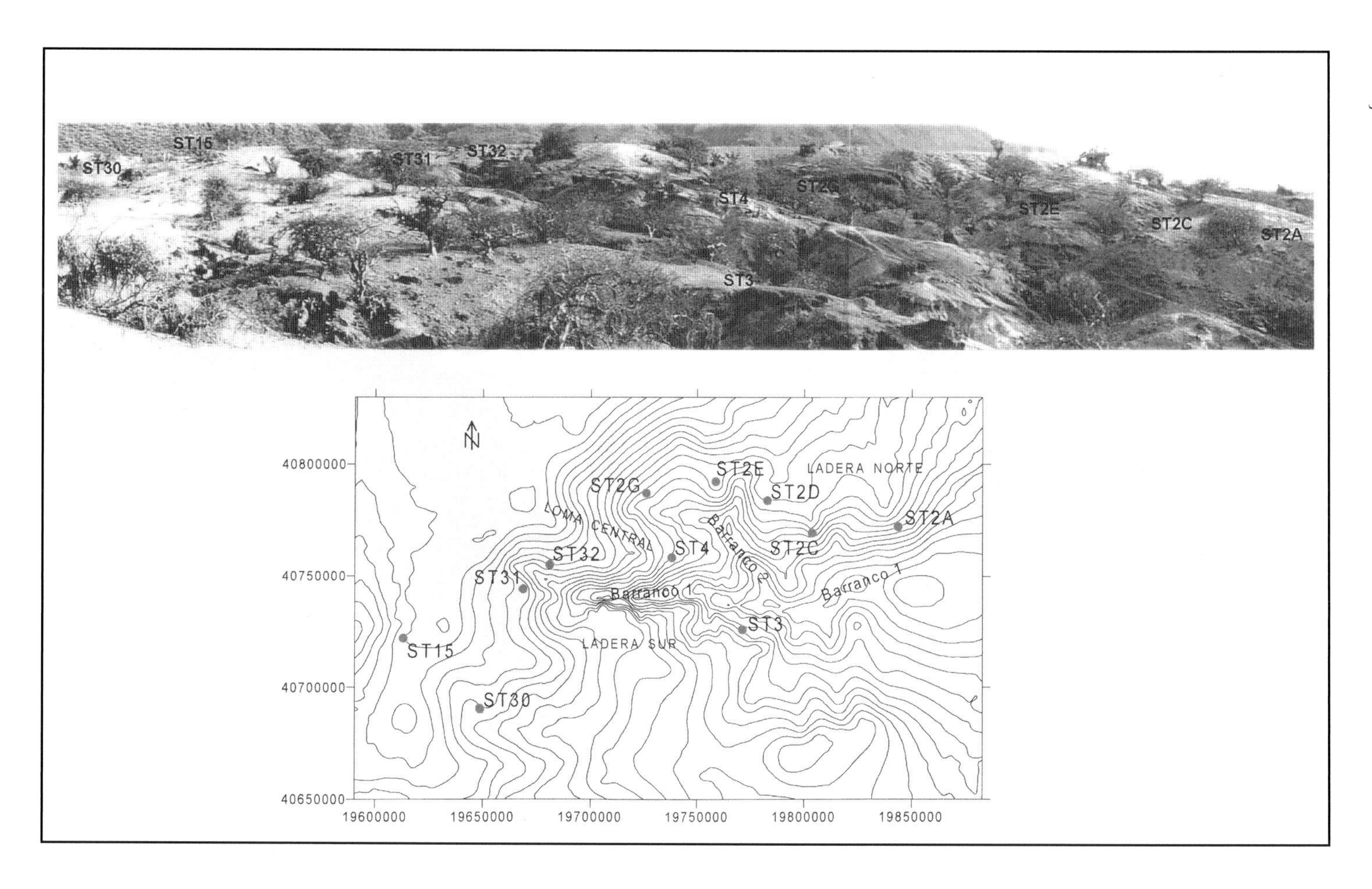

Figure 4.8 Distribution of the sites in the ST Site Complex (de la Torre, 2002). See Figure A.3 in the Appendix for a color version.

there seems to be a lack of significant transport by erosion. The densities of the material observed on the surface are similar to those documented through excavation in the same localities. Second, in the entire Maritanane area, no archaeological materials have been retrieved above the horizon documented on T1 in the USC member. Third, there is no difference whatsoever in raw material types, technology or morphology of the artifacts collected on the surface from those unearthed through excavation. Finally, most of the materials (both faunal and lithic) not affected by chemical dissolution or etching by humidity processes and soil pH appear fresh, without any traces of erosion from exposure. Despite these observations, to further ensure that the archaeological remains on the surface belonged to the same archaeological horizon identified on T1, only those remains observed on the upper parts of the slopes, directly under T1, were included in this study. The ST6 site was not included in the present analysis, because even though faunal remains and stone tools have been documented in it, none of it has been collected so far.

Archaeological materials included in this study therefore come from three sources: surface, *in situ* on the exposed tuff paleosurface, and *in situ* through excavation. Test excavations were carried out at all sites (except ST6, ST31, and ST32) to establish relationships between materials unearthed and those adhered to the exposed paleosurface. These relationships were expressed in terms of density of archaeological remains, taphonomic variables (i.e., preservation stages of fauna, types and sizes of bones represented, orientation, abrasion, polishing, taxonomy), and technology and raw material types used in the production of the lithic artifacts. The sedimentological and taphonomic analyses of the carbonate surfaces of the exposed tuff proved that the archaeological materials were adhered to the tuff in the past and were not adhered as a result of recent re-elaboration and carbonation of sediments. In some cases, anatomical parts of the same animal were found both in the horizon covered by sediments and in the horizon on the exposed paleosurface. The carbonate sediment on top of the tuff was indistinguishable from that documented under the sedimentary deposits. The same density of materials and the same taphonomic results were obtained in both buried and exposed archaeological contexts, further supporting their common provenience.

In 1995, an extensive excavation was initiated at ST4 because this site contained the largest cluster of archaeofauna and showed better cortical preservation on the bones than was found at the other sites. Bone preservation in the ST Site Complex in general is not very good. A significant amount of the bone specimens have cortical surfaces damaged by carbonate crusts, water etching and weathering. A smaller amount of specimens show better preservation, especially around ST4. Only the bones with intact, well-preserved cortical surfaces were used for the analysis of bone surface modifications.

The distribution of artifacts and faunal remains was plotted by using a laser theodolite. Only those archaeological materials appearing encrusted on the sediments overlying the tuff or which were exposed *in situ* during excavation were plotted. Surface materials from other sites were documented and collected but were not added to the laser theodolite data base, since we wanted to keep the spatial information as accurate as possible. Information was downloaded daily from the theodolite into a computer to check for possible errors or missing areas in the topography. A detailed topographic map of the main gully and the surrounding area of the ST Site Complex was made to plot archaeological materials more accurately. All the main geological features, especially

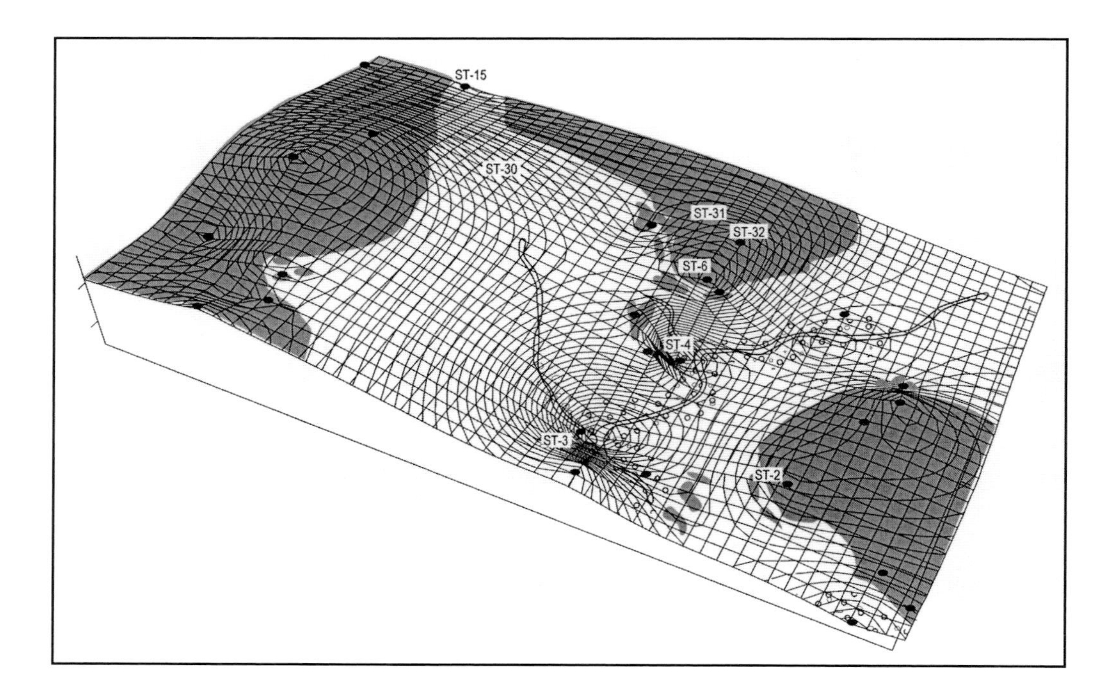

Figure 4.9 Location of the archaeological sites in the ST Site Complex on the paleotopography of the area, showing the paleoecdogical landmarks. The shaded area shows zones where T1 has been preserved. White areas show erosive processes created by channels.

the network of faults in the gully, were also recorded. This enabled a three-dimensional reconstruction of the paleotopography of the area, featuring the main characteristics of the paleolandscape and the site distribution therein (Figure 4.9).

Site descriptions

ST2

This site is found at the top of a small slope in the northeast of the ST Site Complex in a fine- and coarse-grained sandy matrix, directly above the tuff (T1) and about 2.8 m above the Main Tuff. The archaeological remains appear on the top of the T1 surface, covered by 10 cm of greenish sands and clays and a thin sand fill in a channel structure overlying clayish lacustrine and distal alluvial sediments. In a part of the exposed outcrop, T1 appears eroded due to a small channel. The erosion is very straight-edged and could be conditioned by a fault which follows a 105°-107° E direction.

ST3

This site is situated in the eastern margin of the ST Site Complex. Archaeological remains are mainly found at the base of a sand deposit in a former river channel. The base of the channel has a bio-precipitated laminated carbonate, where some faunal specimens and stone tools are found. Other archaeological remains appear on the non-cemented sides of this carbonate and even underneath it. The channel runs west-east and was formed in brown sands with rootcasts. Under these sands, coarse-grained sands containing basaltic blocks from the tuff are found. The site is about 1.5 m above the Main Tuff.

ST4

This site is located in the middle of the ST Site Complex, on top of a slope which runs west-east. It is comprised of coarse-grained sands in a channel-fill deposit formed in a clay matrix, which might have followed a northwest-southeast direction. The channel includes blocks from the tuff (T1) and has a carbonate layer on the basal erosive surface. At the base, volcanic blocks have been retrieved which have a diameter of approximately 0.5 m and which are covered by algae-deposited carbonate. The base of the channel is situated only 0.4 m above the top of the Main Tuff. The proximity of the channel to the Main Tuff could confuse the true stratigraphic location of the channel by placing it in the channel deposits found at the bottom of the USC. However, the existence of the carbonate level at the bottom of the channel, the upper contact with the greenish sands and the presence of basaltic blocks from T1 prove that the formation of the channel took place after the deposition of T1, which was eroded by the channel. This site was strongly altered by tectonic movements, affecting the position of the archaeological materials, which range from 0.2 m to 0.4 m in depth due to faulting.

ST6

This site is found on the same slope as ST4, about 20 m to the west. Archaeological remains are found on the T1 surface in the cemented greenish sands. A few meters away T2 can be observed, which was a little displaced by tectonic dynamics, but is stratigraphically above the archaeological deposit. In this zone, T1 is 0.4 m thick. It is not easy to determine the position of the site with respect to the Main Tuff because it was highly modified by tectonic faulting, but it ranges between 2.5 m and 3.5 m above the Main Tuff.

ST15

This site is the westernmost locality in the ST Site Complex. It is situated on a slope about 1.8 m above the Main Tuff. The archaeological level is situated in heavily cemented sandy sediments and lies directly on the surface of T1, which appears highly altered. At the same point, an erosional front can also be observed, which cuts the tuff at the level of formation of a small stream channel with coarse-grained sands. The channel fill appears laminated and its base is close to the top of the Main Tuff. A few meters south of the site, the tuff is continuously exposed with 1.2 m of underlying sand sediments and 0.4 m of clays with carbonate nodules above the Main Tuff.

ST30

This site is situated on a small platform formed in between two small gullies. It is the southernmost site in the ST Site Complex. Its surface is practically horizontal. T1, whose surface is highly weathered, can be observed to the south of the site. To the north, the greenish sands with rootcasts are found. The facies where archaeological materials appear is composed of cemented sands. The archaeological level corresponds to the level of the paleosurface where the other sites are located. From the southern outcrop of the locality it can be deduced that the site is situated 5 m above the Main Tuff.

ST31

This site is located at the beginning of the gully that separates ST4 and ST3. The archaeological horizon is located on the surface of T1 in a sandy matrix and under the fill deposit of a shallow paleochannel (0.3 m thick). T2 overlies this channel. The base of the sequence is covered with sediments. Therefore, the distance between ST 31 and the Main Tuff is uncertain. Under T1, there are a few cm of greenish sands.

ST32

This site is similar to ST31 and both sites are separated by an erosional gully. The site lies directly on the tuff surface; however, there is more than 1 m of sedimentary sequence between T1 and T2, which could indicate that there was a posterior subsidence movement or that T2 was deposited on an elevated paleorelief. The site is 3.5 m above the Main Tuff.

The shared stratigraphic position of all the archaeological sites of the ST Site Complex suggests that all the sites were deposited on the same paleosurface, i.e., on the surface of the T1 tuff. This would make it a complex of sites or a mega-site spanning ca. 3,500 m^2 of exposed outcrops. The eruption of a volcanic cone to the south of the lake created the deposition of T1 in an alluvial fan environment. Sedimentation eventually stopped, and then erosion proceeded to weather, crack and shatter the tuff. Next the lake briefly covered the area, forming a laminar carbonate layer in the lowest areas because of the action of algae and stromatolites. Afterwards sedimentation by a deltaic system followed the regression of the lake, and sands and clays were deposited on the carbonate and tuff surface. In the initial phase of this deltaic process, while some large and several small and shallower channels were forming, hominids occupied the area and created the ST Site Complex.

The paleorelief of the area (Figure 4.9) has been obtained by measuring the vertical distances of sites to the Main Tuff (after considering recent tectonic changes but not those that occurred after the deposition of the Main Tuff and before the formation of T1). The data were obtained along an extended area where the outcrops have allowed us observe the thicknesses of different stratigraphic units. The most elevated zone is situated to the south and west. A channel, which originated in the northwest, goes across ST4 and ST3. Other smaller streams joined this main channel. The direction of relief is north-south, similar to those fault directions currently observable, which indicates the reactivation of the faulting. According to these data, the most depressed parts of the ST Site Complex correspond to the zones where T1 has been eroded and channel deposits are well-represented. This could indicate that T1 and even the older deposits were eroded by the channels. The most elevated areas are situated on the well-preserved tuff (T1).

The position of the deposits supports the paleotopographic reconstruction, with a minor relief exposed showing a difference of 4 m between the lowest and the highest point. The larger channel would flow southeast or south through the zones where ST4 and ST3 are located. Smaller shallow channels appear nearby, also eroding T1. The overall impression is that hydraulic energy did not play a major role in the final configuration of archaeological sites. The side channels are too small and the situation of ST4 and ST3 on the edge of the channel supports the contentions that most of the materials were not transported by water. Judging by their size, if seasonality was as marked as it is in modern savannas, most of the streams in the areas where archaeological materials were deposited might have been dry or low at the time hominids were foraging in the area.

Zooarchaeological analysis of the ST Site Complex[1]

Data for this analysis were obtained through the study of the bone assemblage collected on the surface of ST and, in the case of ST4, also including the fauna collected in a test trench excavation carried out in 1995, 1996, 2000 and 2001 (see Chapter 6). The total number of faunal specimens recovered from the ST Site Complex is summarized in Table 4.3. The higher number of

Table 4.3 Total number of faunal specimens found in the ST site complex

Site	ST4	ST2	ST3	ST 15	ST30/31/32
Macro Mammals					
Identifiable sp.	296	129	39	58	55
Skull	54	15	9	8	6
Axial	53	26	8	17	5
Appendicular	189	88	22	33	44
Non-identifiable sp.					
Axial	118	167	22	18	26
Appendicular	71	43	48	54	22
Total Macro Mammals	485	339	109	130	103
Chelonia	43	51	36	19	4
Reptilia (non-Chelonia)	13	7	2	0	3
Aves	0	0	0	0	1
Fish	0	0	0	0	212
Crustaceae	6	0	0	0	0
Total	547	397	147	149	323

Identifiable specimens are those identified to element type and carcass size. Non identifable specimens refer to specimens identified to anatomical section but not to element type or carcass size.

specimens documented in ST4 is due to the more prolonged and extensive excavation at this site. For this reason, the ST4 faunal data are presented separately from those of the other sites, where only test trenches have been carried out. The taxa identified in the ST Site Complex are: *Ceratotherium simum*, *Hipparion* cf. *cornelianum*, *Equus* sp., *Sivatherium maurusium*, *Giraffa* cf. *pygmaea*, *Syncerus* sp., *Tragelaphus strepsiceros*, *Megalotragus kattwinkeli*, *Connochaetes taurinus*, *Damaliscus niro*, *Hippotragus* sp., *Aepyceros melampus*, *Sylvicapra* sp., *Antidorcas* cf. *recki*, *Gazella* sp, *Kolpochoerus* sp., *Metridichoerus compactus*, *Hippopotamus* sp., *Parmularius angusticornis*, and *Theropithecus* sp. (Geraads 1987; our fieldwork).

A taphonomic study of the fauna based on specimen size distribution, skeletal part representation and bone abrasion and polishing was conducted to evaluate the influence of water in the bone accumulations. Most specimens do not show the abraded and polished surfaces and edges typical of water transport (Table 4.4). More than 95% of the specimens do not seem to have undergone any significant transport beyond local rearrangement. The only exception seems to be the ST 3 assemblage, in which as many as 19% of the specimens show polishing on some of their edges and moderate abrasion on their surfaces, both typical of water-modified assemblages. Here, either hydraulic transport or local water erosion seem to have been operating. Some of the ST3 specimens are located in a coarse-grained sand deposit in a river channel, indicating a high-energy context.

The distribution of specimen sizes also indicates lack of selection by transport (Table 4.4, Figure 4.10). Most ST Site Complex sites show a distribution of specimen sizes similar to experimental assemblages that have not undergone any water transport, and also similar to some archaeological sites in low-energy depositional environments for which non-significant hydraulic transport is inferred (Blumenschine 1988, 1995; Selvaggio 1994; Capaldo 1995,

Table 4.4 Distribution of the faunal sample in three taphonomic categories: Abrasion, etching, and specimen size

	ST4	ST2	ST3	ST15	ST30/31/32
Polishing/Abrasion	22/485 (4.5)	11/339 (8.8)	21/109 (19.2)	4/149 (2.6)	4/103 (3.8)
Cortical Loss/Etching	387/485 (80)	303/339 (89)	106/109 (97)	130/149 (87)	88/103 (85)
Specimen Size %					
2-3 cms	88/485 (18)	51/339 (15)	11/109 (10)	31/149 (21)	15/103 (15)
3.1-4 cms	93/485 (19)	47/339 (14)	16/109 (15)	25/149 (17)	19/103 (19)
4.1-5 cms	106/485 (22)	58/339 (17)	11/109 (10)	19/149 (13)	21/103 (21)
5.1-6 cms	78/485 (16)	51/339 (15)	13/109 (12)	16/149 (11)	13/103 (12)
6.1-7 cms	38/485 (8)	35/339 (10)	11/109 (10)	13/149 (9)	10/103 (10)
7.1-8 cms	19/485 (4)	30/330 (9)	11/109 (10)	10/149 (7)	10/103 (10)
8.1-9 cms	10/485 (2.5)	24/339 (7)	14/109 (13)	10/149 (7)	3/103 (3)
9.1-10 cms	10/485 (2.5)	18/339 (5)	8/109 (7)	7/149 (5)	4/103 (4)
>10 cms	43/485 (9)	25/339 (8)	14/109 (13)	18/149 (12)	8/103 (7)

Numbers in numerator are for the number of specimens in each category. Numbers in the denominator indicate the total sample size. Numbers in brackets indicate the corresponding percentages.

1997). The high percentage of small-sized specimens (2-5 cm in length) indicates that most of the faunal assemblage was deposited by non-hydraulic agents in the same area from which it was recovered. Further support comes from the types of elements represented. Axial, cranial and appendicular elements appear together in the same sites (see below), despite the differences in density factors.

The spatial distribution of bone remains, always associated with stone tools, also suggests a strong hominid involvement with carcass manipulation in the ST Site Complex. Furthermore, over two-thirds of the appendicular specimens show fractures created while the bone was fresh. A significant percentage of bones also bear cut marks and percussion marks made by hominids (see below).

Skeletal part representation in the ST Site Complex fauna

Tables 4.5-4.8 present skeletal part information for all the ST sites. From the 1,544 identifiable mammal bone specimens (NISP), a total Minimum Number of Elements (MNE) of 306, belonging to different anatomical parts, has been documented. MNE identification for long limb bones was made within each carcass size group by examining bone specimens individually and using the following criteria: animal size, cortical thickness according to bone section, overlap of homologous parts, differences in size and morphology. There is a small number of limb bone epiphyses, so most of the diagnostic criteria were applied in identifying the abundant shaft fragments. Specimens were compared against one another to make sure they belonged to different elements before using them as valid MNE indicators. These estimates very likely underestimate the total number of bones originally present in the bone accumulation at the ST sites. However, they offer much more accurate information than would be obtained if the MNE estimate were based solely on the identifiable epiphyseal fragments (e.g., Bunn and Kroll 1986;

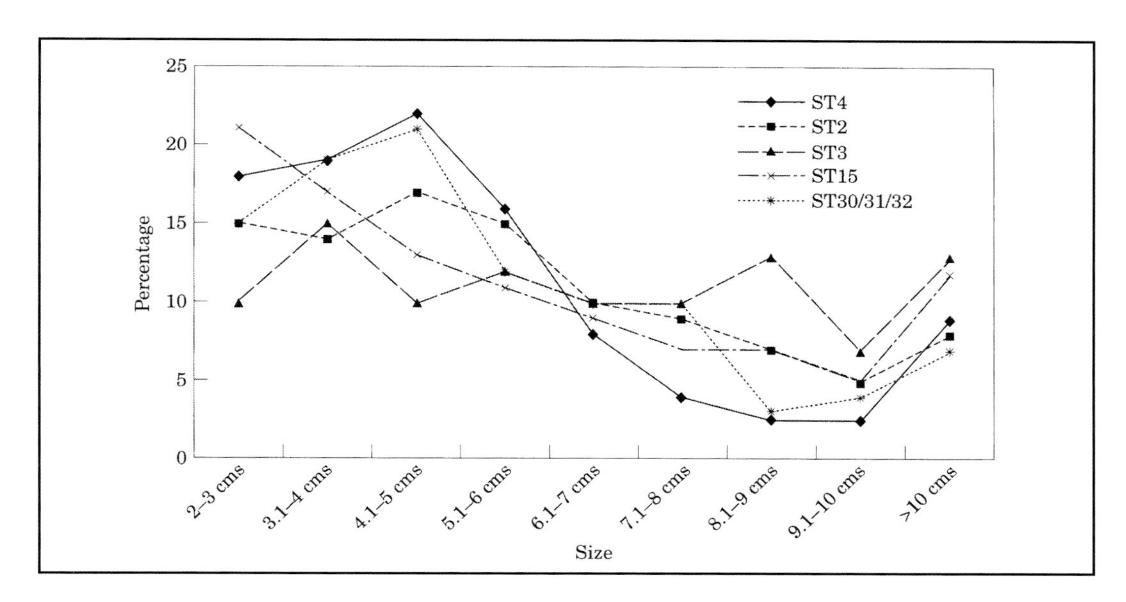

Figure 4.10 Size distribution of bone specimens from the ST Site Complex. All sites show abundant amounts of small remains, indicating a lack of transport.

Table 4.5 Number of elements (MNE) identified in the ST site complex

Site	ST4	ST4	ST4	ST3	ST3	ST3	ST2	ST2	ST2	ST15	ST15	ST15	ST*	ST*	ST*
Size	1, 2	3	4, 5, 6	1, 2	3	4, 5, 6	1, 2	3	4, 5, 6	1,2	3	4, 5, 6	1, 2	3	4, 5, 6
SK	6	7	5	2	2	3	1	4	3	4	4	3	4	2	3
CV	4	1	2	0	1	0	4	2	1	2	1	0	0	1	0
TV	2	2	0	0	0	0	1	2	0	0	1	0	0	0	0
LV	0	0	1	0	1	0	1	0	0	0	0	0	0	0	0
PL1/2	2	3	2	0	0	0	0	1	1	1	1	0	0	0	1
SC	3	3	3	0	0	0	1	1	2	1	1	0	2	1	0
RB	7	3	5	2	0	0	4	3	1	3	1	0	5	4	2
HM	5	5	6	2	1	1	1	2	2	2	1	0	2	1	1
RU	5	6	2	0	0	0	2	2	0	0	2	1	2	1	1
C	6	4	3	0	0	0	1	2	0	0	0	0	1	1	1
MC	1	2	3	0	1	0	2	3	0	0	0	0	1	1	0
FM	2	2	3	0	0	0	0	0	0	2	1	0	0	0	0
TB	3	3	3	1	1	0	1	2	0	2	1	0	1	0	0
T	0	0	3	0	1	0	2	3	2	0	0	2	0	0	0
MT	5	2	2	0	0	0	2	2	1	1	1	0	0	1	0
PH	1	1	0	0	0	0	0	0	0	0	0	0	3	0	1

*MNE values were obtained lumping together ST30, 31 and 32.
SK: skull; CV: cervical vertebrae; TV: thorathic vertebrae; LV: lumbar vertebrae; PL: pelvis; SC: scapula; RB: rib; HM: humerus; RU: radio-ulna; C: carpals; MC: metacarpal; FM: femur; TB: tibia, T: tibia; T: tarsals; MT: metatarsal; PH: phalanges.

Table 4.6 Analysis of epiphyseal and shaft remains and percentage of change

	ST4			ST2			ST30/1/2			ST15			ST3		
Carcass Size	1,2	3	4, 5, 6	1, 2	3	4, 5, 6	1, 2	3	4, 5, 6	1, 2	3	4, 5, 6	1, 2	3	4, 5, 6
Epiphysis (NISP)	10	10	17	3	5	1	1	0	1	2	1	0	0	0	0
Shaft (NISP)	124	98	38	44	58	29	28	24	14	31	47	9	35	28	7
Expected epiphyses*	42	40	38	16	22	6	12	8	4	14	12	2	6	6	2
Epiphysis shaft ratio	0.08	0.1	0.44	0.06	0.08	0.03	0.03	0	0.07	0.06	0.02	0	0	0	0
% change	76.1	75	55.2	81	77.2	83.3	91.6	100	75	85.7	91.6	100	100	100	100

* According to MNE

Table 4.7 MAU and %MAU values for the ST4 faunal assemblage

	MAU (ST4)			% MAU (ST4)			% MAU (ST4)*		
Carcass Size	1, 2	3	4, 5, 6	1, 2	3	4, 5, 6	1, 2	3	4, 5, 6
SK	6	7	5	100	100	100	-	-	-
CV	0.8	0.14	0.28	13.3	2	5.6	26.6	4.6	9.3
TV	0.13	0.13	0	2.1	1.8	0	4.3	4.3	0
LV	0	0	0.2	0	0	4	0	0	6.6
PL	1	1.5	1	16.6	21.4	20	33.3	50	33.3
SC	1.5	1.5	1.5	25	21.4	30	50	50	50
RB	0.25	0.1	0.17	4.1	1.4	3.4	8.3	3.3	5.6
HM	2.5	2.5	3	41.6	35.7	60	83.3	83.3	100
RU	2.5	3	1	41.6	42.8	20	83.3	100	33.3
C	0.08	0.16	0.16	1.3	2.2	3.2	2.6	5.3	5.3
MC	3	2	0.5	50	28.5	10	100	66.6	16.6
FM	1	1	1.5	16.6	14.2	30	33.3	33.3	50
TB	1.5	1.5	1.5	25	21.4	30	50	50	50
T	0	0	0.25	0	0	5	0	0	8.3
MT	2.5	1	1	41.6	14.2	20	83.3	50	33.3
PH	0.08	0.16	0	1.3	2.2	0	2.6	5.3	0

*Removing the skull

Table 4.8 MAU and %MAU values of the ST3, ST2, ST15, St30, ST31, and ST32 sites

	MAU (ST complex)			% MAU (ST complex)			% MAU (ST complex)*		
Carcass Size	1, 2	3	4, 5, 6	1, 2	3	4, 5, 6	1, 2	3	4, 5, 6
SK	11	12	12	100	100	100	-	-	-
CV	0.85	0.7	0.14	7.7	5.8	1.1	24.2	28	5.6
TV	0.06	0.2	0	0.5	1.6	0	1.7	8	0
LV	0.16	0.16	0	1.4	1.3	0	4.5	6.4	0
PL	0.5	1	1	4.5	8.3	8.3	14.2	40	40
SC	2	1.5	1	18.1	12.5	8.3	57.1	60	40
RB	0.5	0.33	0.1	4.5	2.7	1.1	14.2	13.2	4
HM	3.5	2.5	2	31.8	20.8	16.6	100	100	100
RU	2	2.5	1	18.1	20.8	8.3	57.1	100	50
C	0.33	0.5	0.16	3	4.1	1.3	9.4	20	6.4
MC	1.5	2.5	0	13.6	20.8	0	42.8	100	0
FM	1	0.5	0	9	4.1	0	28.5	20	0
TB	2.5	2	0	22.7	16.6	0.1	71.4	80	0
T	0.25	0.5	0.5	2.2	4.1	4.1	7.1	20	20
MT	1.5	2	0.5	13.6	16.6	4.1	42.8	80	20
PH	0.12	0	0.04	1	0	0.3	3.4	0	1.6

*Removing the skull

Marean and Kim 1998). The process of MNE identification using shaft specimens is time-consuming compared to the relatively straightforward identification when using epiphyseal fragments, but usually provides higher estimates for long limb elements which more closely approximate the original amount. This is especially true for assemblages that have undergone post-depositional carnivore ravaging (Blumenschine and Marean 1993; Table 4.6, Figure 4.11).

By transforming the MNE obtained in the different ST sites into Minimum Animal Units (MAU), an overall underrepresentation of animal units according to the MNI (Minimum Number of Individuals) can be observed in the bone assemblage. When the percentage of these animal units (%MAU) is calculated, the axial parts of the skeleton (vertebrae and ribs) and the small compact limb bones (carpals, tarsals and phalanges) appear underrepresented. This situation is documented in all the sites irrespective of animal size. Pelves and scapulae are moderately represented. Long limb bones are dominant, when excluding skull remains. Upper limb bones (ULB) (humeri and femora) and intermediate limb bones (ILB) (radio-ulnae and tibiae) are represented by similar numbers of elements, although ULB seem to be more abundant in most animal size categories. Metapodials are comparatively underrepresented. Curiously, forelimbs seem to be more abundant than hindlimbs. This situation was documented in all the ST sites (Tables 4.7–4.8).

Carnivore modification of the ST Site Complex fauna

Prior to any behavioral interpretation of the ST Site Complex archaeofaunas, a thorough taphonomic analysis is required. Experiments with carnivores, both in captivity (Marean et al. 1992) as well as in the wild (Capaldo 1995, 1998) show that the bones most likely to be destroyed by hyenas, when they have access to bone

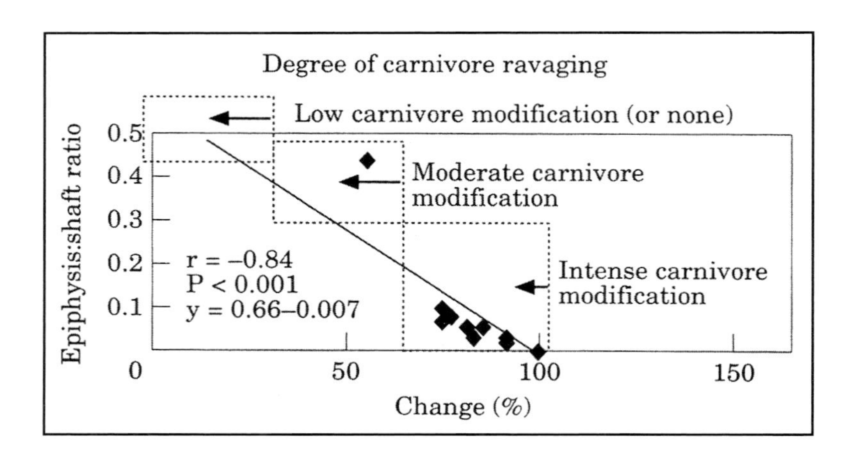

Figure 4.11 Distribution of data from all the ST sites for the epiphyseal fragment:shaft fragment ratio and the percentage of change, suggesting intense post-depositional ravaging by carnivores.

remains discarded by humans, are the vertebrae, ribs, pelves and scapulae, followed by the small compact limb bones. Capaldo's (1995, 1998) research on bone modification patterns by carnivores in complete skeletons demonstrates that carnivore ravaging can result in the deletion of more than 95% of axial elements. His study also shows that carnivores may introduce a 85%–95% bias in the recovery of skeletal parts, even when tooth marks, the most telling signs of their intervention, are relatively infrequent: as low as 5%–10% of bone specimens being tooth-marked in experimental assemblages (Capaldo 1995). The most underrepresented bones in the ST faunal sample are vertebrae, ribs and compact limb bones; this would therefore suggest a significant carnivore impact on the original ST faunal assemblages.

Experimental work with carnivores has also shown that the ratio of epiphyseal fragments to shaft fragments can be used to identify carnivore involvement in the formation of bone assemblages and to measure the intensity of their ravaging (Blumenschine and Marean 1993). The grease contained in fresh limb epiphyses make these bone portions very attractive to scavenging bone-crushing carnivores (such as hyenas). This results in low epiphyseal fragment:shaft fragment ratios. Blumenschine and Marean (1993) documented a negative relationship between the post-ravaged epiphyseal fragment:shaft fragment ratio and the observed percentage change in the original number of epiphyseal fragments when carnivores have access to limb bone elements. This type of comparative analysis was also applied to the ST Site Complex. In Table 4.6, a stark contrast in representation of epiphyses versus diaphyses can be observed; epiphyses are clearly underrepresented. To obtain the percentage of change in the original number of epiphyseal fragments, the total MNE was used (Table 4.6, Figure 4.11), assuming that each element was originally represented by two epiphyses (proximal and distal). The percentage of change is obtained by dividing the difference of the total number of epiphyses ideally represented by the MNE, minus the actual number of epiphyses observed, by the total number of epiphyses ideally represented by the MNE; this value is then multiplied by 100 to obtain a percentage value. The total number of epiphyseal fragments in the bone assemblage was probably higher than our estimate, but using the total MNE gives a conservative, minimum estimate which allows some basic interpretations.

When plotting the epiphyseal fragment:shaft fragment ratio against the percentage of change, most of the different carcass size groups in all the sites cluster around the same values (Figure 4.11). According to previous experimental studies (Marean et al. 1992; Marean 1998; Blumenschine 1988; Blumenschine and Marean 1993: Capaldo 1995, 1997, 1998) the clustering of the ST data as shown in Figure 4.11 would be indicative of intense carnivore ravaging. The relationship between the two variables is negative and highly significant.

Carnivore intervention was also documented by the percentage of epiphyseal fragments that were conspicuously tooth-marked (50%). Failure to find higher rates of tooth-marking in these bone portions may have been due to poor preservation of the cortical surfaces of most of the epiphyseal fragments. In summary, the low representation of ribs, vertebrae, compact limb bones, the low epiphyseal fragment:shaft fragment ratio, and the significant tooth-marking of epiphyseal fragments support the hypothesis of significant carnivore involvement in the formation of the faunal assemblages at the ST Site Complex.

The utility of skeletal part profiles to reconstruct hominid behavior at ST Site Complex

If the skeletal part profiles are to be used for comparative analyses it becomes necessary to create proper taphonomic filters to avoid comparing frameworks that are unrelated and are taphonomically biased in different ways. O'Connell (1997) recently proposed that Plio-Pleistocene sites from areas such as Olduvai could represent near-kill locations like those documented among the modern Hadza hunter-gatherers. No comparative analysis has been made in this regard, but the dense concentration of remains in the Olduvai sites and the high density of individuals represented contrast with the low density of individuals clustered around the near-kill spots documented by O'Connell et al. (1992) among the Hadza and the very widespread nature of these bone assemblages. However, O'Connell et al.'s study did not consider the time-averaging processes involved in the formation of the Plio-Pleistocene archaeofaunas. Future comparative work should either support or disprove this hypothesis as a good explanatory framework for the Olduvai sites. Nevertheless, the low density of remains, MAU and MNI represented in the ST Site Complex, as well as the widespread nature of the bone assemblage, grant tentative support to the idea that these sites might in fact represent a near-kill location which hominids repeatedly visited to obtain carcasses and butcher them.

O'Connell et al. (1992) have published the MAU represented at these near-kill locations among the Hadza (Table 4.9). MAU information regarding carcass transport among the Hadza is also available (O'Connell et al. 1988, 1990; Bunn et al. 1988; Monahan 1998; Table 4.10). To use this information for comparative purposes, it is necessary to transform the data by adding a taphonomic filter. Such a taphonomic filter should be based on the percentage of elements lost after carnivore ravaging of human-made accumulations, as we infer in the ST Site Complex (Table 4.6). Failure to do so produces correlations between the ST faunal data and the Hadza faunal information (both from near-kill locations and from transported assemblages) that are neither positive nor meaningful (Table 4.11). A taphonomic filter is a modification of the percentages of some of the bones that undergo the most drastic deletion after post-depositional ravaging by

Table 4.9 %MAU per body size class at Hadza kill sites (1)

Carcass size	2, 3	4, 5, 6	2, 3*	4, 5, 6*
SK	78.5	100	95.6	100
CV	35.7	92.8	2.1	5
TV	35.7	78.5	2.1	4.2
LV	32.1	85.7	1.7	4.2
PL	39.2	78.5	23.9	39
SC	100	85.7	56.5	71.4
RB	46.4	71.4	6.5	4.2
HM	64.2	64.2	78.2	64.2
RU	71.4	78.5	86.9	78.5
C	78.5	92.8	4.7	5
MC	78.5	92.8	95.6	92.5
FM	75	85.7	91.3	85.5
TB	78.5	85.7	95.6	85.7
T	82.1	100	4.5	5
MT	82.1	100	100	100
PH	53.5	85.7	5.2	8.5

(1) O'Connell et al. (1988, 1990) as shown in Monahan (1998).
* With taphonomic filter (see text).

Table 4.10 %MAU per body size class in Hadza transport assemblages (1)

Carcass size	2, 3	4, 5, 6	2, 3*	4, 5, 6*
SK	75	25	78.6	25
CV	93.7	50	4.9	2.5
TV	96.8	57.5	1.6	5
LV	100	50	4.9	2.5
PL	96.8	50	50.8	25
SC	95.3	62.5	100	62.5
RB	49.3	45	2.2	2.5
HM	81.2	100	85.2	100
RU	71.8	87.5	75.4	87.5
C	67.1	0	3.2	0
MC	67.1	25	70.4	25
FM	73.4	50	77	50
TB	68.7	50	72.1	50
T	67.1	0	3.2	0
MT	62.5	25	65.5	25
PH	87.5	0	4.9	0

(1) O'Connell et al. (1988, 1990) as shown in Monahan (1998).
* With taphonomic filter (see text).

Table 4.11 Correlations (Pearson's coefficient) between the ST archaeofauna and the Hadza data set

% **MAU Hadza near kill locations** (1)

Site	carcass size	r	X2	P
ST4	small	0.25	0.06	0.085
ST4	middle	0.17	0.02	0.115
ST4	large	-0.44	0.19	0.099
ST *	small	0.21	0.04	0.069
ST *	middle	0.28	0.07	0.053
ST *	large	-0.52	0.27	0.044

% **MAU Hadza transport assembages** (2)

Site	carcass size	r	X2	P
ST4	small	-0.23	0.05	0.404
ST4	middle	-0.12	0.01	0.661
ST4	large	0.66	0.43	0.007
ST *	small	-0.15	0.02	0.572
ST *	middle	-0.25	0.06	0.350.
ST *	large	0.62	0.39	0.012

* ST2, ST3, ST15, ST30, ST31, and ST32
(1) Table 4.7 First two columns, without taphonomic filter.
(2) Table 4.8 First two columns, without taphonomic filter.

carnivores. Marean et al. (1992) and Capaldo (1995, 1998) have clearly demonstrated that axial bones (ribs and vertebrae) and small compact limb bones are almost deleted completely by hyenas, with averages of modification always between 90% and 100% as mentioned above. If the Hadza data are modified with this perspective to simulate carnivore post-depositional ravaging, both faunal datasets (Hadza accumulations and archaeological assemblages) can be compared more equitably.

Cranial MNE and MAU in archaeological assemblages are usually overrepresented compared to postcranial remains. Differential preservation, usually due to the density of these elements (especially teeth), accounts for their overrepresentation. For this reason, several individuals appear represented by isolated teeth. Given contrasting representation of cranial and postcranial remains, and the uncertainty about linking cranial representation to human behavior, it was decided to base skeletal part comparative analyses on the postcranial skeletal elements only. The reason for doing so is that %MAU representation is very sensitive to the inclusion or exclusion of cranial elements (Bunn and Kroll 1986).

The %MAU from the Hadza near-kill assemblages, taphonomically modified, was compared to the ST Site Complex %MAU. Most comparisons, especially those of the small and medium-sized carcasses, show a positive and significant correlation (Figure 4.12). Overall, the data obtained at the ST sites seem to fit the predicted model for skeletal part representation based on the Hadza near-kill locations. The small carcass sizes match the percentages obtained in Hadza near-kill locations by 77% ($r2=0.77$) at the ST4 site and by 60% ($r2=0.60$) at the other ST Site Complex sites. The medium-sized carcasses show similar values: $r2=0.68$ at the ST4 site, and $r2=0.69$ at the other ST Site Complex sites.

However, when comparing the ST site information to the %MAU data from Hadza transported assemblages, a similarly positive and significant correlation is obtained (Figure 4.13). Small fauna correlates to a percentage of 67% ($r2=0.67$) at ST4 and 75% ($r2=0.75$) at the other ST Site Complex sites. The medium-sized fauna shows a similar pattern: at ST4, $r2=0.67$ and at the other ST Site Complex sites, $r2=0.64$.

This is a good example of equifinality (Gifford 1991): that is, different processes leading to the same end product. Based on skeletal part profiles alone, there is no way significant differences can be found between the near-kill location and transport models. This case supports multiple researchers' assertions regarding the limited utility of skeletal part profiles for hominid behavioral inferences when bone assemblages have undergone strong carnivore attrition and modification (Blumenschine 1988, 1995; Blumenschine and Marean 1993; Marean et al. 1992; Marean 1998; Capaldo 1995, 1998).

Faunal density in the ST Site Complex

One of the features that makes the ST Site Complex different from the Olduvai sites is the density of remains and individuals represented. A total of 1,544 faunal specimens have been recovered over 3,500 m^2 of the exposed paleosurface in the ST Site Complex. This provides a density value of approximately 0.5 bone specimens per m^2. If we select any of the Olduvai sites, bone density is much higher. For instance, at FLK Zinj (also known as FLK Level 22), more than 40,000 bone specimens were recovered (Potts 1988) from 290 m^2, giving an average density of 137 bone specimens per m^2. Even if we restrict the area to the immediate surroundings near clusters of faunal remains, Olduvai Bed I sites still have much higher densities of remains than the ST sites (Figure 4.14, Table 4.12).

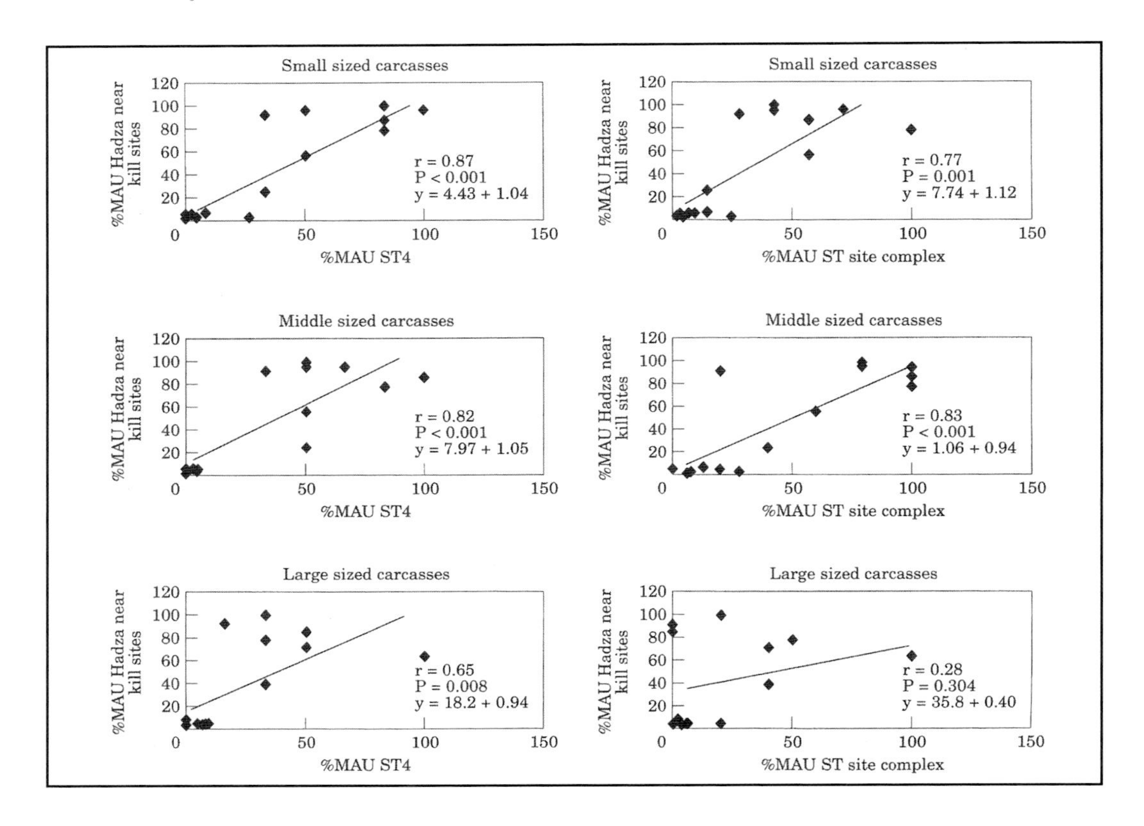

Figure 4.12 Relationship between skeletal parts represented at the ST sites and those documented in Hadza near-kill sites according to carcass size. The archaeological sample has been divided into two comparative units: ST4 and the remainder of the ST Site Complex sites.

The same applies to the MNI figures (Figure 4.14, Table 4.12). Olduvai sites have an average above 30 MNI, whereas ST sites show less than half that amount. Even if we compare the excavated assemblage from the ST4 site with assemblages from Olduvai sites, the density at ST4 is still less than that of any Olduvai archaeological site, although it is closer to one of them, FLK North 6. Density estimates provided here are preliminary, and most sites await further excavation to validate the comparison with other localities.

O'Connell et al. (1992) state that "multiple kill sites at intercept locations may be scattered over areas of up to several thousand square meters or more depending on the characteristics of the location itself." In these types of sites, bones from individual prey animals are often left in discrete and widely separated concentrations. Such concentrations have also been documented at some ST sites. Several elements of an equid were clustered in ST2A; articulated elements from an impala were found in ST32; bones from a large bovid were clustered in ST30 and elements belonging to most of the anatomical areas of the skeleton of a *Giraffa* sp. were found in ST4 (see Chapter 6).

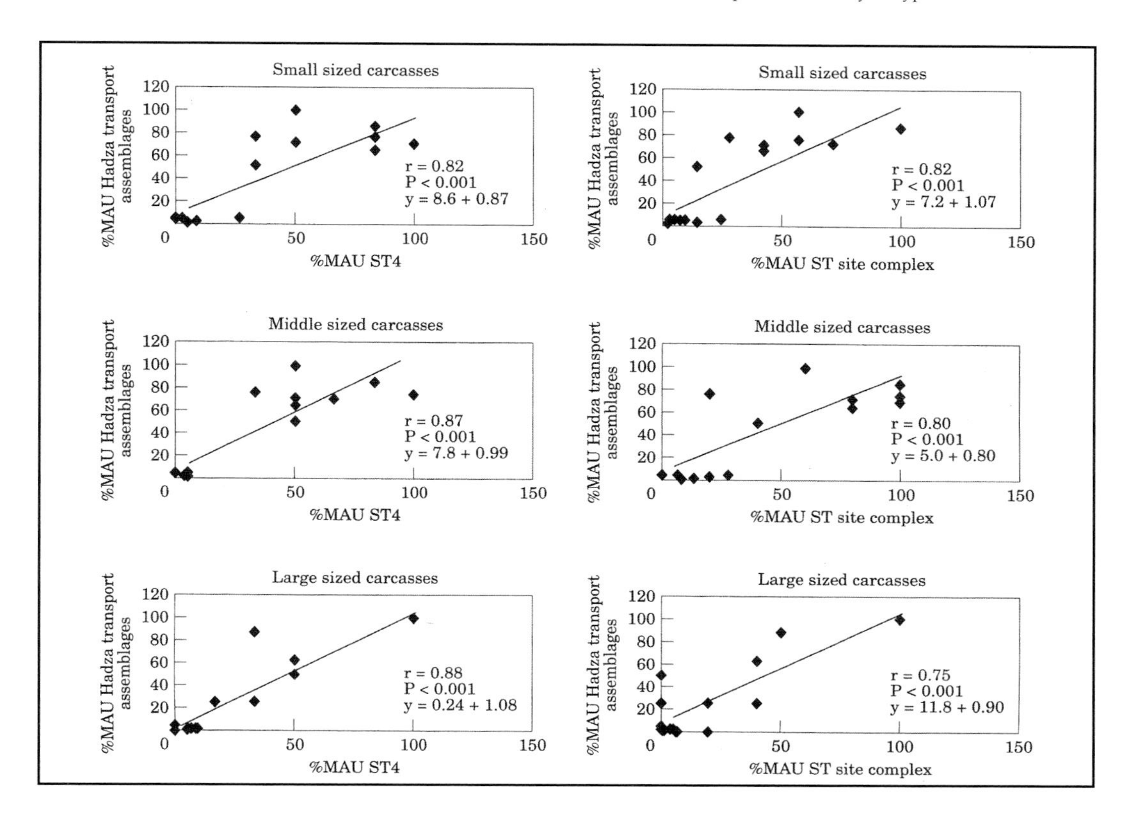

Figure 4.13 Relationship between skeletal parts represented at the ST sites and those documented in Hadza transported assemblages according to carcass size. The archaeological sample has been divided into two comparative units: ST4 and the remainder of the ST Site Complex sites.

Bone weathering in the ST Site Complex

Table 4.13 summarizes the weathering stages identified in the ST archaeofauna. More than two-thirds of the bone specimens in the archaeological sites could not be used in this analysis, since their original cortical layers have been lost due to etching from being under water or in humid soil for a long time. The cortex has also disappeared in a large sample of bones because they were covered by a carbonate matrix, resulting from the diagenesis of the sediments making up the paleosurface and overlying the tuff surface.

Bunn and Kroll (1986) have pointed out that bone weathering data do not effectively measure accumulation time but rather measure burial time. Behrensmeyer (1978) clearly showed that different elements from the same carcass can exhibit different weathering stages. Given the structural differences among axial, cranial and appendicular bones, weathering analyses in the ST Site Complex focused only on shaft fragments from long limb bones. From the sample of bones that had not been affected by diagenesis, specimens from the ST4 site were the most well-preserved, with stages 0 and 1 being

Table 4.12 Distribution of appendicular specimens according to weathering stages and cortical loss

Sites		ST4	ST2	ST3	ST15	ST30/1/2
Appendicular specimens		260	131	70	87	66
Affected by water/carbonate		126	74	51	63	50
Weathering stages	0	109	10	3	18	7
	1	19	30	10	0	0
	2	6	8	0	6	0
	3	0	9	6	0	9

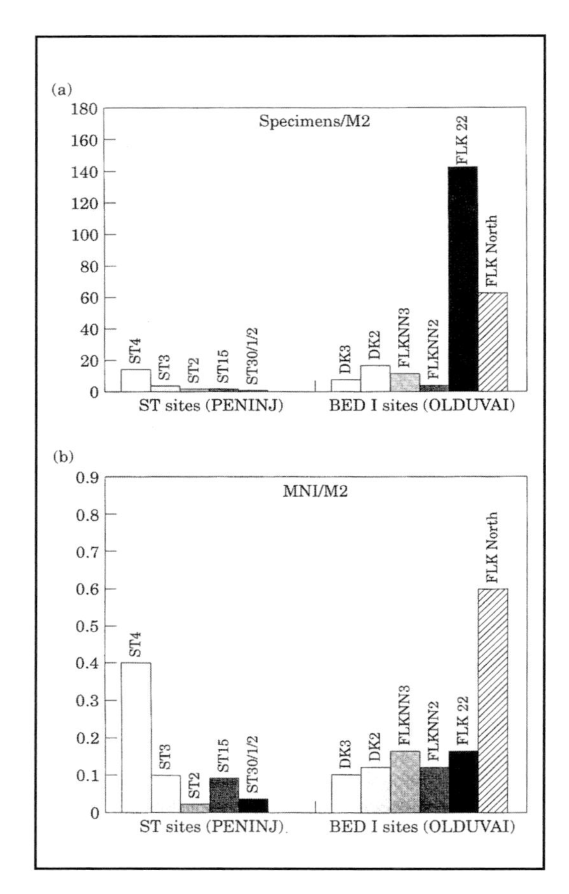

Figure 4.14 Comparison of bone density in the ST sites and Olduvai Bed I sites, according to NISP/m^2 *(A) and MNI/m*^{2} *(B).*

predominant. ST2 and ST 3 showed a wider variety of weathering stages, with stage 1 being predominant and with some specimens showing stages 2 and 3. ST15 showed mostly fresh (stage 0) bone surfaces with only 25% of the specimens showing stage 2 features. The ST30-31-32 complex showed a curious distribution. Most of the bones from this group of sites which were scored for weathering came from ST30. Bone specimens range between fresh (stage 0; 43%) and stage 3 (57%). In this case, it can be clearly observed that the fresh bones belong to small fauna and the more weathered specimens are from larger fauna, especially the postcrania of a member of the bovini tribe. This would rule out differential preservation in bones from the same carcass and would support an interpretation of weathering in this site as an indicator of the time of the total bone accumulation.

Most of the bone assemblage from ST4 and ST15 would have been deposited in a relatively short time period, whereas ST2 and ST30-31-32 seem to have spanned a longer period, probably involving several occupational episodes. An alternative explanation would be that the bones representing weathering stages 2 and 3 are a small part of the sample, which could be part of a background scatter which was deposited prior to the hominid presence in the area. Future research may grant further support to either interpretation.

Table 4.13 Densities of remains and MNI in the ST sites and Olduvai Bed I sites

SITES PENINJ*	m^2	specimens/m^2	MNI/m^2
ST4	40	547 (13.6)	16 (0.4)
ST3	50	147 (2.9)	7 (0.1)
ST2	500	397 (0.79)	12 (0.02)
ST15	100	149 (1.4)	9 (0.09)
ST30/1/2	400	323 (0.8)	15 (0.03)
SITES OLDUVAI Bed I**			
DK3	345	2,433 (7)	36 (0.10)
DK2	345	5422 (15.5)	42 (0.12)
FLKNN3	209	2261 (10.8)	34 (0.16)
FLKNN2	186	478 (2.5)	23 (0.12)
FLK22	290	40,172 (138)	36 (0.12)
FLK NORTH 6	37	2258 (61)	23 (0.60)

* Area restricted to the main bone clusters.
** Based on Potts (1988).
Numbers in brackets indicate density of bone specimens or MNI per 1 m^2.

Bone surface modifications

Over the past 15 years, the study of cut marks, percussion marks, and tooth marks has become increasingly relevant in the discussion of hominid and carnivore interactions with fossil faunal assemblages. Blumenschine (1988, 1995) used the percentage and distribution of tooth marks according to bone portion (long limb shafts) as good indicators of the order of access to carcass remains by carnivores. Early access creates high rates of tooth-marked shaft specimens, whereas secondary access to demarrowed bones results in low percentages of tooth-marked shaft specimens. Domínguez-Rodrigo (1997b, 1999a) also developed a methodology regarding cut mark distribution according to bone section and element type to differentiate between primary and secondary access to carcasses by hominids. Domínguez-Rodrigo (1999a) also conducted a study on flesh distribution and availability in large carnivore kills to compare with cut mark distribution patterns on the overall anatomy of medium-sized carcasses.

Most of the analysis that follows is based on long limb bones for two reasons: they are the most common skeletal part represented in many faunal assemblages, and most of the analytical methodology developed to differentiate hominid and carnivore interaction has been based on bone surface modifications on these elements. Long limb bones are divided into three categories as was described earlier in this chapter (Domínguez-Rodrigo 1997a): upper limb bones (ULB; humerus and femur), intermediate limb bones (ILB; radius-ulna and tibia), and lower limb bones (LLB; metapodials). Blumenschine's (1988) analytical method for quantifying percussion and tooth marks is based on bone portions (epiphyses, near-epiphyses and diaphyses), with diaphyses being the most diagnostic. Data on percussion and tooth marks will be presented for limb bone shafts only. Most marks appear on Bunn's (1982) size 2 and size 3 carcasses. In this analysis, both carcass sizes will be lumped together, and will be considered as smaller mammals. Larger mammals comprise carcass Bunn's

Table 4.14 Distribution and percentages of tooth marks and percussion marks*

TOOTH MARKS	smaller mammals	larger mammals	Total
Upper limb bones	2/46 (4.3)	1/12 (8.3)	3/58 (5.1)
Intermediate limb bones	1/51 (1.9)	1/11 (9)	2/62 (3.2)
Lower limb bones	2/28 (7.1)	0/4 (0)	2/32 (6.2)
Total	5/125 (4)	2/27 (7.4)	7/152 (4.6)
PERCUSSION MARKS	smaller mammals	larger mammals	Total
Upper limb bones	12/46 (26)	3/12 (25)	15/58 (25.6)
Intermediate limb bones	11/51 (21.5)	3/11 (27.2)	14/62 (22.5)
Lower limb bones	1/28 (3.5)	0/4 (0)	1/32 (3.1)
Total	24/125 (19.2)	6/27 (22.2)	30/152 (19.7)

* Shaft specimens. Numerator indicates the number of marked specimens. Denominator shows the number of specimens in each category. Numbers in brackets are for percentages.

Table 4.15 Cut mark patterns and distribution in the ST archaeofauna

	smaller mammals	larger mammals
HUM PSH	0/0 (0)	0/0 (0)
HUM MSH	4/10 (40)	2/4 (5)
HUM DSH	0/2 (0)	0/2 (0)
RAD PSH	2/3 (66.6)	0/2 (0)
RAD MSH	2/12 (16.6)	1/4 (25)
RAD DSH	0/1 (0)	0/0 (0)
MC PSH	1/3 (33.3)	0/0 (0)
MC MSH	0/4 (0)	0/3 (0)
MC DSH	0/2 (0)	0/2 (0)
FEM PSH	0/1 (0)	0/0 (0)
FEM MSH	2/5 (20)	1/2 (50)
FEM DHS	0/2 (0)	0/2 (0)
TIB PSH	0/2 (0)	0/1 (0)
TIB MSH	2/8 (25)	0/3 (0)
TIB DSH	0/2 (0)	0/0 (0)
MT PSH	0/1 (0)	0/0 (0)
MT MSH	0/3 (0)	0/1 (0)
MT DSH	0/3 (0)	0/0 (0)
TOTAL	13/64 (20.3)	4/26 (15.3)
cut marked ULB*	3	2
cut marked ILB*	3	1
cut marked LLB*	0	0
ULB:ILB RATIO		
ULB	9/20 (45)	5/10 (50)
ILB	9/28 (32.1)	2/10 (20)
MSH	10/41 (24.3)	4/17 (23.5)
ENDS	2/21 (9.5)	0/9 (0)
MEATY MSH %	10/13 (76.9)	4/4 (100)
NICMSP:NISP	13/64 (20.3)	4/26 (15.3)
NCMMSSP:NICMSP	10/13 (76.9)	4/4 (100)

* Non-identifiable to element type. All shaft specimens.

Table 4.16 Cut marked specimens from the ST archaeofauna*

	smaller mammals	larger mammals
VERTEBRA	3	0
RIBS	1	2
PELVIS	1	1
SKULL	1	0
SCAPULA	2	1
CALCANEUM	0	1
PHALANGE	1	0

*Non-long limb bone specimens.

(1982) size 4 through size 6. Bone surface analysis has been applied to all specimens scored as weathering stage 0 (Behrensmeyer 1978), whose cortical layers have not been affected by subaerial weathering or other diagenetic processes such as soil pH and water etching.

Results from the analysis of bone surface modifications and their anatomical distributions are presented in Tables 4.14–4.16. Tooth marks have been observed in almost one out of two epiphyses from limb bones and one out of three rib fragments. Tooth marks are fairly infrequent (4.6%) on limb shaft fragments, with a moderately higher percentage in large-sized carcasses (Table 4.15). Metapodials appear tooth-marked at a higher rate than meat-bearing limb bones.

Percussion marks occur on all limb bone types. There is a marked contrast between their occurrence on meat-bearing bones and on metapodials. Percussion mark rates are broadly similar in both carcass size categories. As can be seen in Figure 4.15, both percussion marks and tooth marks at the ST Site Complex support the hypothesis that hominids had primary access to carcass resources, both meat (low tooth mark percentages) and marrow (see percussion mark percentages). Tooth mark occurrence on shaft fragments appears in the lower range of variation of the hominid-carnivore dual-patterned

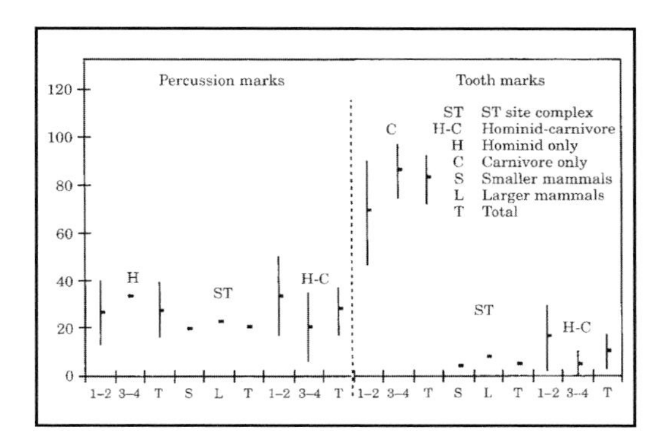

Figure 4.15 Distribution of tooth-marked and percussion-marked specimens in the ST Site Complex bone assemblage and in experimental assemblages simulating hominid-carnivore interactions. Data from experimental models have been obtained from Blumenschine (1988, 1995) and Capaldo (1995, 1997).

experimental model (Blumenschine 1988, 1995: Capaldo 1995, 1998).

Cut mark patterns also support this behavioral inference (Tables 4.15–4.16). From the 154 bone fragments showing weathering stage 0, several were smaller than 4–5 cm and were hard to identify to element (e.g., humerus or femur), even when limb section (e.g., upper limb bones) could be ascertained in several cases. Intermediate and lower limb bones were also

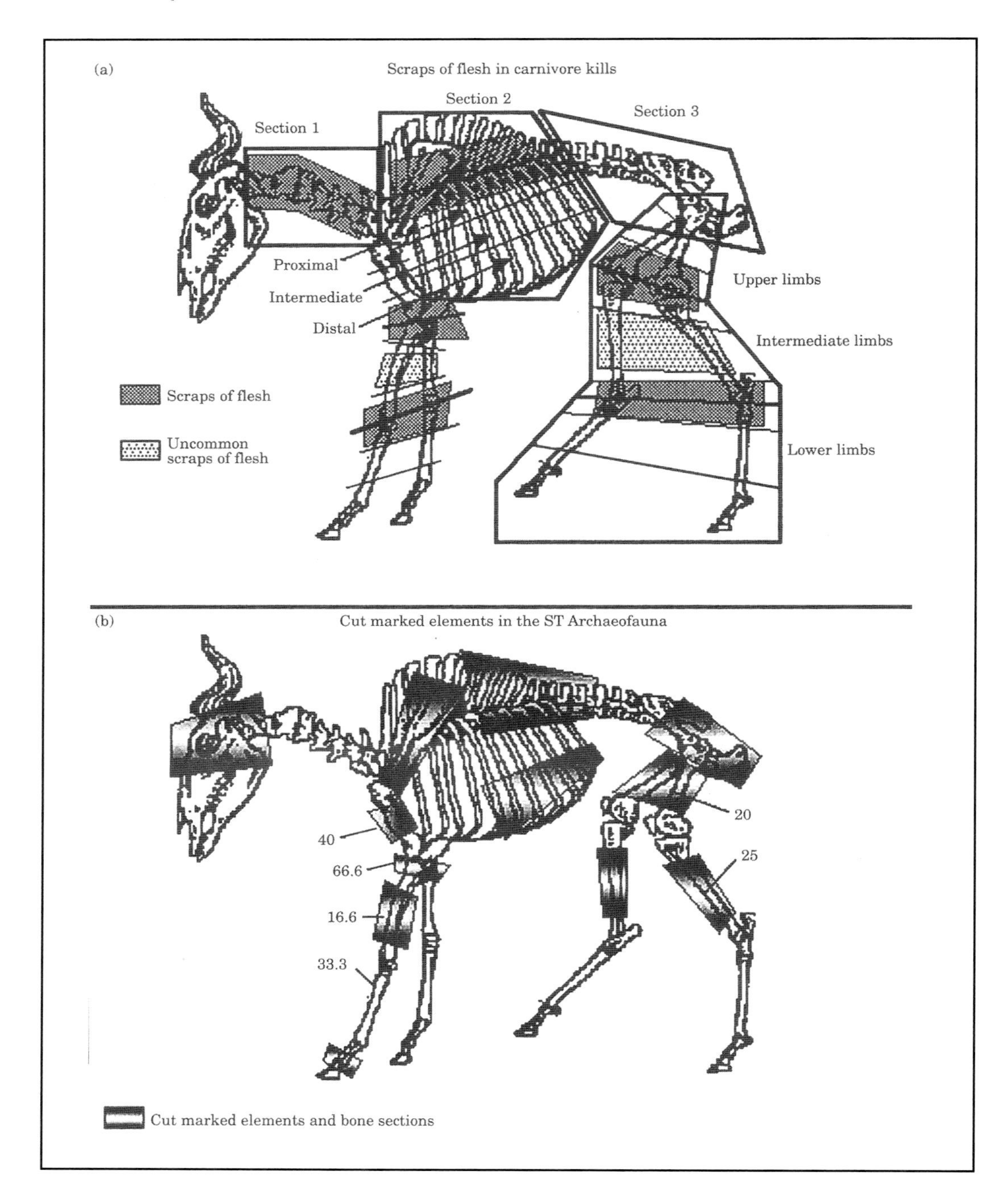

Figure 4.16 Anatomical distribution of scraps of flesh on carcasses abandoned at carnivore kills: a) Domínguez-Rodrigo 1999b; b) cut mark anatomical distribution on carcasses in the ST Site Complex bone assemblage. The comparison suggests that hominids had primary access to carcasses and carnivore intervention was secondary.

difficult to distinguish from one another in several specimens smaller than 5 cm, due to their similar cortical thickness and small size. A total of 90 shaft fragments with weathering stage 0 were classified to element type. The resulting cut mark distribution shows a high proportion of cut-marked upper limb bone shaft specimens, followed by intermediate and lower limb bone fragments. Both in smaller mammals and larger mammals, the cut mark patterns are broadly similar. Furthermore, midshaft specimens are cut-marked at a much higher rate than limb bone ends. Only two epiphyses (both proximal radii from size 3 carcasses) bear cut marks on the near-epiphyseal section. The remainder of the cut marks have been documented on midshaft sections (Table 4.15, Figure 4.16).

Furthermore, a large majority (>75%) of meat-bearing midshaft specimens are cut-marked in both carcass sizes. This is indicative of intensive defleshing (Domínguez-Rodrigo 1999a). Overall, the bone surface modification data reflect primary hominid access to fleshed carcasses as modeled in hominid-carnivore experimental scenarios (Domínguez-Rodrigo 1997a, 1997b, 1999a). The main difference observed between these experiments and the ST archaeofauna – which for the analysis of cut marks comes mostly from the well-preserved bone assemblage excavated at ST4 – is the scarcity of cut-marked limb ends. Therefore, the only carcass-processing activity that can be reconstructed with the currently available information is carcass defleshing, but not disarticulation. This could reflect either hominid behavior or sampling bias. Cut marks have been observed in all the ST assemblages (except ST3), although in smaller amounts than in the ST4 assemblage. Other analytical indicators support our inferences, based on cut mark percentages and distribution, for access to fleshed carcasses by hominids. The cut-marked ULB: cut-marked ILB ratio shows a positive relationship with ULB showing slightly higher percentages of cut-marked fragments. The ratio of NCMMSSP (number of cut-marked midshaft specimens) to NICMSP (number of total identified cut-marked specimens), divided by the ratio of NICMSP to NISP also indicates this (Domínguez-Rodrigo 1997a; Table 4.15).

In addition, specimens from other anatomical regions are also cut-marked (Table 4.16). If all the data from Tables 4.15–4.16 are combined as shown in Figure 4.15, it becomes clear that the cut-marked elements from all carcass sizes indicate defleshing. For medium-sized carcasses, a study of distribution of flesh at carnivore kills (Domínguez-Rodrigo 1999a) also suggests that hominids at the ST Site Complex were not restricted in access to largely defleshed carcasses such as those found in those sites (Blumenschine 1986).

Discussion and conclusions

The paleosurface exposed at Maritanane (Type Section), totalling an area close to 0.6 km^2, offers a unique opportunity to assess hominid behavior across a paleolandscape. Although small scatters of archaeological remains have been discovered in places along this paleolandscape, most archaeological materials are concentrated in the ST Site Complex. This area comprises a small part of the exposed paleosurface. Thus, it seems that hominids were specifically selecting that particular area for activities related specifically to carcass processing. The abundance of stone tools and bone remains with hominid-made modifications supports this interpretation.

The detailed geological analysis of the area shows that the ST sites were created in an alluvial setting in a deltaic environment at the intersection of several river channels. Although isotopic analyses should be conducted to confirm it, the abundance of fossil rootcasts from plants bigger than grasses suggests some degree of closed vegetation. A landscape taphonomic analysis of the ST Site Complex and other areas where the paleosurface is exposed suggests that in the immediate vicinity of the sites there was a low degree of carnivore competition, compared to the other areas of the exposed paleolandscape (Domínguez-Rodrigo 2001). Despite this, carnivore post-depositional intervention in the bone assemblages created by hominids was very significant, as seen above. This carnivore disturbance is explained by secondary access to carcasses by carnivores, in a situation in which different predatory taxa were not in a high-stress situation. A high-stress situation, involving overlapping use of space, would have been reflected in an even greater deletion of bone remains (Blumenschine 1989). If carnivore competition was related to landscape type in the Early Pleistocene as it is today in modern African savannas (Blumenschine 1986; Tappen 1992, 1995; Domínguez-Rodrigo 1996), then the ST Site Complex might have been located in a moderately bushy environment close to very open habitats (Domínguez-Rodrigo et al. 2001a).

The wide distribution of the archaeological materials at the ST sites over 3500 m^2 contrasts with the discrete, high-density sites documented at Koobi Fora and Olduvai Gorge. The existence of spatially differentiated bone clusters, each belonging to the same individual, the different degrees of weathering on limb bone shafts, the overall low densities of materials (both of stones and bones) and the distribution of artifact types (abundance of flakes, scarcity of core types and manuports) all support the interpretation that the ST Site Complex represents an overlapping set of *loci* in the alluvial landscape of Maritanane that hominids created by repeatedly visiting the area. Carcasses may have been obtained in or near that alluvial setting. Remains bearing cut marks and belonging to large-sized animals such as rhinoceros and *Giraffa* sp. suggest that hominids obtained carcasses close by. The skeletal part representation of the *Giraffa* sp. remains at ST4 (see Chapter 6) indicate that they belonged to the same individual, which was complete when processed by hominids. If the alluvial area was not very open, the abundance of antilopini and alcelaphini would be indicative of their transport by hominids from nearby, more open areas. Regardless, most game could have been obtained very close to the ST sites given the overall open nature of the landscape (Domínguez-Rodrigo et al. 2001a).

Most likely, carcasses were fully fleshed when transported to sites, as indicated by the percentages and distribution of cut-marked and tooth-marked bone specimens. Based on a landscape taphonomy study in the area (Domínguez-Rodrigo 2001), there was a high degree of carnivore competition in Maritanane during the formation of the paleosurface. This contradicts the widely accepted scavenging scenarios proposed to account for early hominid behavior. A prolonged stay in the same spot by hominids would have created dense accumulations of materials. This is observed in Olduvai Bed I and Bed II sites, where hominids also accumulated substantial amounts of stones as the result of repeated transport of raw materials to the same places. Most Oldowan tool types appear in the Olduvai assemblages, whereas these types are virtually absent in Peninj. Manuports are accumulated because their utility is foreseen in the future if the use of a predetermined spot is either frequent or prolonged. The absence of this lithic element at Peninj indicates that, behaviorally, sites at Olduvai and Peninj may reflect different patterns. Potts (1994) already stressed that not all the Plio-Pleistocene sites should be the result of the same behavior. Hominids at the ST sites came with cores, flaked them, and used the resulting tools. This is more suggestive of a restricted, short-term use of the space. Hominids at the ST Site Complex may have used the area as a carcass obtaining and butchering arena. This would support O'Connell's (1997) "near-kill location" model, although it is unknown whether hominids butchered carcasses and then transported their products to other loci or if everything was consumed on the spot.

Monahan (1998) has highlighted the possible biological and ecological differences between modern hunter-gatherers and Plio-Pleistocene hominids. According to him, predation risks from large carnivores at fresh carcasses would have limited the amount of meat-stripping, bone discard and the time spent at kill/butchering places. This would have made hominids transport and discard bones and associated edible tissues together. Monahan introduces a "weight minimizing" model and a "food maximizing" model. The former would create a significant "schlepp effect," since hominids would select light anatomical parts (limbs) which are easier to transport over the more nutritious but heavier axial parts. The latter model implies that hominids would simply have focused on those parts with the highest caloric yields and abandoned the rest, depending on the number of carriers. Monahan (1998:421) assumes that since early hominids lacked the bow and arrow technology of modern hunter-gatherers, which are effectively employed to drive carnivores away from carcasses, and because Plio-Pleistocene carnivores were larger than modern ones, hominids were at a higher risk at kills and had little or no processing time.

In theory, this may be true, but in practice, this model is highly speculative. Modern humans - e.g., Mwalangulu (Kenya), Maasai pastoralists (Tanzania) - can deter large carnivores from their kills without bows and arrows or any projectile weapon. One of us (MDR) has witnessed a few times how both groups kept lions at bay armed only with sticks. If hominids had clubs, branches or spears they may have been able to do the same, especially bearing in mind that they were probably much stronger than modern humans (Wolpoff 1999). Monahan´s assumption also obviates two facts. First, hominids may have exploited a mid-day temporal niche during which carnivore activity (and therefore predation risk) is marginal (Domínguez-Rodrigo 1994a). Second, hominids were obtaining their animal food in alluvial environments (such as those documented in Olduvai, Koobi Fora or Peninj), where predation risk is very low because those habitats show the lowest degree of carnivore overlap in the use of the space (Blumenschine 1986; Domínguez-Rodrigo 2001). Furthermore, the wider diversity of Plio-Pleistocene carnivores does not necessarily mean a more intense presence of large carnivores on the landscape at any given time. This carnivore diversity is an interpretation based on the amalgamation of diachronic and geographically diverse paleontological deposits. No archaeological area in East Africa has provided us with synchronic taxa representing the whole carnivore range documented during the Plio-Pleistocene. Furthermore, due to time-averaging in the fossil record, it is impossible to ascertain if carnivore diversity represents time-averaged, intermittent use of the space by predators (without interaction), or simultaneous, overlapping use of space by inter-specific competitors.

The ST Site Complex at Peninj suggests that hominids were obtaining and transporting fleshed carcasses to certain areas and then leaving these places after carcass processing. If the

ST sites were "near-kill locations," hominids could have been involved in pre-consumption carcass preparation, as modern hunter-gatherers are (O'Connell 1997). This would be supported if some sort of reciprocity (i.e., food-sharing) existed at that time. Alternatively, complete carcass consumption could also have been carried out in those places. The current resolution of the archaeological record does not allow us to confidently support either option. Either way, the behavior reconstructed implies a substantial degree of complexity, planning and dynamic interaction with the environment to obtain carcasses, a behavior in which predatory strategies on small and medium-sized animals should seriously be considered. The recent discovery of Acacia wood residues on handaxes from an Acheulian site roughly contemporary with the ST Site Complex (Domínguez-Rodrigo et al. 2001b) would support the interpretation that those hominids may have already crafted rudimentary spears with which hunting had become a feasible carcass obtainment strategy.

Endnote

1 This section was previously published in Journal of Archaeological Science, vol. 29, Domínguez-Rodrigo et al. (2002), pp. 639-665, Copyright 2002. Reproduced with permission by Elsevier.

5

Isotopic Ecology and Diets of Fossil Fauna from the T-1 (Type Section, Maritanane) Paleosurface

Nikolaas J. van der Merwe

Introduction

The diets and habitat of the fossil fauna from Type Section (Maritanane) at Peninj have been assessed by means of stable carbon and oxygen isotope ratios ($\delta^{13}C$ and $\delta^{18}O$) in the tooth enamel of 40 specimens. Of these, 32 specimens were identified to the level of genus or better by the excavators, while eight specimens were bovids of indeterminate species. The carbon isotope ratios show that all these animals were grazers or mixed feeders. The oxygen isotopes indicate that they had access to permanent water sources that were not seriously subject to isotopic enrichment by evaporation. The habitat that provided the appropriate resources for this faunal population was apparently an open grassland savanna with few trees, in the vicinity of the extant Lake Natron. The process by which these conclusions have been reached is described in this chapter.

Stable light isotopes and tooth enamel

The assessment of prehistoric diets and environments by means of the isotopic analysis of bone has developed over the past thirty years and is widely used in archaeology (Vogel and van der Merwe 1977; van der Merwe and Vogel 1978; for reviews see van der Merwe 1982, Schoeninger and Moore 1992; Katzenberg 2000). Of particular importance has been the determination of $^{13}C/^{12}C$ ratios ($\delta^{13}C$ values) in bone collagen, which provide a measure of C_3 and C_4 plants in the diet. Accordingly, the early isotope applications involved the introduction and spread of maize (a C_4 plant) in C_3 biomes such as the North American Woodlands and the tropical forests of South America (van der Merwe et al. 1981). Stable nitrogen isotope ratios ($\delta^{15}N$ values) can also be measured in collagen; these provide an indication of trophic level in well-watered environments and can be used to assess the meat intake.

Collagen in bone has a limited lifetime, especially in wet and warm environments where organic materials deteriorate rapidly. The oldest hominin collagen that has been successfully analyzed for carbon and nitrogen isotopes were Neanderthal specimens from cold, dry caves (Bocherens et al. 1999; Richards et al. 2000). The faunal specimens from Maritanane are some 1.3 million years old and contain no collagen, but stable carbon and oxygen isotopes can be measured in the mineral phase of their fossilized skeletons. The isotopic analysis of fossils has been developed over the past twenty years and has had a major impact in paleontology. The first measurements were done on fossil bone (Sullivan and Krueger 1981, 1983), but tooth enamel has proved to be the best material for analysis (Lee-Thorp and van der Merwe 1987, 1991; Lee-Thorp et al. 2000). Tooth enamel is a biological apatite, which contains about 3% carbonate. Being highly crystalline, the tooth enamel is resistant to alteration by carbonates in ground water. With appropriate pretreatment, the carbon and oxygen isotopes in tooth enamel carbonate can be measured to provide dietary information. The stable carbon isotope ratio provides an average of the distinctive carbon isotope ratios of plants at the base of the foodweb of an individual. This isotopic signal is acquired when herbivores eat plants and is passed along the foodchain to carnivores and omnivores. The oxygen isotope ratio provides a measure of the body water of an animal, whether acquired from plant water or by drinking, and is attenuated by its thermophysiology. Oxygen isotopes can contribute to dietary and environmental reconstructions (Quade et al. 1992; Bocherens et al. 1996; Cerling et al. 1997; Sponheimer and Lee-Thorp 1999), but are not as well understood as carbon isotopes in the foodchain.

Carbon and oxygen isotope ratios are measured at the same time in a mass spectrometer, the sample gas being CO_2 produced from tooth enamel carbonate. The isotope ratios of the samples are compared with those of the international VPDB carbonate standard and the results are reported in the "delta notation" in parts per thousand (per mil or ‰). In the case of the carbon isotope ratio, for example, this is calculated as follows:

$$\delta^{13}C = \frac{{}^{13}C/{}^{12}C\ \text{sample} - {}^{13}C/{}^{12}C\ \text{standard}}{{}^{13}C/{}^{12}C\ \text{standard}} \times 1000\ ‰$$

To interpret carbon isotope results in dietary terms, it is necessary to determine the C_3 and C_4 "end members" for a given time and place. Typically, the C_3 end member can be determined by measuring the tooth enamel $\delta^{13}C$ values of dedicated browsers, such as giraffes, which eat the C_3 leaves of trees and shrubs. The C_4 end member can likewise be obtained from dedicated grazers, e.g., alcelaphines like the wildebeest. At the moment, browsing herbivores of the South African interior have mean tooth enamel $\delta^{13}C$ values of about -14.5‰, while grazers have mean values of about -0.5‰. These values can be altered by climatic or atmospheric conditions. The "Industrial Effect" of the past 200 years, for example, increased the CO_2 content of the atmosphere substantially and made its $\delta^{13}C$ value (and hence that of all plants) more negative by 1.5‰. Increased humidity may make the $\delta^{13}C$ values of C_3 plants (but not C_4) more negative by as much as 2‰ (Tieszen 1991). Tropical forests provide extreme examples in this regard, with C_3 plant values as much as 10‰ more negative than those of plants growing in the open. This is the result of high humidity, low light, and recycled CO_2 from rotting leaf litter (van der Merwe and Medina 1989). By contrast, the $\delta^{13}C$ values of C_3 plants may increase (become less negative) in response to aridity and bright sunlight (Ehleringer et al. 1986; Ehleringer and Cooper 1988), while C_4 plants may respond with slightly more negative values, due to the increased prevalence of enzymatic C_4 subtypes that are adapted to such conditions.

To determine the ratio of C_3 and C_4 plants in the diets of herbivores is simply a matter of interpolating between the C_3 end member (100% C_3 diet) and C_4 end member (100% C_4 diet) for a given ecosystem.

Oxygen isotope ratios in tooth enamel carbonate can contribute to an understanding of the local habitat. The primary source of oxygen in biological apatite is water (from drinking or from food) and from oxygen bound in food. The $\delta^{18}O$ values of liquid water vary with latitude, altitude and temperature (i.e., evaporation), while leaf water $\delta^{18}O$ values vary with aridity (evapo-transpiration). The $\delta^{18}O$ value of tooth enamel carbonate also varies with the water output of the animal (sweat and urine). (For review see Lee-Thorp et al. 2003).

The integrity of the $\delta^{13}C$ and $\delta^{18}O$ values of tooth enamel carbonate is best judged by observing the patterns of values between different animal species in the same ecosystem. A considerable database of such values for the Plio-Pleistocene is available by now, both published and unpublished. As can be expected, browsing herbivores (C_3 feeders) invariably have distinctively more negative $\delta^{13}C$ values than grazing herbivores (C_4 feeders). More subtly, however, there is also an observable pattern between different species of grazers: wherever they have been compared, specimens of *Damaliscus* sp. (e.g., the tsessebe) have more positive $\delta^{13}C$ values than *Connochaetes* sp. (wildebeest), which, in turn, have more positive values than *Equus* sp. (zebra). This pattern results from different amounts of C_3 plants like forbs that are included in the diets of *Damaliscus* sp., *Connochaetes* sp. and *Equus* sp. In the case of oxygen isotopes, the $\delta^{18}O$ values of hippos are invariably more negative than those of grazers like the equids. They sweat less, because they live in the water during the day, and therefore do not lose as much of the light ^{16}O isotope as land animals.

Materials and methods

Tooth specimens of 40 mammalian fossils (34 from Maritanane; 6 from the escarpments for comparison) were provided for analysis by Manuel Domínguez-Rodrigo. The stable carbon and oxygen isotope ratios of their enamel carbonate were measured in the Stable Light Isotope Facility of the Archaeology Department at the University of Cape Town. The procedures used were developed in this facility and have been described in detail elsewhere (Lee-Thorp et al. 1997; Sponheimer 1999; Luyt 2001).

Only 1 mg of pretreated tooth enamel powder is required for a single determination of carbon and oxygen isotope ratios. To allow for duplicate measurements, about 3 mg of enamel powder was drilled under magnification from each tooth specimen, using a diamond-tipped dental burr of 1 mm diameter in a low-powered, slow-turning hand drill. Where possible, the powder was ground from broken enamel edges, instead of drilling a hole. Care was taken not to drill into dentine or to heat the enamel. Since 3 mg of enamel powder equals about two sugar grains, the sampling damage was minimal and the specimens were returned to the excavator.

The enamel powder was gathered on smooth weighing paper and poured into a small centrifuge vial with a snap lid, in which all the subsequent chemical pretreatment was carried out. The powder was pretreated with 1.5–2% sodium hypochlorite for 30 minutes (to remove organic materials and humic acids), rinsed, and then reacted with 0.1 M acetic acid for 15 minutes (to remove readily dissolved carbonates). After washing and drying, 1 mg of the enamel powder was weighed into an individual reaction vessel of a Kiel II autocarbonate device. Each powder sample was reacted at 70°C with 100% phosphoric acid; the resulting CO_2 gas was cryogenically distilled and its isotope ratios measured in a Finnegan MAT 252 ratio mass spectrometer. The $\delta^{13}C$ and $\delta^{18}O$ values were calibrated against the international VPDB carbonate standard by using a calibration curve established from NBS standards 18 and 19 and by inserting secondary laboratory standards of "Lincoln Limestone" and "Carrara Z marble" at regular intervals in the autocarbonate device. The precision of replicate analyses was better than 0.1‰.

Results

The $\delta^{13}C$ and $\delta^{18}O$ results are listed in Table 5.1 and plotted in Figures 5.1–5.2.

Carbon isotopes

Compared to most early hominin sites in South and East Africa, the collection of fossil fauna from Peninj is unusual for its lack of browsing animals. Of some 1,500 identifiable specimens reported by the excavators (Domínguez-Rodrigo et al. 2002), only one specimen was that of a browser: a partial

Table 5.1 $\delta^{13}C$ and $\delta^{18}O$ values for tooth enamel of fossil fauna from Peninj. Precision 0.1‰

Specimen No.	Taxon	$\delta^{13}C_{VPDB}$	$\delta^{18}O_{VPDB}$
ST2 53	*Ceratotherium simum*	1.9	-3.9
ST37A 39	*Hippotragus equinus*	1.7	-2.6
ST4 U0 21	*Damaliscus niro*	1.5	-1.0
ST38A 309	*Damaliscus niro*	1.8	0.8
		$\bar{X}$ (n = 2) = 1.6 (0.3)	-0.1 (0.1)
ST3A 27	*Redunca* cf *redunca*	1. 2	-1.9
ST38A 24	*Redunca* cf. *redunca*	1.7	-1.1
		$\bar{X}$ (n = 2)= 1.4 (0.4)	-1.5 (0.6)
ST4 U1 308	*Megalotragus kattwinkeli*	0.9	-1.4
ST30D 33	*Connochaetes taurinus*	1.6	-1.2
ST30D 4	*Connochaetes taurinus*	-0.0	-0.7
		$\bar{X}$ (n = 2) = 0.8 (1.2)	-1.0 (0.4)
ST4 45	*Theropithecus* sp.	-0.4	-3.0
ST15	*Theropithecus* sp.	-0.5	-2.0
ST4 44	*Theropithecus* sp.	-0.3	-3.7
		$\bar{X}$ (n = 3) = -0.4 (0.1)	-2.9 (0.8)
ES 5	*Equus* sp.	-0.1	-0.9
ES 5	*Equus* sp.	0.1	-0.5
EN 17	*Equus* sp.	-0.2	-1.9
ES4 145	*Equus* sp.	-0.3	-0.9
ES4 128	*Equus* sp.	-1.4	-1.2
ST4 53	*Equus* sp.	0.6	-1.6
ST4 132	*Equus* sp.	-0.5	-0.4
ST4 U0 24	*Equus* sp.	-0.9	-1.6
ST4 O2 23	*Equus* sp.	-0.4	-0.9
		$\bar{X}$ (n = 9) = -0.4 (0.6)	-1.2 (0.5)
EN2A 90	*Metridiochoerus compactus*	-0.7	-1.7
ST4 160	*Metridiochoerus compactus*	-0.8	-0.5
		$\bar{X}$(n = 2) = -0.8 (0.04)	-1.1 (0.8)
ST4 U2 10	*Kobus* cf. *kob*	-1.0	-0.3
ST0A 18	*Gazella* sp.	-1.0	-2.4
ST0 209	*Hippopotamus* sp.	-2.6	-5.7
ST4 216	*Hippopotamus* sp.	1.0	-2.7
ST4 173	*Hippopotamus* cf. *gorgops*	-2.1	-2.2
		$\bar{X}$ (n = 3) = -1.2 (2.0)	-3.6 (1.9)
ST4C 471	*Kolpochoerus* sp.	-1.8	2.2
ST4 197	*Kolpochoerus* sp.	0.5	-4.0
ST4 135	*Kolpochoerus* cf. *limnetes*	-1.9	-2.4
		$\bar{X}$ (n = 3) = -1.2 (1.4)	-2.9 (1.0)
ST53A 20	Bovidae, indeterminate	-4.6	-1.4
ST3	tooth fragment	2.8	-1.7
ST4 125	premolar	2.7	0.1
ST4 surface	lower molar	1.9	-0.9
ST4 72	tooth fragment	1.6	-1.4
ST4 U0 22	incisor	1.1	0.0
ST1	mandible, large	0.9	-1.5
ST4 U2 37	premolar	-0.1	0.5
ST2A 10A	upper M3	-0.6	-4.0

skeleton of *Giraffa* cf. *pygmaea*. A tooth sample of this specimen was not available for analysis.

The dominance of grazers in the Peninj collection is clearly demonstrated by the $\delta^{13}C$ values of the 40 tooth samples that were analyzed. With the exception of one specimen of *Taurotragus* sp., a mixed feeder related to the eland, the collection is comprised of grazing animals. This includes three specimens of *Theropithecus* sp., a distant relative of the extant grazing gelada baboon of Ethiopia. The $\delta^{13}C$ values range from -4.6‰ at the negative end (*Taurotragus* sp.) to 1.9‰ at the positive end (*Ceratotherium simum*, the white rhino).

It is not possible to establish the C_3 end member of this collection. The C_4 end member can be placed at about 2‰; a rough approximation of the C_3 end member is about -12‰. This means that *Taurotragus* sp. had about 50% C_3 and C_4 plants in its diet, while *Damaliscus niro*, *Hippotragus equinus* (roan), and *Ceratotherium simum* had essentially pure C_4 diets. The integrity of the $\delta^{13}C$ measurements can be judged from the fact that the C_3 dietary component of *Equus* sp. was higher than that of *Connochaetes* sp., which in turn was higher than that of *Damaliscus* sp.

The C_3 dietary component of *Hippopotamus* sp. (n = 3) was approximately 23%. Hippos are grazers who feed on C_4 grasses at night, but they may also include C_3 water-edge plants like *Typha* sp. (bullrush) in their diet. In the case of *Taurotragus* sp., the C_3 component is more likely to have been the leaves of trees and shrubs, so the landscape was not totally devoid of such plants.

The faunal community and carbon isotope ecology of Maritanane appears to have been similar to that of the modern Serengeti Plain. In both cases, the landscape was formed on a layer of volcanic tuff, which could not easily be penetrated by the roots of trees. The result was an open savanna dominated by C_4 grasses. It is important to note that the $\delta^{13}C$ values of modern Serengeti fauna, when corrected for the Industrial Effect by adding 1.5‰ to the results, are essentially identical to those of the Peninj fauna (van der Merwe, unpublished data). The plant community of Peninj was apparently closely similar to that of the modern Serengeti.

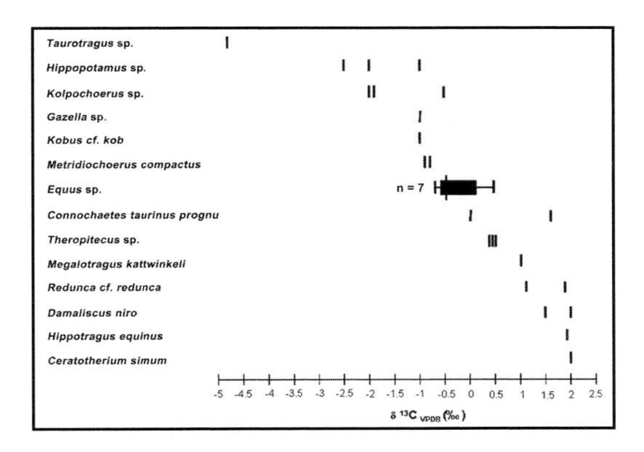

Figure 5.1 $\delta^{13}C$ values for tooth enamel of fossil fauna from Maritanane (34 teeth) and the escarpments (six teeth). The values for unidentified bovids are not depicted. In the box-and-whisker plot for Equids, the central vertical line depicts the mean, the black box depicts 25%-75% of the range and the whiskers denote 10%-90% of the range.

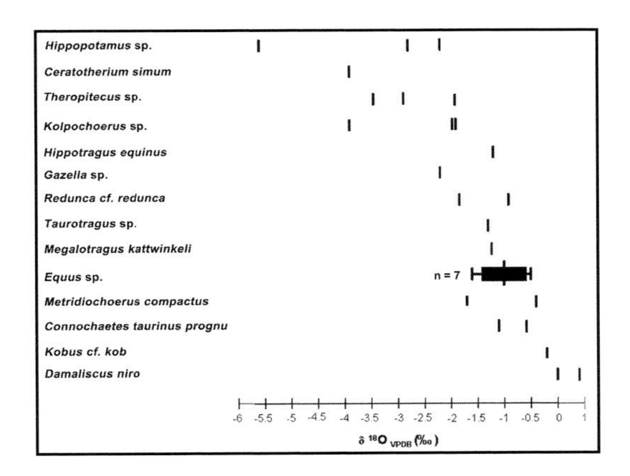

Figure 5.2 $\delta^{18}O$ values for tooth enamel of fossil fauna from Maritanane (34 teeth) and the escarpments (six teeth). The values for unidentified bovids are not depicted. In the box-and-whisker plot for Equids, the central vertical line depicts the mean, the black box depicts 25%-75% of the range and the whiskers denote 10%-90% of the range.

Oxygen isotopes

The $\delta^{18}O$ values for tooth enamel from Peninj are distinctively negative, relative to the VPDB standard. They vary from -5.7‰ (*Hippopotamus* sp.) to 0.8‰ (*Damaliscus niro*). Essentially all the animals in the collection are grazers, which take in water by drinking. The obvious source of this water was Lake Natron and the river flowing into it from the Kenya Highlands. The rain feeding this source had lost much of the heavy ^{18}O isotope by the time it fell on the highlands, hence the negative $\delta^{18}O$ values. The variability in $\delta^{18}O$ values that can be observed in the data is due to the thermophysiology of different species: the values for water-dwelling hippos are more negative than those for grazers such as *Equus* sp., *Connochaetes* sp., and *Damaliscus* sp. (in that order).

The $\delta^{18}O$ values for modern animals from the Serengeti show a strong contrast to those from Peninj: they are essentially all positive, relative to VPDB (van der Merwe, unpublished data). The major rivers that flow through the Serengeti (e.g., the Mara River) originate in the Kenya Highlands, as does the Njiru River that supplies Lake Natron. The contrasting oxygen isotope values of rain in the Kenya Highlands at 1.5 Ma and the present suggest strongly that the water vapor source for the rain changed during this time period from the Atlantic Ocean to the Indian Ocean. The Atlantic Ocean is a much longer distance from the Kenya Highlands than the Indian Ocean, which would result in increasingly negative oxygen isotope values for rain as the clouds traveled across the Congo basin, losing ^{18}O along the way. This hypothesis has important implications for paleoclimate studies and requires further research.

Conclusions

The stable carbon isotope ratios for 40 tooth enamel specimens from Peninj at ca. 1.3 Ma show that the fauna from this time and place were essentially all grazers with diets of C_4 grasses. The exceptions in this collection are *Taurotragus* sp., a mixed feeder with a diet of about 50% C_4 grasses and 50% C_3 plants (trees and shrubs). *Hippopotamus* sp. had about 23% C_3 plants in their diet, on average. This C_3 component was probably obtained from water-edge plants such as *Typha* sp. (bullrush). A single specimen of *Giraffa* cf. *pygmaea*, presumably a browser, was collected at Peninj, but its teeth were not available for analysis. The dominance of grazers at Peninj resembles the modern faunal community of the Serengeti Plain. When the $\delta^{13}C$ values for modern Serengeti fauna are corrected for the Industrial Effect (by adding 1.5‰), they are essentially identical to those of the Peninj collection. Based on the prevalence of grazing animals in the Peninj collection (and, indeed, the near absence of browsers and mixed feeders), it can be concluded that the landscape was an open savanna, essentially grassland with very few trees. It likely resembled the modern Serengeti Plain.

The stable oxygen isotope ratios for Peninj fauna were essentially all negative, relative to VPDB. The animals, mostly grazers, were dependent for their water on the open water source of Lake Natron and the river that fed it from the Kenya Highlands. The rivers that flow through the Serengeti Plain today also rise in the Kenya Highlands, but the stable oxygen isotope ratios for animals in the modern Serengeti are positive, relative to VPDB. This observation strongly suggests that the water vapor source for the rain in the Kenya Highlands changed from the Atlantic Ocean at 1.5 Ma to the Indian Ocean in modern times. With less distance to travel from the Indian Ocean today, the clouds lose less of the heavy ^{18}O isotope along the way than when they traveled all the way across the Congo basin. This hypothesis deserves further investigation, since it has important implications for paleoclimatic interpretations.

6

Archaeological Evidence of Carcass-Processing Spots Created by Lower Pleistocene Hominids from the ST4 Site

Manuel Domínguez-Rodrigo, Fernando Diez-Martín, Luis Alcalá, Luis Luque, Rebeca Barba, Rafael Mora, Ignacio de la Torre, and Pastory Bushozi

Introduction to the ST4 site

The ST4 site is one of the sites of the ST Site Complex in Type Section (Maritanane), described in Chapter 4. The wealth of archaeological and paleontological localities in this area has been described in the previous chapters. ST4 is located in the intermediate zone of the ST Site Complex, on top of a slope running west-east. It lies in a channel cutting through the paleosurface overlying Tuff 1 (T1) of the Upper Humbu Formation, which has been described in previous chapters. The site's sediments are moderate and coarse-grained sands in a channel fill formed in a clay matrix which might have followed a north-west-southeast direction. This site was altered by tectonic movements. The paleoenvironmental reconstruction of the area suggests a deltaic environment close to a lacustrine floodplain.

The ST4 site was extensively excavated beginning in 1995, with excavations continuing during 1996, 2000, and 2001 and finishing in 2004. The dates for the Upper Humbu Formation are bracketed between the Main Tuff at the base and the base of the Moinik Formation at the top. The Main Tuff is traditionally dated to 1.6 million years ago; however recent dating efforts place it at 1.3-1.2 Ma (see Chapter 2). The Intra-Moinik basalt is dated to 1.34 million years ago. Given the stratigraphic position of the paleosurface containing the ST4 site, overlying the Main Tuff (above the T1 tuff), a maximum estimate of 1.6 million years and a minimum estimate of 1.2 million years (based on recent dating by A. Deino, in progress) is inferred for the site.

Depositional history of ST4

Fossil bones and stone tools at the ST4 site appear evenly distributed through a deposit that spans over 1.5 m in depth, although the densest part of the deposit is concentrated in approximately 70 cm of depth. The distribution of materials in the deposit seems to be the result of several occupational episodes, as suggested by the different weathering stages of limb shaft specimens and the differential distribution of materials throughout the deposit. The number of occupations is unknown, since no discrete archaeological horizons are discernible. Furthermore, a number of materials seem to have been locally rearranged by channel dynamics, thereby losing their original spatial properties. This rearrangement seems to be uneven through the deposit, minimally affecting its lower part. Beyond this, however, hydraulic disturbance does not seem to have been significant, despite the medium-sized sand matrix in most of the channel. The high percentage of small-sized bone and stone tool specimens, together with the fact that most of them show no abrasion or polishing typical of water transport, supports the local or nearby deposition of the archaeological materials. The presence of all size categories and bones belonging to all the anatomical sections for all carcass sizes also grants support to this assertion. Biometric data on bones from the same animal

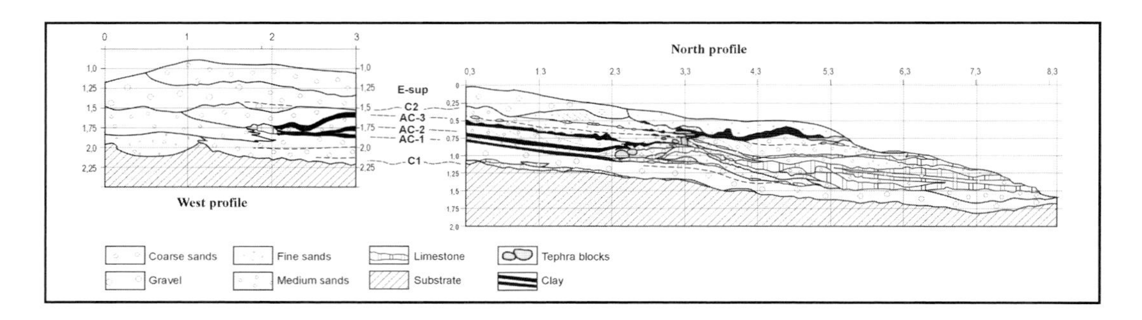

Figure 6.1 Stratigraphy of the west and north profiles of the ST4 excavation.

sizes and taxonomic groups also indicate that bones possibly belonging to the same individuals are occasionally found distributed in different depths of the deposit, thus supporting the idea that water disturbance may have resulted in the rearrangement, but not the sorting out, of most of the archaeological materials.

The distribution of materials over a deep deposit has been documented to be common in modern settlements that occur in sandy depositional contexts (Gifford 1977). Materials deposited by modern populations in loose sand from river/stream channels or lacustrine beaches have been documented to undergo sub-soil mobility under the effects of gravity and trampling of the overlying sand. In these cases, downward movements of materials spanning depths similar to that reported for the ST4 site have been documented (Gifford 1977). More recently, Roche et al. (1999) have also shown how the flaked stone products belonging to the same cores in the 2.3 million-year-old site of Lokalelei 2 appeared distributed through a larger deposit almost 1 m deep, in a sandy context similar to that reported here.

The main geological features of the site are defined by the size of the channel and its lithology. The channel is approximately 10 m wide and less than 150 cm deep. The channel was formed by eroding the T1 tuff and it almost reaches the top of the Main Tuff that divides the Upper from the Lower Humbu Formation. Some lava boulders from T1 were incorporated into the sides of the channel. At the bottom, carbonate was precipitated. Above the carbonate, medium-sized sands are found which grade into finer-grained sands towards the upper part of the channel. The carbonate becomes thicker in the center of the channel and thinner on the sides. In the central part of the channel, a series of erosional events took place, eroding the original channel and creating smaller ones (showing at least five cycles) with medium- to coarse-grained sands. However, these smaller channels did not significantly affect the materials contained on the sides of the previous larger channel.

The stratigraphy can be observed in Figure 6.1. The substratum of the channel shows an irregular outline, creating a fill composed of gravels and sands, with an erosive outline on the underlying clay base, which also shows some scattered carbonate patches. The lower unit (C1) is comprised of an erosive surface near the base of the channel, overlying some gravels and underlying coarse-grained sands. Laterally, several carbonate nodules can be observed. Overlying this unit, a level (AC-1) composed of clays in the west and carbonate in the east can be observed. Lava blocks occur in small scatters. This level is tilting towards the east. It is interbedded with a small deposit of coarse-grained sands. Another unit (AC-2) of clay in the west and south and carbonate in the east overlies the previous unit. This stratum still contains small numbers of blocks and tuff rocks. The lithology is as described for the previous unit. Above this stratum lies AC-3, a subhorizontal unit composed of clays in the west and carbonate in the east, laterally eroded, overlying some gravels and underlying some coarse-grained sands. After this unit, an erosive episode took place (C2), leaving some horizontal patches of clay, which appear eroded laterally. This is the beginning of another unit (E-sup) representing a series of erosive episodes filling the channel with sands and gravels.

Figure 6.2 Excavation of ST4 on the edge of the gully (southern side of excavation).

Despite these geological levels, it became difficult to differentiate discrete archaeological levels, since bones showed a continuous vertical distribution in the deposit. As noted above, biometrics showed that bones probably belonging to the same individual could be found in the same square at different depths, sometimes in different geological units. This is also seen in the fact that 11.4% of the bones showed a high degree of vertical tilting. This percentage was even higher when considering elements longer than 10 cm (46.7%). This could be typical of a sandy sediment in which the lack of compaction enables archaeological materials to migrate vertically due to various reasons: differential gravity according to differing weights, variability in substratum resistance due to moisture contents, and the differential vertical migration created by trampling in certain areas, etc. This, together with the fact that the lithological features that enabled us to define at least five different erosive/depositional episodes were not continuously observable across an horizontal plane, but rather appeared in patches, prevented us from clearly establishing different archaeological levels, although we know that the archaeological deposit was formed through multiple occupations of the site by hominids. This can even be observed in the most clear discrete archaeological horizon situated at the base of the channel (see below). Therefore, in our preliminary taphonomic analysis presented below, we will study the distributions of various bone modifications by diverse agents taking into account the vertical distributions of the materials throughout the deposit, which may help in assessing what parts of the deposit were due to the action of non-hominid agents or more disturbed by natural processes.

The excavated area of ST4 measures 10 m x 6 m. The testing area was initially carried out on the western edge, which forms the top of a ridge that divides two small gullies (Figure 6.2). If the global distribution of the archaeological materials is considered, the deposit seems to be horizontally concentrated and vertically continuously dispersed at the west side of the excavation (the first 5 m on the Y-axis, to the west wall of the

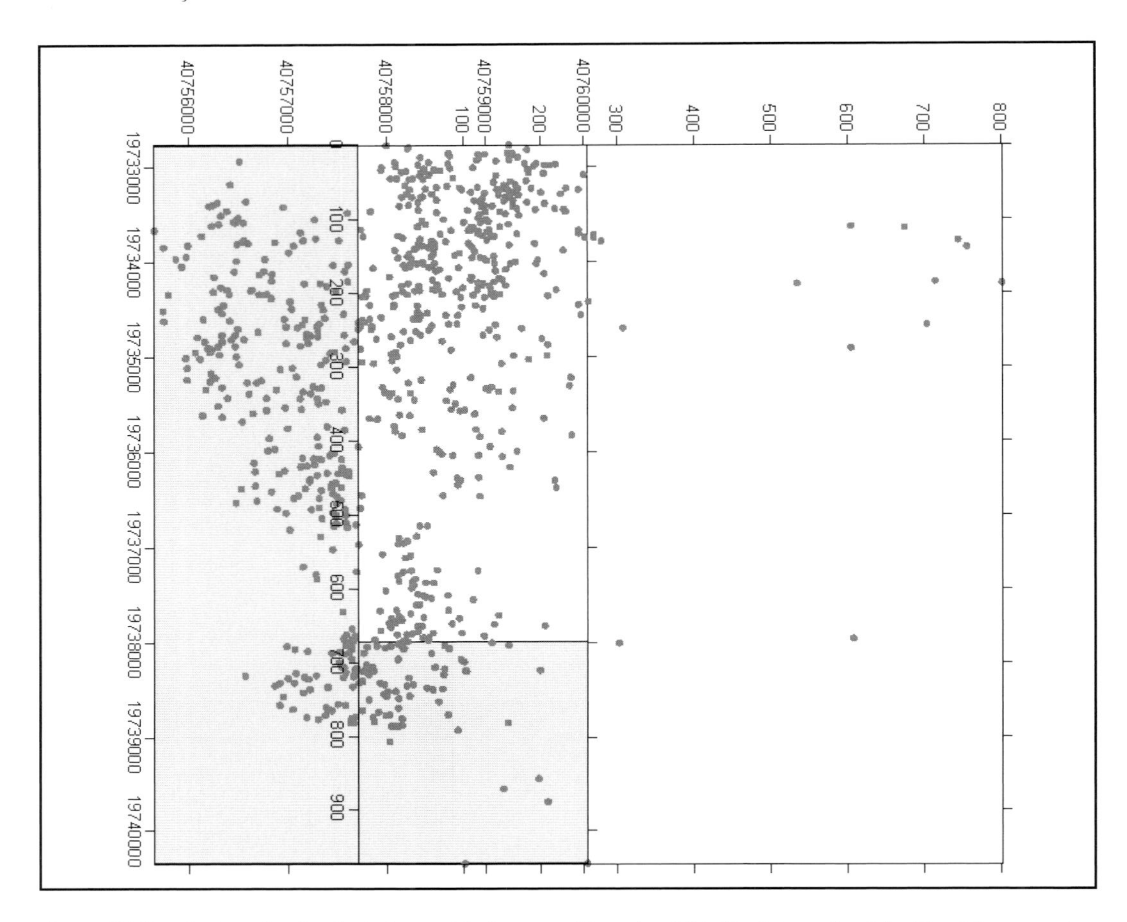

Figure 6.3 Distribution of archaeological materials (stones and bones) at ST4. North is on the right side of the excavation. The materials appear concentrated towards the south. The shaded area corresponds to the 1995–2000 excavations. The unshaded area was excavated in 2001–2004.

excavation; Figure 6.3). Fossils appeared more spaced out both horizontally and vertically as one moves towards the north side of the excavation (Figures 6.4–6.5). The easternmost half of the site shows a clear scarcity of remains and various depositional intervals as seen vertically, which may be related to the riverine cycle of multiple-channel processes (Figure 6.5c–d). This part of the site is the innermost distribution of remains inside the channel. The bulk of the deposit lies at the edge of the channel, in a lower depositional environment and in direct contact with the alluvial plain. Both lithics and bones are spatially associated and distributed in relatively similar proportions in each area (Figures 6.6–6.8). They are mostly concentrated in 70 cm of depth of deposit.

Bone preservation, animal size distribution and specimen size distribution show no specific spatial distribution when lumping all the data from the different levels together, and this is especially the case in the lower level (see below).

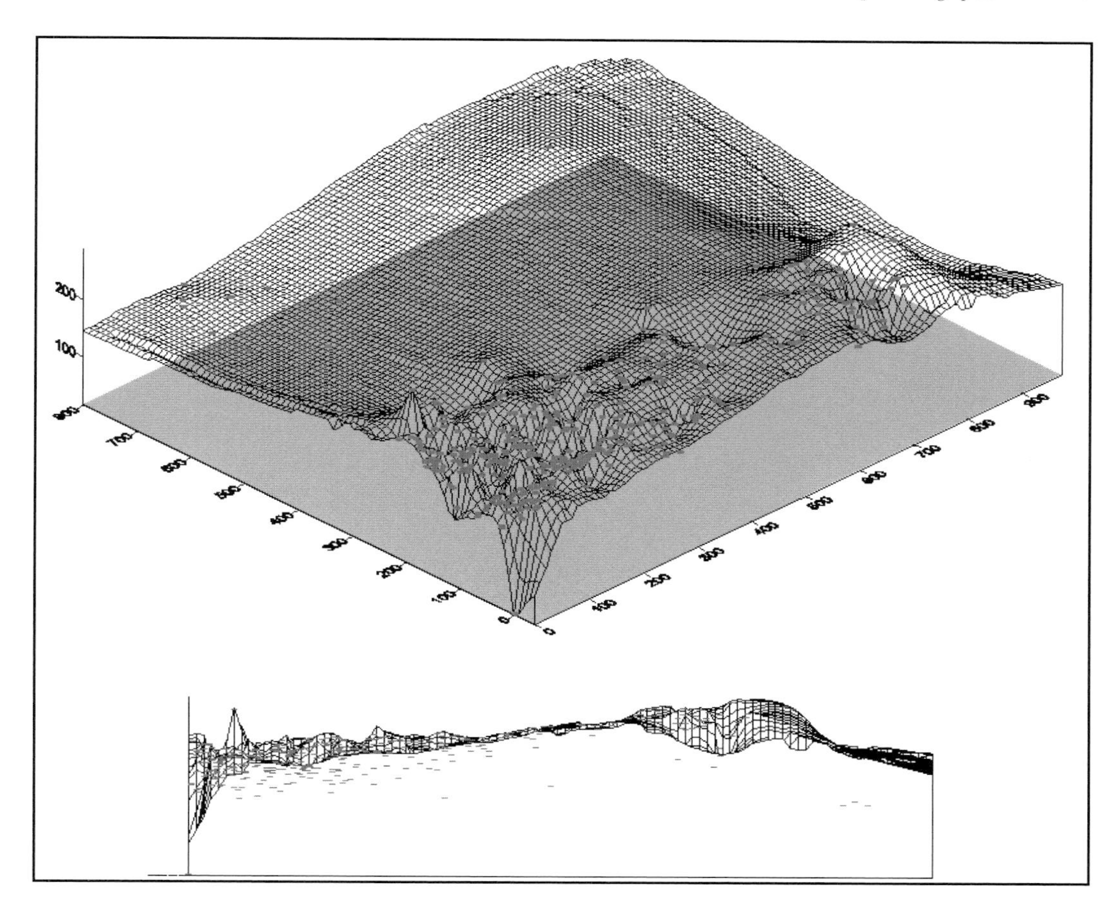

Figure 6.4 Isometric reconstruction of the vertical and horizontal distribution of materials at ST4.

All the specimen sizes are represented but a bias towards the larger ones can be observed. Specimens smaller than 3 cm are substantially low, especially when compared with specimens larger than 10 cm which appear overrepresented (Figure 6.9). Almost 25% of the specimens show some degree of diagenetic breakage, either in one plane or more. This means that the smaller specimens should be slightly overrepresented due to fossil diagenesis. However, when size distribution is examined using only those specimens with exclusively green fractures, the frequencies are mostly similar to those obtained with all bones (Figure 6.9). This indicates some taphonomic post-depositional size-sorting process. This can also be observed when only long limb shaft fragments are used. Small specimens are underrepresented and the largest ones are overrepresented (Figure 6.10).

In experimental assemblages, there is a negative relationship between the frequency of fragments and their size. Small specimens are several times more represented than the larger ones (Blumenschine 1995). In ST4, and contrary to what is observed in sites such as FLK Zinj at Olduvai, the larger specimens are highly

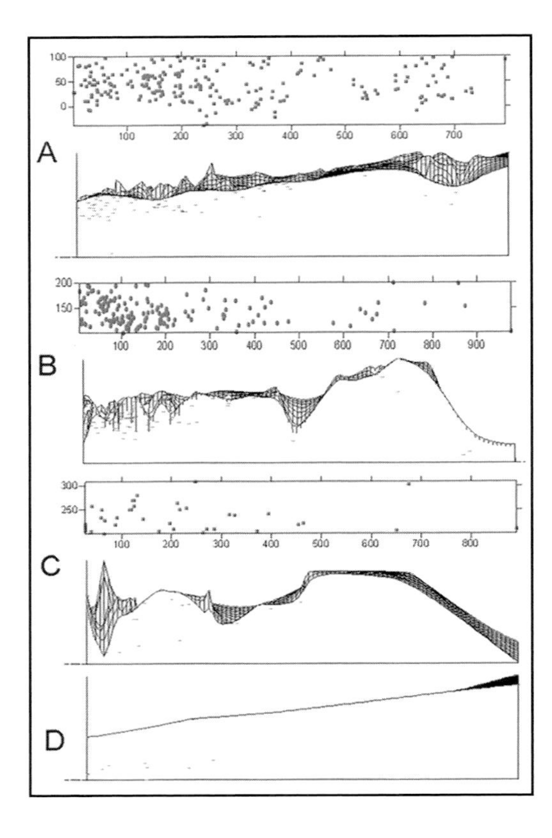

Figure 6.5 Vertical and horizontal distribution of materials at ST4 per 1 m transect; from the south of the excavation (A) to the north (D).

represented and the smaller ones are not. However, small-sized specimens measuring 4–6 cm are represented in proportions similar to what has experimentally indicated low post-depositional disturbance (Blumenschine 1995). Therefore, this suggests that the assemblage underwent some post-depositional modification due to low- to moderate energy hydraulic processes which varied in between cycles. Also, the greater representation of large specimens when compared to experimental assemblages or sites like FLK Zinj is also due to the fact that a substantial number of bones from large-sized fauna (giraffes and hippos) are present. This low bias is also be supported by the presence of all anatomical parts from various carcass sizes which would not be present if the site were the result of transport by physical agents (see below).

This interpretation is also supported by the very low frequency (1.2%) of bone which has been polished and abraded due to water transport. Water in the channel hardly brought any bone into the site, but seems to have taken away some of the smallest fragments. The few polished specimens showing some degree of transport are small fragments (<4 cm). This suggests that the agent that transported these few specimens into the site also deleted a portion of the same size population from it. Further support for this assertion comes from the analysis of bone orientation. When all bones are plotted irrespective of size, no distinctive orientation can be observed. Only a slightly preferred northwest-southeast orientation can be observed (Figure 6.11). This becomes the dominant orientation when only large specimens (>6 cm) are considered. Long bones show a clear preferred orientation, perpendicular to the inferred axis of the channel, which is the direction in which the energy must have flowed (Figure 6.12). Therefore, water must have rearranged the original distribution of the archaeological materials in most of the deposit, except in the lowermost part of it, where it seems to have affected the distribution of bones and stones in a minimal way (Figure 6.13).

This can be clearly observed when the orientation of bones is analyzed by depositional sections. Figure 6.13 shows the distribution of bone orientation in three artificial sections in the archaeological deposit, which were created using arbitrary sections measuring approximately 30 cm in depth each. It is clear that bones in the upper 30 cm of the deposit show a slight but not remarkable orientation (Figure 6.13a), whereas

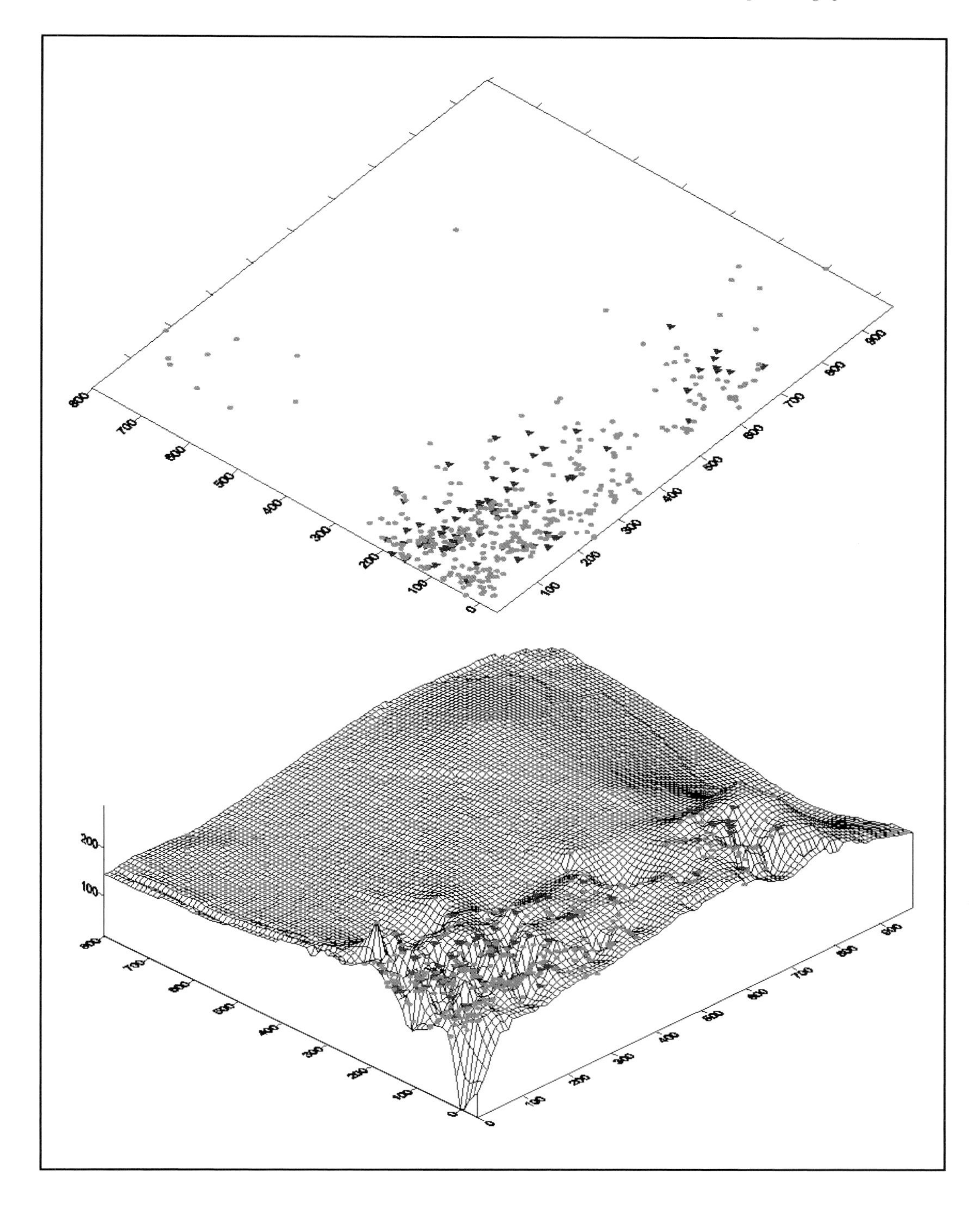

Figure 6.6 Spatial distribution of bones (light circles) and stone tools (dark triangles) at ST4. See Figure A.4 in the Appendix for a color version.

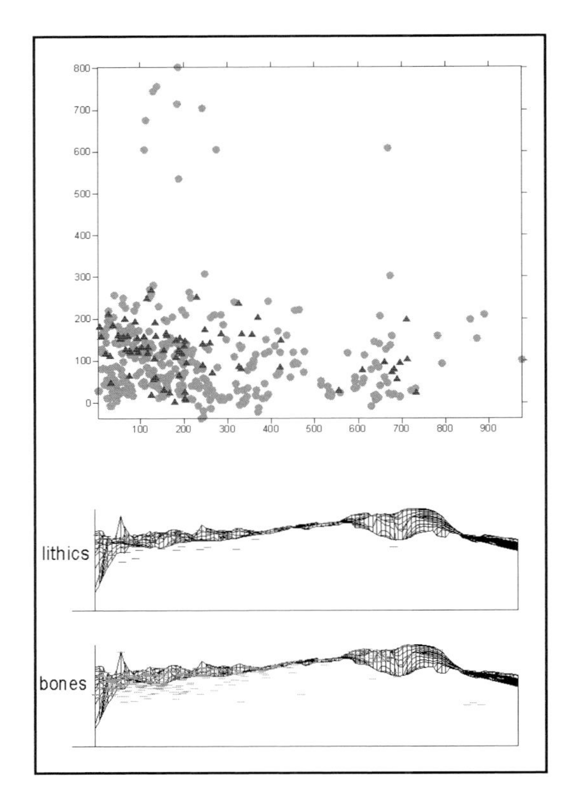

Figure 6.7 North-south view of the vertical and horizontal distribution of bones (light circles) and stone tools (dark triangles) at ST4. See also Figure A.5 in Appendix for color version.

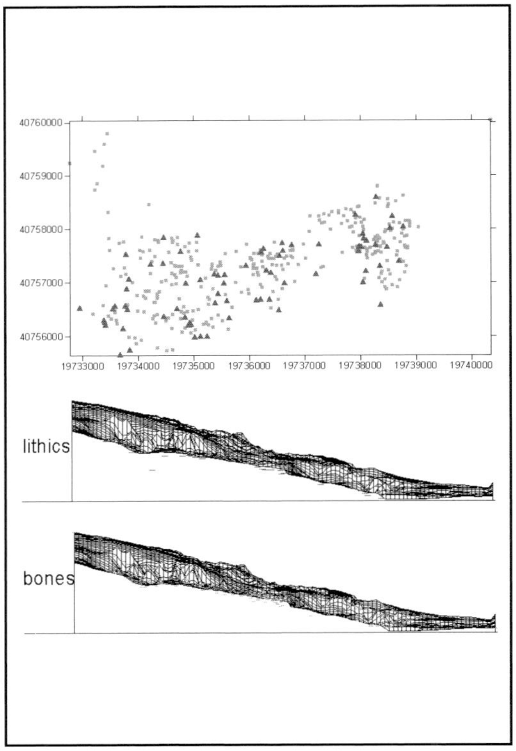

Figure 6.8 West-east view of the vertical and horizontal distribution of bones (light circles) and stone tools (dark triangles) at ST4. See also Figure A.6 in Appendix for color version.

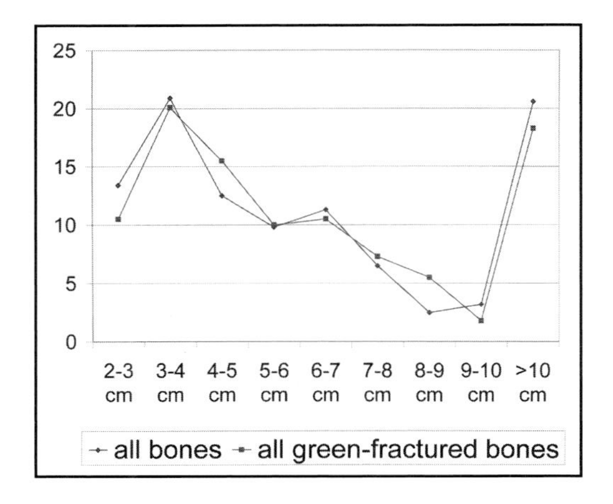

Figure 6.9 Distribution of bone specimen sizes, using all bones and using only green-fractured specimens.

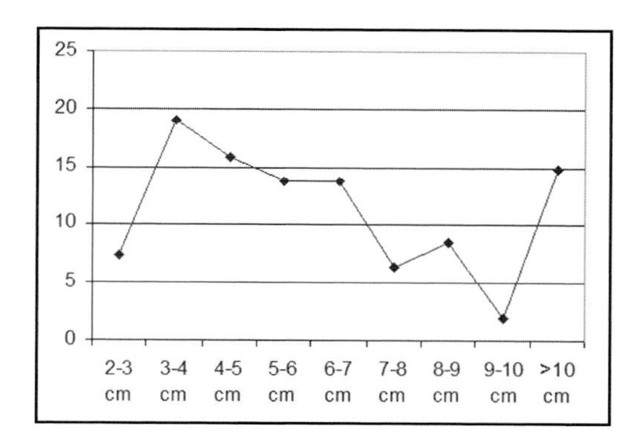

Figure 6.10 Distribution of specimen sizes using long limb shaft specimens, which have similar densities.

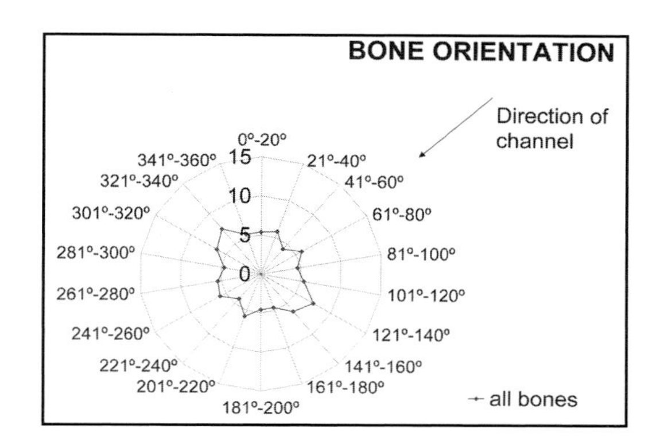

Figure 6.11 Orientation of all bone specimens at ST4.

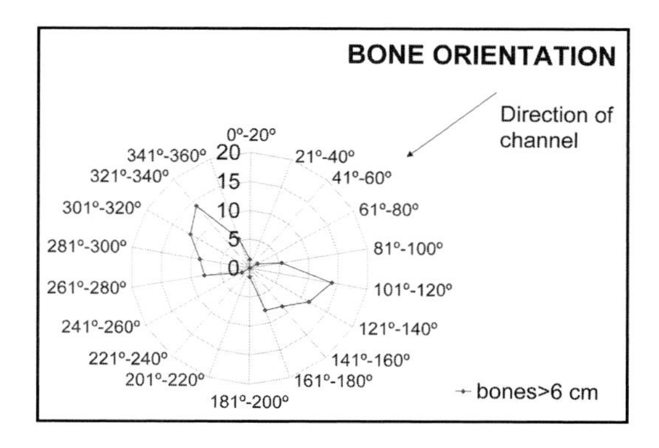

Figure 6.12 Orientation of all bone specimens larger than 6 cm at ST4.

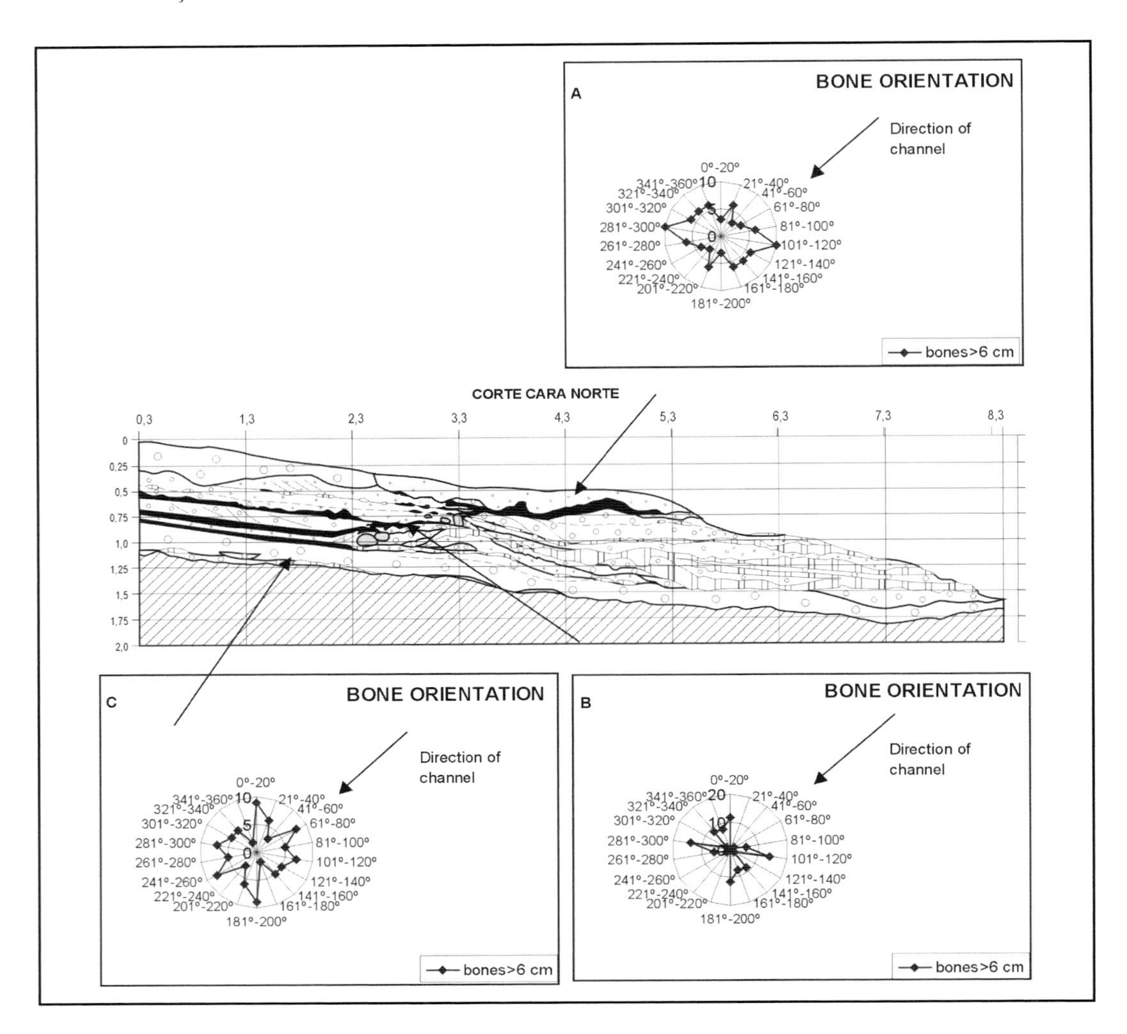

Figure 6.13 Orientation of all bone specimens larger than 6 cm in the top, middle, and lower sections of the deposit at ST4.

bones in the middle section shows a sharper bi-directional orientation (Figure 6.13b), indicating stronger modification by hydraulic energy. It is in this middle section that we found the smallest number of small-sized bone specimens, which clearly indicated that the small channel episode observed in the mid-section of the stratigraphy of the main channel must have been responsible for significant bone loss (Figure 6.1). The lower section of the deposit, representing the shallow phase of the main channel, shows no bone orientation at all (Figure 6.13c). In this lower part of the deposit, materials have preserved a higher degree of integrity, and clear spatial associations between anatomical elements of differential individuals and stone tools can be observed (see below). From a depositional point of view, these taphonomic approaches (specimen size and orientation, bone polishing and abrasion) suggest that ST4 is not a hydraulic jumble. Rather, it seems to be an archaeological deposit retrieved in the same location in which it was formed, by the edge of a channel; some of the smaller materials were likely taken away because of water energy, and the other materials were probably rearranged to some extent (both horizontally and vertically) with respect to their original positions, especially in the upper 60 cm of the deposit.

Taphonomic integrity of the ST4 assemblage

Marean et al. (2004) suggested a method for determining the degree of reliability of any given faunal collection to be properly understood from a zooarchaeological point of view. This method placed special emphasis on various aspects of shaft fragments which are usually neglected by archaeologists both in curating as well as in analyzing any faunal assemblage. When limb bones are broken by humans or carnivores they typically produce numerous shaft fragments and some articular ends, and the latter are often consumed by the carnivores (e.g., Blumenschine 1988; Binford et al. 1988; Bunn 1986, 1991; Marean and Spencer). Generally the articular ends of broken limb bones have a section of shaft attached that is complete in circumference (Bunn's Type 3; Bunn 1982), while the isolated shaft fragments may show more than half the circumference (Bunn's Type 2) or less than half the circumference (Bunn's Type 1). Experiments have shown that when bones are broken either by humans or carnivores, the percentages of each of these types may vary, but the proportion of Types 3 and 2 to Type 1 range from 0.44 to 0.10; that is, Type 1 specimens are dominant. Any assemblage where the ratio is >1 clearly indicates a severe taphonomic or analytical bias (Marean et al. 2004). The ST4 limb assemblage shows a ratio of 0.33 (Figure 6.14), within the range of variation of experimental assemblages, although slightly higher than human-carnivore assemblages, probably because a significant number of small specimens (mostly Type 1) have been transported away from the site by water. This shows that the assemblage contains a high degree of integrity in terms of analytical bias and that it can reliably be used to determine skeletal part profile estimates.

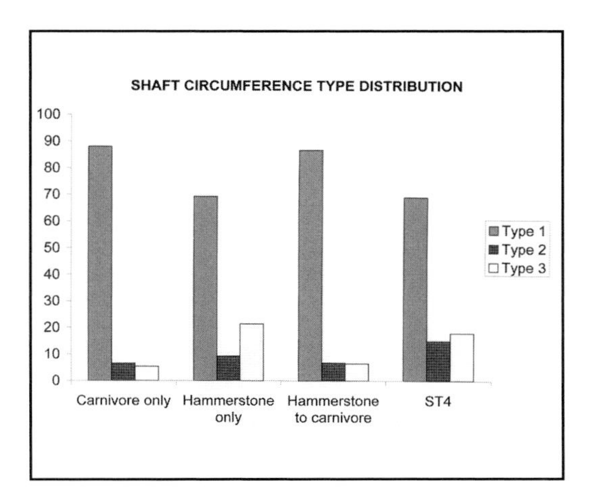

Figure 6.14 Distribution of the different shaft circumference types for long bone fragments in experimental assemblages and at ST4. Data for experimental assemblages are from Marean and Spencer 1991 and Marean et al. 2004.

The second approach in the method suggested by Marean et al. (2004) was to measure the ratio between epiphyses and shafts. Every study of modern processes that affect the relative survival of bone shaft and epiphyseal portions provides evidence that limb bone shafts survive at rates higher than epiphyses (Pickering et al. 2003; Marean et al. 2004). Not a single study identifies the loss of shaft portions at rates greater than ends, nor at rates equal to ends. Thus, we have a universal pattern documented by modern observations that limb bone shaft portions survive carnivore ravaging at rates higher than epiphyses. We also know that limb bone shaft portions are denser than epiphyses and lack nutrition that is bound in the bone (such as grease), providing a cause and effect link between ravaging by carnivores and differential survivorship of portions within limb bones. The average of shaft fragments per epiphyseal fragment ranges from 3 (Blumenschine 1988; Marean et al. 2004) or 5–7 (Capaldo 1995) to 10 (Blumenschine 1988) in modern experiments on bone breakage, with or without carnivore intervention. As can be seen in Table 6.1, the proportion of shafts to epiphyses is 2.9, partly due to the fact that most epiphyses were diagenetically fragmented into several specimens. During excavation, especially with the larger fauna, fragments would be unearthed as single specimens but then would break into various pieces after removal, despite attempts to consolidate them. Therefore, the number of epiphyseal fragments is artificially inflated. Another consideration is that a significant amount of small shaft fragments have been deleted from the record due to transportation away from the site. This would also have decreased the original shaft sample. Despite this, shafts outnumber epiphyses. This suggests that the bias in the collection introduced by physical agents is small and that there is no analytical bias in the collecting process.

Zooarchaeological analysis of ST4

The ST bone assemblage is currently composed of 735 macro-mammal remains.[1] Given the detritic context of the deposit, no micromammal remains were found, although a few fragments of crab and fish were discovered (Table 6.2). Crocodiles and turtles are well-represented: a total of 113 turtle remains were discovered. A large array of mammal species are documented by teeth and/or horns: *Ceratotherium simum*, *Equus* sp., *Elephas recki*, *Sivatherium maurusium*, *Giraffa* cf. *pygmaea*, *Syncerus* sp., *Tragelaphus strepsiceros*, *Megalotragus kattwinkeli*, *Connochaetes taurinus*, *Damaliscus niro*, *Parmularius angusticornis*, *Kobus* cf. *cob*, *Hippotragus gigas*, *Aepyceros melapus*, *Sylvicapra* sp., *Antidorcas* cf. recki, *Gazella* sp., *Kolpochoerus* sp., *Metridiochoerus compactus*, and *Theropithecus* sp. If using horns and teeth, a minimum of 28 individuals (MNI) have been documented. However, post-cranial elements document a substantially smaller number of individuals. A total of 13 individuals (5 small carcasses, 5 medium-sized carcasses, and 3 large carcasses) account for most of the post-cranial elements recorded at ST4. The appendicular and cranial anatomical areas of small and medium-sized carcasses seem to be well-represented (Table 6.3). Only one large carcass (belonging to a pygmy giraffe) appears represented by most anatomical areas. Some of the partial skeletons of 11 carcasses (four *Antidorcas* cf. recki, one *Gazella*, three *Parmularius* sp., one *Connochaetes taurinus*, one *Metridiochoerus compactus*, and one *Giraffa* cf. *pygmaea*) show traces of hominid modification in the form of cut marks, percussion marks and percussion notches.

Appendicular element identification was carried out using a combined approach of epiphyseal and shaft element quantification as described in Barba and Domínguez-Rodrigo (2005). As can be seen in Table 6.3, MNE estimates based on shaft fragments are systematically more numerous than MNE estimates based on epiphyseal specimens. This difference is consistent across all elements except metapodials, and is not restricted to elements from small carcasses. Humeri and femora of small and medium-sized carcasses would be almost non-existent had the MNE estimate been based only on epiphyseal fragments. Only in large carcasses (Bunn's sizes 4–6; Bunn 1982) do both shafts and epiphyses

Table 6.1 Distribution of skeletal part profiles (NISP) at ST4 according to carcass size (using Bunn 1982)

Size	1 and 2	3	4 and 5
HORN	5	5	6
SK	2	5	0
MAND	3	5	2
TTH	23	38	10
CV	3	1	2
TV	2	3	7
LV	0	1	1
V	8	9	1
PL	3	6	8
SC	8	8	9
RB	27	10	12
HUM prox.	1	3	5
HUM MSH	5	2	1
HUM dist.	3	3	4
RAD prox.	2	2	1
RAD MSH	6	16	3
RAD dist.	2	4	0
MC prox.	1	0	4
MC MSH	1	1	2
MC dist.	3	7	7
FEM prox.	1	1	1
FEM MSH	6	5	4
FEM dist.	0	0	0
TIB prox.	0	2	2
TIB MSH	9	20	1
TIB dist.	1	7	0
MT prox.	1	3	0
MT MSH	7	3	0
MT dist.	3	2	1
CARPAL	2	0	8
TARSAL	2	5	6
PHALANX	3	1	0
ULB	21	17	10
ILB	12	15	23
LLB	23	16	2
INDET	23	21	10
TOTAL	222	250	153

ends	77
shafts	227
Diagenetic breakage	151/617 (24.4%)
Polishing/abrasion	7/617 (1.2%)
Circumference types of long limb bones	
type 1	200/304 (66%)
type 2	36/304 (11.8%)
type 3	68/304 (21.2%)

yield similar MNE estimates. Had the MNE estimates been based on epiphyseal ends and only those shafts with the most easily identifiable landmarks, the ST4 assemblage would have been presented as a lower limb- and skull-dominated assemblage typical of scavenging behaviors (as is the case in Stiner 1994). In contrast, the shaft-based approach has allowed the identification of those elements that would be least visible to the traditional analyst's eye, these being meaty long bones among which humeri and femora are predominant. The poor preservation of upper limb bone epiphyses relative to those of metapodials is probably related to the former being less dense and more desirable for scavenging carnivores given their grease contents (Marean et al. 1992; Capaldo 1995; Pickering et al. 2003).

Table 6.2 Faunal remains at ST4

Macromammals	
Identifiable species	**632**
skull	59
axial	197
appendicular	376
Non-identifiable species	**103**
axial	71
appendicular	32
Total macromammals	**735**
Chelonia	113
Reptilia (non-Chelonia)	13
Aves	0
Fish	16
Crustaceae	6
Total	**883**

Table 6.3 Distribution of skeletal part profiles (MNE) at ST4 according to carcass size (Bunn 1982)

	size 1, 2	size 3	size 4, 5, 6
horns	4	8	3
skull	2	3	6
atlas	0	0	3
axis	0	0	2
CV	0	4	1
TV	0	5	3
LV	0	3	0
pelvis	2	5	2
scapula	0	5	5
ribs	5	8	10
humerus ep.	3	3	5
humerus ep.+ shafts	6	7	5
radius ep.	2	4	1
radius ep.+ shafts	3	6	1
metacarpal ep.	3	7	7
metacarpal ep.+ shafts	3	8	7
carpals	3	7	11
femur ep.	1	1	1
femur ep.+ shafts	5	5	2
tibia ep.	1	7	2
tibia ep.+ shafts	6	8	3
metatarsal ep.	1	6	1
metatarsal ep.+ shafts	1	6	1
tarsals	1	6	4
phalanges	3	5	1

Further support for this observation can be obtained from the representation of axial elements. Their relative scarcity in the assemblage suggests carnivore post-depositional ravaging, as experimentally observed in human-to-carnivore assemblages (Capaldo 1995). However, the very presence of axial elements is important, because given the evidence of hominid exploitation of the appendicular skeletons of almost a dozen carcasses, they suggest complete carcass transport. In fact, a few axial elements also bear cut marks. Large and medium-sized carcasses might have been transported fairly complete to the site, given the presence of axial elements from these carcass sizes in the assemblage. Small carcasses very likely underwent the same process, but given their lower bone density, carnivores very likely deleted their axial bones from the assemblage.

Small carcass bones do not seem to correlate with meat consumption. Correlation between element representation (%MAU) and the general utility index (GUI) yielded a negative ($r = -0.17$) and non-significant ($p = 0.60$) relationship. However, given that carnivore intervention in the assemblage has been widely documented (see below), correlating bone representation with carcass nutritional value is not appropriate. Carnivore scavenging affects primarily axial bones, which have a high nutritional value. When we use only the long limb bones, which can be clearly reconstructed even after carnivore ravaging by using element identification through shaft quantification, the results are different (Figure 6.15). In this case the correlation between the appendicular bones represented and their nutritional value is positive ($r = 0.32$), although still not significant ($p = 0.53$), probably because of the small sample size. Bones from medium-size carcasses also do not correlate with meat consumption when considering all skeletal elements ($r = -0.078$, $p = 0.782$). However, if using only long limb bones, the correlation is positive ($r = 0.69$, $p = 0.146$), as was the case with small carcasses (Figure 6.15). Once again, by using the more dense, ravaging-resistant skeletal elements, taphonomic biases introduced by carnivores can be overcome and meat exploitation can be inferred.

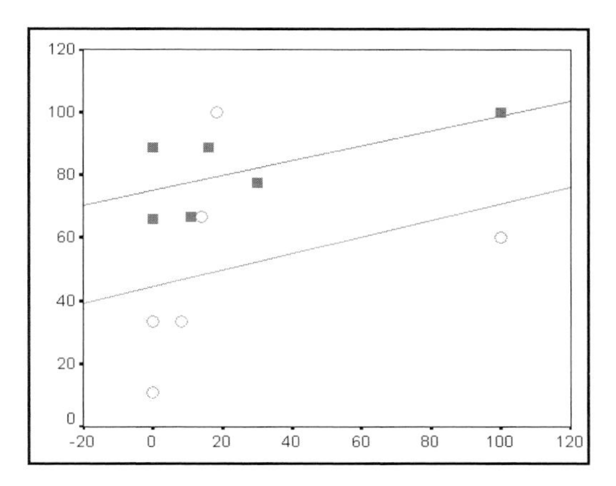

Figure 6.15 Correlation between long limb bone representation at ST4 (%MAU y axis) and their nutritional value (GUI x axis). Utility indices for small and large carcasses have been derived from data on meat and marrow values in gazelle and wildebeest carcasses, as published by Blumenschine and Caro (1984). Open circles represent small carcasses, closed squares represent large carcasses.

Marean and Cleghorn (2003) identified a threshold of bone mineral density above which bone fragments have a much better chance of survival. According to Cleghorn and Marean (2004:56):

> Bones with marrow cavities are attractive to scavengers only up until trabecular portions have been deleted, marrow has been removed, and cortical portions lacking trabecular bone remain. Elements without a substantial cortical portion free of trabeculae, lack this brake on ravaging. As noted

above, there are several elements that are often consumed or completely destroyed despite regions of high density. Marrow bones are less likely than other elements to be completely destroyed by carnivore ravaging. We therefore view the marrow bones – meaning all long bones and the mandible – as a coherent group with respect to survivability. This "high survival" set includes all of the long bones (femur, tibia, humerus, radius-ulna, and metapodials), mandibles (these basically function like long bones due to their dense cortical bone and open medullary cavity), and crania. The cranium is included because teeth and petrosals are both extremely dense and lack nutrient value. They demonstrably survive carnivore ravaging very well. The ulna is included because, in bovids and cervids, this bone often fuses with or is tightly bound to the radius shaft, and has a countable landmark in that area. The "low survival" set includes all vertebrae, ribs, pelves, scapulae (which have thick cortical bone but are difficult to identify and quantify when fragmented), and all tarsals, carpals, and phalanges of size class 1 and 2 ungulates (Brain 1981) since these tend to get swallowed by carnivores. All of these have significant proportions of trabecular bone that, because of high grease content, are especially attractive to scavengers. More importantly, low survival elements lack large areas of dense cortical bone without trabeculae. Although high survival elements are designated as such principally on the basis of their resistance to carnivore ravaging, a survey of maximum density in this group shows gratifyingly little variation as well. This is similar to the leveling effect Stiner attempted to achieve through the ARP. We can test the high survival set against the carnivore ravaging data in much the same way that we tested the ARP. For this test, however, only studies that incorporate long bone shaft fragments into estimates of element survival are appropriate.

This is exactly what can be observed at ST4. Axial elements and compact bones (carpals and tarsals) are underrepresented. They are low survival bones that tend to dissappear after carnivore ravaging takes place. Cranial and especially long limb bones are predominant in the ST4 assemblage. They compose the high survival set which can be more accurately quantified because the shaft approach was used when estimating long limb bone frequency. Carnivore intervention in assemblage modification can also be demonstrated by the presence of tooth marks on several bones (see below). Therefore, bone deletion is attributable to carnivore post-depositional ravaging, since the missing bones are those that tend to disappear first when carnivores intervene. Given that the presence of axials shows that at least some carcasses were transported complete, the low number of these bones together with the low frequency of carpals, tarsals, scapulae, pelves and cancellous epiphyses of long limb bones (especially for small animals) can confidently be attributed to the action of carnivores. Thus, the original skeletal part profiles representing what hominids brought to the site cannot be reproduced because of the bias introduced by carnivores. However, this bias is not sufficient to obliterate the evidence of hominid access to fleshed carcasses, evidence which includes the high presence of upper limb bones and the frequency and distribution of cut marks (see below).

The ST4 assemblage makes an interesting contribution to the debate regarding carcass transport by hominids in that 11 carcasses are represented by several long limb bones from both left and right sides, suggesting that carcass transport could have been complete, as indicated by the presence of axial elements; or if transport was selective, then limbs from the same carcass were transported together. This indicates that the limb elements of the carcass were not previously disarticulated, as would be expected in most carnivore kills. The high presence of humeri and femora for all carcass sizes further suggests that carcasses must have been fleshed when transported. This is demonstrated below in the study of cut mark distribution.

Behavioral significance of tooth and percussion mark distribution at ST4

Blumenschine's (1988) pioneering work on tooth mark percentages and distribution on bone portions (i.e., epiphyseal, near-epiphyseal and midshaft portions of long limb bones) showed that the order of access to carcasses by carnivores (i.e., primary or secondary) could confidently be determined. Carnivore modification of epiphyseal and near-epiphyseal bone portions did not seem to be significantly different when comparing primary and secondary access, since the grease deposits in the cancellous sections of these elements made them appealing both when carcasses were fleshed and defleshed. However, a sharp contrast in tooth mark frequencies was seen when using midshaft portions. When carnivores have access to complete bones, they leave tooth marks on the shafts in the process of breaking them open to obtain the marrow; this results in high tooth mark frequencies, with marks usually occurring on more than 80% of all the broken fragments. When humans access the marrow first by breaking the bones open with hammerstones, carnivores with secondary access will only find dense midshaft fragments devoid of any edible resource. They then leave very few tooth marks: on average in human-first experimental assemblages, carnivores leave between 5% (Blumenschine 1988) and 15% (Capaldo 1995) of bone fragments tooth-marked when dealing with middle-sized carcasses and a similar percentage when dealing with smaller carcasses. This modern referential framework theoretically enables clear-cut differentiation of the order of access to carcasses by carnivores in archaeological bone assemblages, provided that the primary agent of bone breakage and modification fully exploits all the resources available that are suitable for their physiology and dental morphology (carnivores) or technology (hominids).

Regarding hammerstone percussion marks, Blumenschine and Selvaggio (1991) calculated mark frequencies in human-first assemblages which were subsequently modified by hyenas. Percussion mark frequencies on limb bone specimens (irrespective of bone portion) varied by carcass body size (size classes follow Brain [1981]): size class 1 and 2 limb bone assemblages showed percussion mark frequencies of ~30%; size class 3 had frequencies of 21.5%; size class 4 had frequencies of 53.8%. Further, Blumenschine and Selvaggio (1991) established percussion mark to tooth mark ratios in these simulated hammerstone-to-carnivore assemblages that range from 1.5:1 (in size class 3 carcasses) to 2:1 (in size class 2 carcasses). In other words, percussion mark frequency is nearly double in cases in which hyenas act as secondary modifiers of limb bones first demarrowed and abandoned by humans.

Thus, the comprehensive consideration of both cut mark and percussion mark frequencies leads us to suggest that the relative contribution of hominids to the formation of prehistoric bone assemblages could be inferred from calculating those frequencies. If cut mark and percussion mark frequencies are lower than in the modern assemblages in which humans were the primary agents of carcass-processing, this probably indicates a relatively minor overall hominid contribution to the bone accumulation if the site is a palimpsest (Domínguez-Rodrigo et al. 2005). Alternatively, this could also mean that strong taphonomic processes have intervened, biasing the original configuration of the assemblage.

Percussion marks, as well as tooth marks and cut marks, can only be properly identified and evaluated in assemblages in which bone surfaces have been well-preserved, a situation which is the exception rather than the rule. For this reason, any study of bone surface modifications should be carried out on bones that have optimally preserved cortex, mainly on specimens showing very little sub-aerial weathering. The sample collected for this purpose at ST4 was, as discussed above, composed of 151 long limb bone fragments, mostly shafts. The analytical method applied for the study of percussion marks is the one developed by Blumenschine (1988, 1995) in which limb bone fragments are classified as epiphyseal, near-epiphyseal and midshaft portions. The analysis of midshafts was targeted, given the crucial role of these bone portions in determining whether within-bone nutrients were exploited by either hominids or carnivores, and whether carnivores had been primary or secondary agents in carcass obtainment and flesh and marrow consumption.

Tooth marks are very abundant on the few axial remains that have been preserved at ST4. Almost one-third of rib and vertebral remains show at least one tooth mark, primarily on bones from smaller carcasses. This implies a high degree of carnivore activity at the site and the subsequent deletion of most of the axial elements originally accumulated there. On long limb bone shafts, tooth marks are very scarce (Table 6.4). Contrary to what would be expected according to Blumenschine (1988) if carnivores had primary access to carcasses, midshaft fragments are tooth-marked at very low rates (<10%), irrespective of carcass size. This suggests that hominids had primary access. Furthermore, tooth marks are even scarcer (<5%) on upper limb bone midshafts, suggesting that no carnivore took part in the initial defleshing of carcasses, since these parts (especially femora) are the first targeted by carnivores after hunting their prey (Blumenschine 1986).

However, not only were carnivores not involved in the first stages of carcass exploitation at ST4, but they seem to have been relegated to a marginal role in the final stage, after hominids demarrowed the appendicular bones. The few complete bones found at the site were metapodials from size 1 carcasses (two elements) and from a very large carcass (metapodials from a giraffe). The remainder of the bones were broken. Percussion marks occur on 14.8% (in large carcasses) to 23.7% (in small carcasses) of bone fragments, slightly lower than the frequencies observed in modern experiments (probably due to more intensive bone breakage during diagenesis), but very close to them, suggesting that most of the bones were broken through hammerstone percussion by hominids. This is further supported by the percussion mark to tooth mark ratios (see above) at ST4, which are 2.3:1 for small carcasses and 3:1 for larger carcasses, with an average of 2.5:1 if both carcass sizes are lumped. In sum, tooth mark and percussion mark frequencies and distribution are similar to modern experiments replicating primary access by humans to carcasses (Blumenschine and Selvaggio 1991).

Behavioral significance of cut mark distribution at ST4

The arguments that support the analytical utility and explanatory power of cut mark placement are well-established in the literature (Bunn 1982;

Table 6.4 Distribution and percentage of tooth marks and percussion marks*

	Smaller Mammals	Larger Mammals	Total
Tooth marks			
Upper limb bones	1/25 (4)	0/9 (0)	1/34 (2.9)
Intermediate limb bones	5/41 (12.1)	1/7 (14.2)	6/48 (12.5)
Lower limb bones	2/14 (14.2)	0/5 (0)	2/19 (10.5)
Total	**8/80 (10)**	**1/27 (3.7)**	**9/107 (8.4)**
Percussion marks			
Upper limb bones	6/25 (24)	3/9 (33.3)	8/34 (23.5)
Intermediate limb bones	9/41 (21.9)	1/7 (14.2)	10/48 (20.8)
Lower limb bones	4/14 (28.5)	0/5 (0)	4/19 (21)
Total	**19/80 (23.7)**	**4/27 (14.8)**	**23/107 (21.5)**

* Shaft specimens

Numerator indicates the number of marked specimens. Denominator shows the number of specimens in each category. Numbers in brackets are for percentages

Bunn and Kroll 1986; Domínguez-Rodrigo 1997a, 1997b, 1999a, 1999b, 2002; Domínguez-Rodrigo and Pickering 2003; Domínguez-Rodrigo et al. in press; Pickering and Domínguez-Rodrigo 2003). A number of modern cut-marked assemblages are available, derived from both ethnoarchaeological and experimental studies, which indicate that mark frequencies on limb bone specimens are remarkably consistent across these varied datasets, ranging between 15-30 % of all specimens recovered after hammerstone bone breakage (e.g., Bunn 1982 in identifiable bones only; Lupo and O'Connell 2002). Although cut mark frequencies in Domínguez-Rodrigo's experimental butchery dataset are beyond the range cited above, it should be emphasized that those experiments were aimed explicitly at complete flesh removal. Substantial scraps of flesh still adhere to defleshed bones after a more typical episode of human butchery, e.g., in ethnoarchaeological situations where the butchers are hunter-gatherers who usually abandon these adhering flesh scraps. This accounts for the range of 15%-30% of limb bone fragments being cut-marked in modern assemblages. Indeed, in the one experiment by Domínguez-Rodrigo (1997a-b) in which typical butchery was conducted, cut mark frequencies of 29% agree with the expected values for typical butchery scenarios.

However, obtaining these values in archaeological assemblages is rather uncommon. A frequent problem when comparing experimental and fossil assemblages is that results often vary because some taphonomic variables that have intervened in the configuration of fossil assemblages have not been experimentally modelled. Fossil specimens often show poor cortical preservation and diagenetic breakage. Experimental studies have not considered these variables and results therefore differ, irrespective of the hypothesis that is being tested. To overcome this problem, Pickering et al. (2004) have recently created an adjustment index and successfully applied it to the bone assemblage from Swartkrans Member 3. This index combines a bone surface preservation ratio and a diagenetic bone breakage ratio. Briefly, these adjustment ratios are based on the following procedures:

First, bone surface preservation is quantified using a simple formula. Given that bone surface preservation (BSP) is not affected by specimen size, the ratio is obtained as follows:

BSP index = NISP of bones with good cortical preservation/Total NISP

Second, the number of extra fragments resulting from diagenetic breakage is estimated. To quantify this increase, a reliable estimate of bone breakage by specimen size range must be established. An index of diagenetic breakage is obtained which should be subtracted from the sample, the formula being: A.B.I.= X d.b. /2, where A.B.I. is the Adjustment Breakage Index and X d.b. is the NISP with diagenetic breakage.

Given that a bone broken diagenetically will create a minimum of two specimens showing that specific breakage, increase in the number of specimens caused by diagenetic breakage is not directly reflected in the overall diagenetic percentage. For example, consider an original assemblage of 100 bone specimens. Diagenesis breaks 30 of those specimens into at least two fragments per specimen. The resulting number of specimens showing diagenetic breakage is 60, 30 original specimens plus 30 new ones. In the resulting assemblage, would identify 60 NISP with diagenetic breaks out of 130 NISP, giving a diagenetic breakage rate of 46%. But if we were to subtract 46% of the whole assemblage, we would be overestimating diagenesis. By dividing 46% by 2, we obtain a rate of diagenetic breakage of 23% of the 130 fragments, that is, 30 NISP, which is the original number of fragments undergoing diagenetic breakage. The NISP estimate provided by this formula would still be lower than the actual one, since diagenetic breakage of bone specimens often produces more than two fragments.

We may apply this formula to any sample of bones with good cortical preservation, since we would expect that diagenetic breakage occurs randomly, not selectively on bones with good cortical preservation. These indices should be applied whenever differential preservation of cortical surfaces and a substantial amount of diagenetic breakage have been documented. Otherwise, the application of experimental referential frameworks that have not considered these taphonomic processes is of limited use.

For analytical purposes, as shown in Chapter 4, the faunal sample of ST4 was divided into small (sizes 2–3) and large (sizes 4–5) mammals. Most of the cut marks reported previously (Domínguez-Rodrigo et al. 2002) for the ST Site Complex come from ST4, with only three exceptions from ST15 and ST2. Most cut marks in this assemblage are found on size 2 and size 3 carcasses. Most of the large mammal cut-marked sample comes from size 4 remains, and mostly from one individual, a pygmy giraffe. Following the principles described above, only those bones with good preservation (mostly Behrensmeyer's [1978] weathering stage 0) were used. Most of the ST4 bones show poor cortical preservation, as is common in most of the ST Site Complex faunal assemblages. ST4 was selected for excavation not only because it showed the highest density of fauna in the entire Lake Natron basin, but also because bone preservation seemed to be better in some specimens than in any other locality. More than 80% of the total bone assemblage from ST4 shows poor cortex and most of the specimens are affected by etching or, most commonly, by strong carbonation on the surfaces and cortical loss due to a humid depositional environment.

Mark frequencies are reported by anatomical element (Tables 6.5–6.6). However, given that most of the faunal collection is composed of limb bones, detailed anatomical distribution of cut marks is also shown on each appendicular unit (e.g., upper limb bone; humerus, femur) and each bone portion (proximal or distal end, midshaft) as specified in Domínguez-Rodrigo (1997a, 1997b). A total of 139 limb bone fragments, most of them showing weathering stage 0, were classified to element type. Twelve more were classified to anatomical unit (i.e., upper, intermediate, or lower limb bone). In total, a sample comprised of 35 cut-marked bones out of 151 identified long limb fragments showing complete bone surface preservation was obtained. Sixteen additional cut-marked bones from various other anatomical parts (e.g., phalanges or ribs) were also added to the sample. This ranks ST4 the site with the third-highest number of cut-marked bones after FLK Zinj and Swartkrans Member 3, among all Pleistocene archaeological sites older than 1 Ma.

The patterns of cut mark frequency and distribution on long limb bones show a high proportion of cut-marked shaft specimens from upper and intermediate limbs. The overall percentages of cut-marked bones (16% for small mammals and 15.1% for large mammals) are similar to those replicated in human-to-carnivore experiments and those obtained through studies of modern foragers who have access to fleshed carcasses. The fact that they are in the lower part of the range of modern experiments may be related to the amount of diagenetically-broken bone. Given that close to 25% of fragments in the ST4 assemblage were broken diagenetically, the correction index described above would mean that 12.5% of the total sample should de subtracted. This would mean that bones from small mammals were cut-marked at an approximate rate of 18.2%, and larger mammals at 17.2%, which is very similar to frequencies obtained in modern experimental assemblages and from modern hunter-gatherers (Gifford 1977; Bunn 1982, 1983; Lupo and O'Connell 2002). However, this percentage is still not completely accurate because it does not include all the small fragments that post-depositional water flows have taken away from the site.

Table 6.5 Cut mark patterns and distribution in ST4

	Smaller mammals	Larger mammals
HUM PSH	0/1 (0)	0/0 (0)
HUM MSH	4/12 (33.3)	2/5 (40)
HUM DSH	0/3 (0)	0/2 (0)
RAD PSH	2/3 (66.6)	0/2 (0)
RAD MSH	2/20 (10)	1/4 (25)
RAD DSH	0/1 (0)	0/0 (0)
MC PSH	1/4 (25)	0/2 (0)
MC MSH	0/6 (0)	0/4 (0)
MC DSH	1/6 (16.6)	0/3 (0)
FEM PSH	0/1 (0)	0/0 (0)
FEM MSH	3/10 (33.3)	2/4 (50)
FEM DHS	0/2 (0)	0/2 (0)
TIB PSH	0/2 (0)	0/1 (0)
TIB MSH	4/18 (22.2)	0/3 (0)
TIB DSH	0/3 (0)	0/0 (0)
MT PSH	0/2 (0)	0/0 (0)
MT MSH	0/8 (0)	0/1 (0)
MT DSH	0/4 (0)	0/0 (0)
TOTAL	17/106 (16)	5/33 (15.1)
cut marked ULB*	3	3
cut marked ILB*	3	2
cut marked LLB*	1	-
ULB:ILB Ratio		
ULB	10/32 (31.2)	7/17 (41.1)
ILB	11/50 (22)	3/10 (30)
MSH	20/80 (25)	10/21 (47.6)
ENDS	4/32 (12.5)	0/12 (0)
MEATY MSH %	20/56 (35.7)	10/21 (47.6)
NICMSP:NISP	24/113 (21.2)	10/38 (26.3)
NCMMSSP:NICMSP	20/24 (73.3)	10/10 (100)

* Non-identifiable to element type. All shaft specimens.

Table 6.6 Cut marked specimens from the ST4*

	Smaller mammals	Larger mammals
Vertebra	3	0
Ribs	1	3
Skull	1	0
Scapula	2	1
Calcaneum	0	1
Astralgus	1	0
Phalange	1	0

*Non-long limb bone specimens.

Despite these limitations, the frequencies and anatomical locations of cut marks at ST4 remain indicative of primary access to carcasses by hominids. Upper limb bones have the highest cut mark frequencies, followed by intermediate bones. The cut-marked ULB:cut-marked ILB ratio and the higher frequency of cut-marked specimens on midshafts versus ends are indicators similar to those obtained in experiments modeling primary access to fleshed carcasses (Domínguez-Rodrigo 1997a, 1997b, 1999a). Experiments show that shafts usually comprise >75% of the cut-marked sample when humans have primary access to carcasses; by contrast, midshafts are usually cut-marked at very low frequencies (or not at all, especially on upper and intermediate limb bones) when humans have secondary access to carcasses initially defleshed by carnivores. At ST4, the frequency of shafts among cut-marked bone specimens is 73% for small mammals and 100% for larger mammals. The frequencies of cut marks observed at this site and their anatomical placement are indicative of intensive defleshing (Domínguez-Rodrigo 1999a). Shafts, especially those from meaty limb bones, appear too highly cut-marked to support any argument that hominids behaved as passive scavengers (Domínguez-Rodrigo 1997a, 1997b).

The occurrence of cut marks on pelves also supports this interpretation. One of these marks occurs near the neck of the acetabulum, the other two on the iliac blade. In experiments, these areas show no scrap of flesh after felids consume their prey; cut marks in these areas therefore suggest primary access by hominids. So, too, do the cut marks on ribs, some of which are on distal ends, which are sometimes not even present after felid evisceration of a carcass (Domínguez-Rodrigo 1999a). This suggests initial defleshing by hominids. The presence of cut marks on other elements, such as on the medial aspect of calcaneum, the lateral aspect of astragalus and on phalanges, indicates disarticulation. Cut marks on the proximal ends of radii of small mammals are also suggestive of disarticulation. Disarticulation clearly indicates primary access to carcasses by hominids, since carcasses obtained at felid kills do not require dismembering, an energetically costly process, to be successfully exploited; furthermore, bones collected randomly across the landscape would have already been dismembered.

In summary, as can be seen in Table 6.6, cut marks appear on the skull, axial skeleton, on scapulae and pelves and most extensively, on limb bones. This would not be expected in a passive scavenging scenario: the axial skeleton is completely defleshed after felids finish consuming their prey and there are no resources available for a flesh-eating scavenger (Blumenschine 1986; Domínguez-Rodrigo 1999a). The presence of cut marks on axial elements, as well as their significant overall presence at the site (even after intensive carnivore ravaging), are strong indicators that hominids enjoyed primary access to fleshed carcasses.

The lower level of the ST4 archaeological deposit

Under the premise that the closer to the bottom of the deposit, the less disturbed the materials deposited during channel formation should be, intensive excavation of the site was carried out to the base of the channel, to understand the depth of deposit and processes involved in the formation of the deposit.

The lower section of the ST4 deposit at the bottom of the channel, comprising 30 cm of sands, showed the least disturbed materials and clear spatial relationships among the fossils and between these fossils and stone tools (Figure 6.16a). This section tilts downwards towards the central part of the channel, reflecting the paleo-relief of the ground which was higher to the sides of the channel and deeper in the middle. More than half the width of the channel was exposed through excavation.

Several faunal remains were retrieved from the lower section (Figure 6.17a). On the left-hand side of the figure, a cluster of bones belonging to a pygmy giraffe (*Giraffa* cf. *pygmaea*) was found. On the right-hand side, another discrete cluster fossils, associated with stone tools, could also be distinguished. The first cluster, occupying most of the exposed area, consisted of bones belonging to the giraffid front limbs (humeri, radi-ulnae, one unciform and metacarpals) with their associated scapulae, plus seven ribs and five thoracic vertebra, one cervical vertebra, and one pelvis. One upper molar was also identified. Most of the elements from the left limb appear pseudo-articulated within a 2 m radius. These elements suggest that the front legs, at least one hind limb, the skull and the trunk were present at the site (Figure 6.17a). Identification of these elements was carried out mostly during excavation, since several bones were strongly carbonated and affected by extensive diagenetic cracking (Figure 6.17b). The fragments produced by diagenetic breakage were kept together in almost all of the elements identified during excavation, but preservation treatments proved useless for some of them after their excavation. Some specimens turned into powder and others were extremely fragmented. The study of specimens during excavation also allowed clear identification of breakage patterns other than diagenetic. In some cases, carbonate did not cover the whole surface of the bone (especially true of the scapulae), leaving enough of the original surface intact to observe the cortical preservation. Limb bone shafts and scapulae without carbonate showed no traces of subaerial weathering.

Bones belonging to medium-sized bovids (*Parmularius* sp., Bunn´s [1982] size 3a) are mixed with those belonging to the giraffid (Figure 6.17a). Given the presence of two large skull fragments with both horns, two bovid individuals have been identified. However, when considering only post-cranial elements, the presence of just one element of each type and side is indicative of just one bovid individual represented. This is further supported by biometrics: the metapodials represented show the exact same dimensions, suggesting they came from a single individual. The bovid bones belong to all anatomical sections (skull, axial and limbs), although there is a clear underrepresentation of ribs and vertebrae.

Bones in the smaller second cluster to the right included the following elements: the upper sector of the cluster contained two fragments from the giraffid skeleton (one rib and one distal humerus) and some bovid bones; the lower sector of the cluster contained a glenoid cavity of a scapula, a fragment of a femur and a fragment of a humerus, all from a suid. Table 6.7 shows the distribution of bones of each animal represented in the bottom section of the ST4 sequence.

Analysis and results

The distribution of the archaeological materials shows that both the bones and the artifacts are concentrated in two distinct clusters. This is suggestive of, though not sole proof for, a functional behavioral link between both types of archaeological remains (Figure 6.16a).

Despite the intensive diagenetic cracking, most elements show breakage patterns attributable to fracturing while fresh (Figure 16b). Bone surface modifications were sought to discern the agent of breakage (hominids or carnivores). Given that most of the sample was strongly affected by carbonate and that the original bone surfaces of several specimens were lost, only a

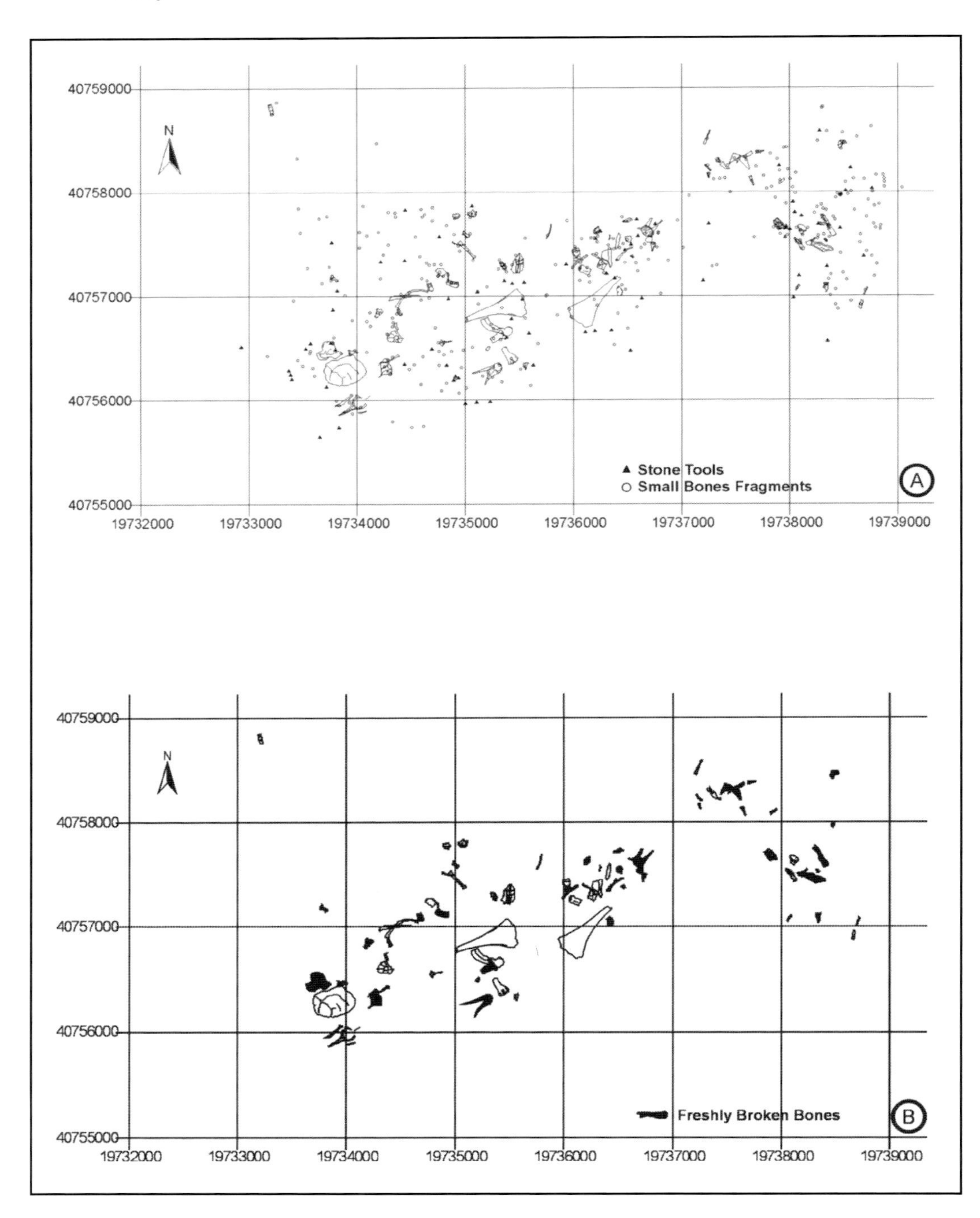

Figure 6.16 A) Spatial distribution of bones and stone tools in the core area of the lower section of ST4; B) distribution of green-fractured bone specimens.

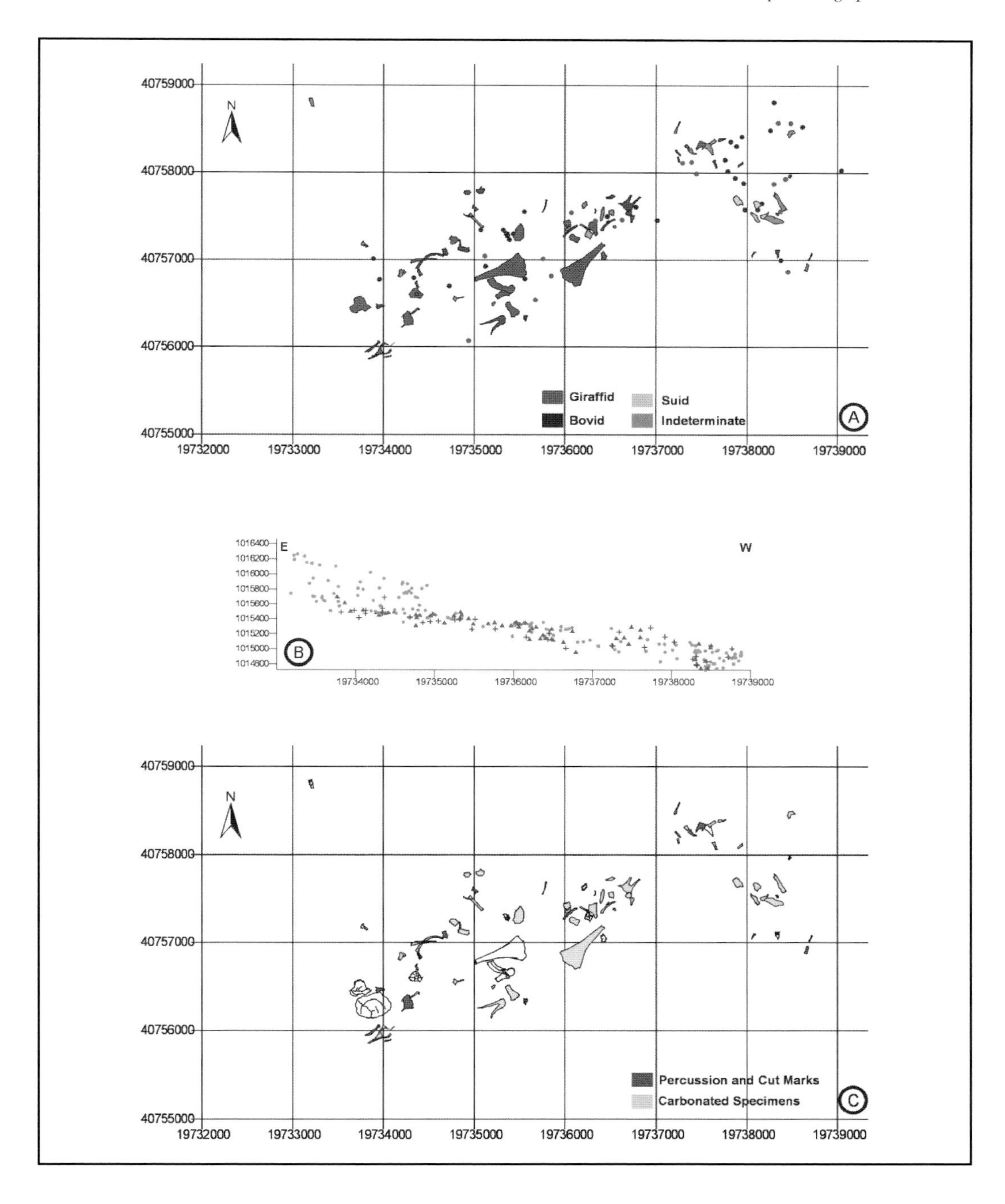

Figure 6.17 A) Spatial distribution of the bones from each of the four carcass types present in the core area of the bottom of ST4 including their vertical distribution (B); C) spatial distribution of carbonate-affected bones and specimens bearing cut marks and percussion marks. See also Figure A.7 in Appendix for color version.

Table 6.7 Element representation for the carcasses present at the bottom of the ST4 deposit

	giraffid			bovid			suid		
element	left		right	left		right	left		right
skull	-	-	-	-	2	-	-	-	-
teeth	-	2	-	-	3	-	-	-	1
mandible	-	-	-	1	-	1	-	-	-
axis	-	-	-	-	-	-	-	-	-
atlas	-	-	-	-	-	-	-	-	-
cervical vertebrae	-	1	-	-	1	-	-	-	-
torathic vertebrae	-	5	-	-	2	-	-	-	-
lumbar vertebrae	-	-	-	-	-	-	-	-	-
ribs	-	6	-	-	3	-	-	-	-
pelves	1	-	-	-	-	1	-	-	-
scapulae	1	-	1	-	-	-	-	-	1
humerus	1	-	1	-	-	1	-	-	1
radius-ulna	1	-	1	1	-	1	-	-	-
carpals	1	-	-	-	-	-	-	-	-
metacarpals	1	-	1	1	-	1	-	-	1
femur	-	-	-	1	-	1	-	-	-
tibia	-	-	-	1	-	-	-	-	-
tarsals	-	-	-	-	-	-	-	-	-
metatarsals	-	-	-	1	-	1	-	-	-

small percentage of bones with well-preserved cortical surfaces were used. Percussion marks and notches were identified in both the giraffid and the bovid elements. Cut marks also appear on bones from both taxa. However, tooth marks were only observed on bovid and suid specimens. The lack of tooth marks on the fresh cortical surfaces of the giraffid specimens, together with the fact that most long limb elements were represented by at least one epiphysis, suggest that no significant carnivore activity took place on the bones belonging to the giraffid.

Further support can be obtained from the study of the axial elements. Most of them are preserved fairly complete (vertebrae) or with most of their sections complete (ribs), fragmented only by diagenesis. The scapulae and pelves are also preserved complete and untouched by carnivores. The latter appeared broken only by diagenesis. The absence of the remaining anatomical elements is therefore very likely due to hominid behavior. Either they were not transported to the site, or if they were, they may be in the unexcavated part of the site. Both humeri show hominid-made marks (cut marks and percussion marks) indicative of defleshing and demarrowing. Defleshing is also observed in two rib specimens bearing cut marks.

The bovid elements show a higher number of percussion marks and notches distributed across all limb bones. It is noted that percussion and cut marks in both taxa are documented in the southwestern end of the largest cluster, which is the least affected by carbonate. Specimens with percussion marks seem to be distributed around a large lava boulder (Figure 6.17b). Scattered fragments of shafts are also documented within the cluster near the epiphyseal fragments of the same elements supporting the spatial integrity of the site.

The presence of tooth marks on several axial elements with good cortical preservation and on the epiphyseal sections of some long limb bones can account for the bias in axial element preservation. The appearance of tooth marks on bovid specimens but not on giraffid ones is surprising. This might be suggestive of both taxa having been transported (or obtained) and deposited in different moments, and excludes carnivores as agents of transportation of the giraffid carcass. No marks have been documented on the suid specimens, with the exception of one proximal humerus. The absence of visible tooth marks is very likely due to the strong carbonate that affected many bone surfaces in the assemblage. Carnivore modification of bovid bones may have affected the original distribution of elements, even if a close association between epiphyses and shaft specimens is still observable. The spatial association between epiphyses and shafts is even closer in the giraffid limb elements (Figure 6.7a). This could be seen as the effect of minor (or nonexistent) post-depositional disturbance by carnivores. In the case of the giraffid (and even in the suid specimens), semi-articulation of elements is observed. As noted above, most elements belonging to the right and left front limbs (scapula, humerus, radius and metacarpal) are found within a two-meter radius. The only two suid specimens (humerus and scapula) are also found spatially associated.

The accumulation of the ST4 bone assemblage

As was mentioned in the beginning of this chapter, the lower part of the deposit at the base of the channel in ST4 showed a fairly intact assemblage in which post-depositional modification was not strong. Remains from several other carcasses appear on the remainder of the overlying deposit. It has been inferred that most of the faunal remains have been accumulated at the site by hominids for the following reasons:

1) The paleosurface that contains the ST Site Complex is horizontally exposed over an area totalling 0.6 km^2. Most of the exposed paleosurface is devoid of fossils, with archaeological sites concentrated on only 5% of this surface. No particular taphonomic bias has been identified which would have caused such a patchy distribution. The ST4 faunal assemblage contains a minimum of 19 individuals (MNI), all of which were recovered from less than 100 m^2 (Domínguez-Rodrigo et al. 2002). This high density in such a reduced area is better explained by transport processes, rather than by he natural deposition of bones on the spot (Potts 1988), especially given the lack of evidence for carnivore involvement with carcass defleshing at the site.

2) Using the same standard taphonomic procedures as were applied to Olduvai[2] (Potts 1988), the high numbers of identifiable specimens (NISP), elements (MNE) and individuals (MNI), together with the limb-dominated skeletal part profile and the taxonomic diversity present in the ST4 assemblage support the hypothesis of transport of faunal remains to the site (Domínguez-Rodrigo et al. 2002). Furthermore, the high MNE:MNI ratio of 7.3 elements (mostly limbs) per individual is also suggestive of carcass transport to the site.

3) All fragment sizes are represented at the site, with an abundance of the smaller specimens, although the smallest ones are underrepresented since they were transported offsite by a moderate water current. As mentioned above, there is a systematic lack of polishing or abrasion on these specimens. The overall even presence of elements belonging to all the anatomical sections of individuals, across all carcass sizes, rules out hydraulic transport as the agent of accumulation (Domínguez-Rodrigo et al. 2002).

4) Carnivores can also be ruled out as the agents of transport because there is a low rate of tooth-marking (<10%) on diaphyseal specimens, which is suggestive of secondary access

by carnivores to bone remains (Blumenschine 1988; Domínguez-Rodrigo 2002). Furthermore, most of the NISP and estimated MNE bear no tooth marks. The low epiphyseal fragment:shaft fragment ratio and the high percentage of change in the expected versus actual ratio of epiphyses according to MNE (Domínguez-Rodrigo et al. 2002) suggests post-depositional ravaging by carnivores of this hominid-made assemblage. The low incidence of carnivore modifications on limb shaft sections and the spatially concentrated nature of the assemblage, together with the presence of articulated and semi-articulated elements, do not support primary transport and processing of remains by carnivores (Marean and Bertino 1994). Most of the carnivore damage is located on the epiphyses and the morphology and size of tooth marks is indicative of hyenid modification (Domínguez-Rodrigo et al. 2002). Since scavenging from hyenids is not supported by any of the criteria discussed in this chapter, the best-supported interpretation is that hominids were primary agents in bone accumulation, with hyenids as secondary bone modifiers.

There is a close spatial association between stone tools and bones at the site, together with a significant number of hominid-modified bone specimens. Although the spatial association of stone tools and bones alone is not sufficient proof of hominid involvement with carcasses, as has been recently argued at Olduvai (Domínguez-Rodrigo et al. 2007), hominid primary access to animal resources at ST4 has been defended here based on the percentage and distribution of cut and percussion marks. The high percentage of cut marks on meat-bearing bones, and more specifically on their diaphyses, excludes the possibility of secondary access by hominids to hyenid-accumulated bone remains. These data exclude the possibilities that the site might have either been a place where animals died from natural causes or a killing arena by carnivores. It could, however, have been a near-kill location or a central-foraging place used by hominids.

Discussion

Manipulation of fairly complete large carcasses by Plio-Pleistocene hominids has only been marginally inferred at some Koobi Fora sites, where bones belonging to the same large-sized individuals (hippopotamus) showing hominid-imparted marks have been found (Isaac 1978; Bunn 1982, 1983). However, large-sized carcasses are underrepresented at early archaeological sites, being a minor (or very often non-existent) component of the faunal assemblages. The sporadic presence of large-sized carcasses in some sites, with elements representing most of the skeleton, suggests a lack of transport and either occasional scavenging by hominids, or the natural deposition of carcasses with which hominids were not involved. The bulk of the fauna discovered in the East African Plio-Pleistocene sites, however, is composed of small and medium-sized carcasses (Bunn 1982). All the analytical data and the experimental research conducted to understand how fauna were accumulated and modified at sites have been based on these carcass sizes (Bunn 1982; Domínguez-Rodrigo 2002). Taphonomic and zooarchaeological research over the past two decades has failed to produce unambiguous and direct evidence of transport and processing of small- and medium-sized carcasses by hominids. Bone transport, however, seems to be supported by taphonomic analyses in few sites.

Hominid involvement in carcass manipulation has been reported for several early sites based on the presence of hominid-modified bones (Domínguez-Rodrigo 2002). The type of strategy used by hominids to obtain and exploit carcasses has been widely debated, and it is at the heart of the hunting/scavenging debate. Primary access to carcasses by hominids would prompt carcass transport, whereas secondary access would more likely mean access only to scattered bone elements. No systematic site excavation has so far provided researchers with direct evidence that hominids were transporting complete or near-complete carcasses rather than a limited number of dismembered bones that could be found at carnivores' kills. It has been assumed that a large number of elements (MNE) per each individual represented (MNI) would very likely reflect carcass transport, although this assumption disregards the possibility that the elements considered may have been obtained from individuals that are not clearly identifiable. In summary, transport and manipulation of small- and medium-sized carcasses has been inferred but not conclusively documented in archaeological contexts.

The area exposed through excavation at the ST4 site (Figure 6.18) shows, together with FLK Zinj at Olduvai (see discussion in Domínguez-Rodrigo et al. 2007), the earliest likely evidence of carcass transport, complete or partial. Hominid transport and modification of an assemblage of bones probably belonging to three different individuals as accumulated in the lower section of ST4, within a spatially restricted area, is documented. The semi-articulated distribution of some anatomical units, with spatially differentiated distribution of taxa, resembles the distribution of bones in Hadza near-kill sites (O'Connell et al. 1992) and might be indicative of the place acting as a butchering spot (Domínguez-Rodrigo et al. 2002). The spatial association of bones and stone tools would support this, as well as the low density of artifacts. Skeletal elements modern foragers' home bases tend to be more widely dispersed and mixed. Semi-articulation of elements in home bases is also marginal or non-existent.

The skeletal part representation and bone surface modification data from ST4 support the hypothesis that lower Pleistocene hominids were transporting and processing fleshed carcasses, rather than just dismembered bones. This gives further support to the hypothesis of carcass obtainment and processing by hominids prior to carnivore intervention. It also discredits interpretations which assume that hominids had only secondary access to marginal carcass resources. Most importantly, it also shows that some sites may have been carcass butchery, preparation and, probably partial consumption spots.

Endnotes

1 This figure is the result of the analysis of the complete bone collection after finishing the excavation of ST4. During this time, 10 boxes filled with fossils from the ST Site Complex (among them, several from ST4) disappeared from the National Museum in Dar es-Salaam, Tanzania. Some of the fossils that disappeared, which do not appear in the current study, were preliminarily analyzed previously (Domínguez-Rodrigo et al. 2002), but given that their study could not be completed, they are not included in the present study.

2 These criteria, however, when recently re-applied to the Olduvai assemblages, have provided new and distinct results suggesting, at most Bed I sites, a marginal role for hominids in site formation (see Domínguez-Rodrigo et al. 2007).

Figure 6.18 ST4 after the completion of excavations.

7

THE TECHNOLOGY OF THE ST SITE COMPLEX

Ignacio de la Torre and Rafael Mora

Methodology

Here we define the techno-typological classification model used in our analysis of the Peninj stone tools; this is important since our technological interpretation and comparisons with other collections depend on the model used. Although in Eastern Africa virtually every author proposes his or her own typology, the best known are those of Leakey (1971:4–8) based on materials from Olduvai and of Isaac et al. (1997:264–268) based on materials from Koobi Fora. In practice, these are very similar. Both are useful, since they make it possible to include any collection in a comprehensible list of categories that can easily be compared, but they do not consider the technical processes involved in any great depth. Here we use a simplified version of these typological lists, following the practice adopted in previous works (de la Torre et al. 2003, 2004). This synthesis and simplification is appropriate to the nature of the ST Site Complex, since our collection does not contain the variety of forms and types identified at Olduvai (Table 7.1). We trust that using a list of categories similar to existing ones will facilitate comparison with the archaeological record of other regions. However, in this work, an evaluation of these categories will be carried out from a technological perspective, in an attempt to recognize the knapping strategies involved in creating each type of object.

The technological information is obtained mainly by analyzing cores and flakes. With regard to the knapping products, the cortical index of the flakes has been calculated by combining the presence of cortex on the striking platform and on the dorsal face of the products (Table 7.2), although in this study this index will only be applied to complete flakes. Combining the cortical characters of the striking platform and the dorsal face is very useful, since it enables combined inferences to be made regarding the exploitation phase of the core and its processes of rotation. Toth's types (1982) are redundant with regard to the characteristics shown in Table 7.2, since the latter include all the possibilities contemplated by Toth. However, since Toth's method of classifying the cortex is so widespread, this attribute will also be included in the analysis of flakes.

Calculating the number of dorsal scars of flakes is a common practice in lithic technology. This variable provides information on the recurrence of knapping on the actual débitage surfaces, and it is, together with the direction of the preliminary detachments, a basic attribute for inferring the methods of exploitation by which flakes are obtained. These diacritical structures have been very commonly used in European archaeology since their definition (Dauvois 1976), but have received little attention in Africa, probably due to the poor quality of the raw materials available, since it is well-known that it is difficult to identify the direction of preliminary detachments in materials other than chert. Therefore, in some cases, it has been impossible to determine the direction of previous scars. In fact, the diacritical schemes described in this chapter (see Figure 7.44) should be regarded as a minimal inventory which only reflects a part of the real variability in flake production.

Table 7.1 Different classifications of the lithic collections in the East African Lower Pleistocene

Leakey (1971)	Isaac et al. (1997)	Present work
Tools	Flaked pieces	Flaked pieces
Choppers	Choppers	Cores
Polyhedrons	Polyhedrons	
Discoids	Discoids, regular	
	Discoids partial	
Proto-bifaces	Discoids, elongate	
Heavy Duty scrapers	Scrapers, core	
Light Duty Scrapers	Scrapers, flake	Small retouched pieces
Burins	Other and misc.	
Awls		
Outils écaillés		
Laterally trimmed flakes		
Sundry tools		
Bifaces	Acheulean forms	
Spheroids/ subspheroids	Pounded pieces	Pounded pieces
Modified battered		
Utilized materials		
Hammerstones	Hammerstones	Hammerstones
Utilized cobbles	Battered cobbles	
Utilized flakes		
Anvils	Anvils	
Debitage	Detached pieces	Detached pieces
Flakes	Whole flakes	Whole flakes
Others	Broken flake	Flake fragments
	Angular fragments	Angular fragments
		Chips / debris
	Core fragments	
Manuports	Unmodified	Unmodified material

Table 7.2 Cortex in the whole flakes

Dorsal face	Striking platform	
	Cortical	Non-cortical
Cortical		
Cortex > 50%		
Cortex < 50%		
Non-cortical		

The detailed study of cores is critical to understand the various methods of reduction. In previous studies (de la Torre et al. 2003, 2004; de la Torre and Mora 2004), we proposed a core classification based on considering cores as geometric volumes in which at least six schematic surfaces have been differentiated. Flaking on these surfaces and the resulting interaction between them result in unifacial, bifacial, trifacial and multifacial systems. The direction of flaking makes it possible to distinguish unidirectional, bi-directional - not bipolar, as we mistakenly referred to it in a previous work (de la Torre et al. 2003) - and centripetal strategies. The angle formed by the intersection of the different surfaces exploited can be described as simple or abrupt. Taking all these attributes into account, in previous studies the existence of seven different methods was proposed (de la Torre and Mora 2004; de la Torre et al. 2003, 2004). We introduce a small modification here, given that we no longer consider the polyhedral system as a structured method (*sensu* Texier and Roche 1995) to be present in Peninj, and the examples ascribed to this method are now considered irregular multifacials. In the same way, a new method, the peripheral bifacial system (de la Torre 2005), is included. This was not identified in the previously published materials but has been detected in those obtained from the most recent excavations. Considering all these attributes together, the exploitation strategies followed by hominids and documented in the ST site lithic assemblage are below (Figure 7.1).

Type 1. Unifacial simple partial exploitation. It is represented by unifacial choppers. Flaking takes place on a surface generated by the natural or cortical plane. The striking platform and the flaking surface adopt an acute angle; that is, the edge appears only on part of the perimeter of the core.

Type 2. Unifacial centripetal/peripheral exploitation. This consists of the exploitation of the horizontal plane from both the sagittal and transversal planes. Flaking is carried out from unprepared striking surfaces. It is differentiated from Type 1 in the development of the edge, which now occupies the entire perimeter of the core. In addition, the only exploitation (flaking) surface is generated through radial flaking.

Type 3. Unifacial abrupt exploitation. This can also be defined as the exploitation of the transversal and/or sagittal plane from one or two of the horizontal planes. Thus, from natural or prepared striking platforms, parallel and longitudinal flakes are obtained. The flaking surface forms a straight percussion angle with respect to the striking platform.

Type 4. Bifacial partial exploitation. This is the strategy documented for chopping tools or bifacial choppers (Leakey 1971). The negative scars of flaking on one plane are used as the striking surface to flake the adjacent plane. A configuration edge is obtained this way with a simple angle. The edge occupies only a specific area of the piece, not its entire perimeter.

Type 5. Bifacial hierarchical centripetal exploitation. The geometric volume of these cores is divided into two asymmetrical convex surfaces which share an intersection plane. The surfaces are hierarchical; the subordinate surface acts as a preparation plane to obtain the radial flakes that characterize the main surface. Additionally, the striking surface is oriented with respect to the flaking surface in a way in which the edge created by the intersection of both surfaces is perpendicular to the knapping axis of the centripetal flaking.

Type 6. Bifacial peripheral exploitation. These cores have two asymmetric exploitation surfaces, one

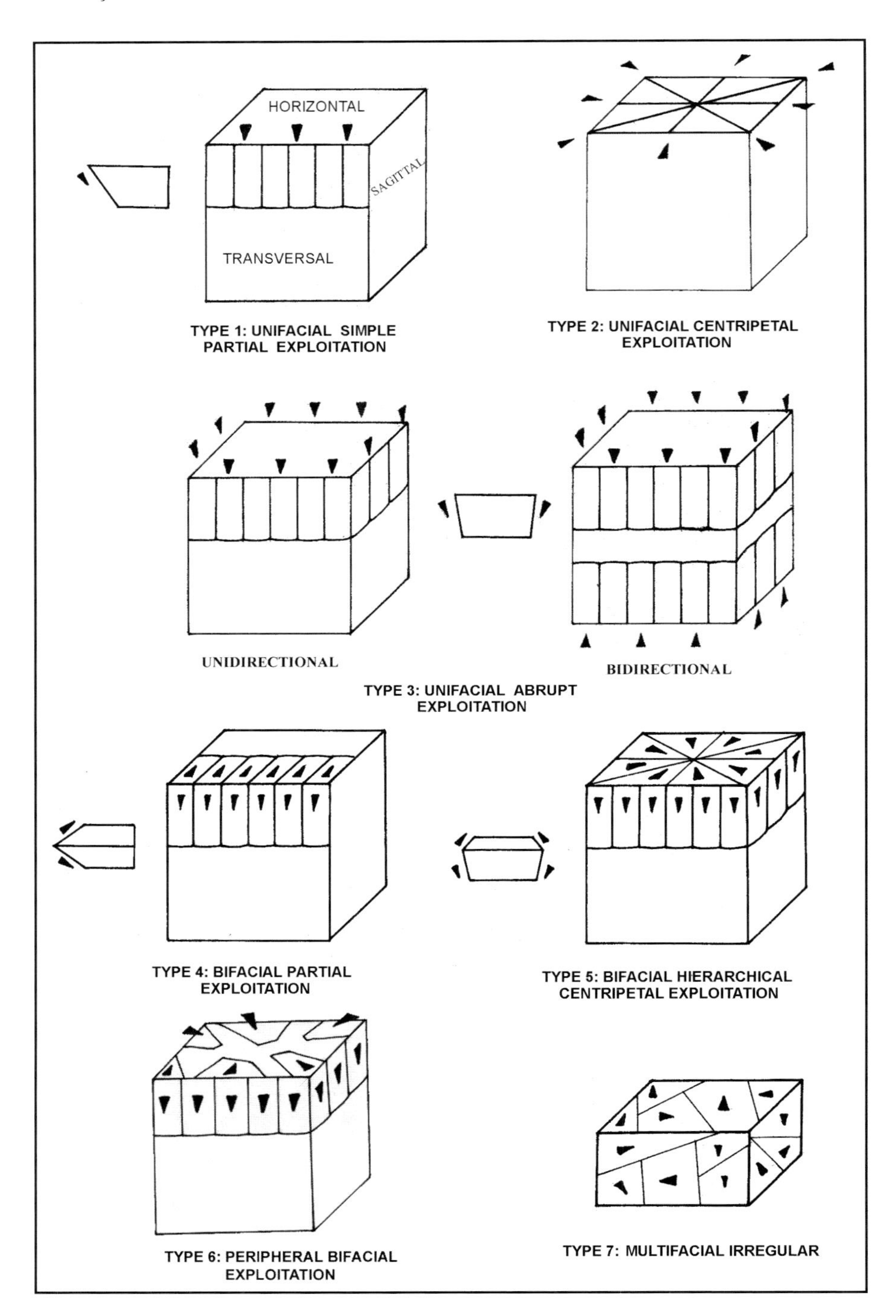

Figure 7.1 Ideal schemes of the exploitation methods documented in the ST Site Complex.

of which acts as preparation plane for the extractions from the main surface. In principle this method is similar to the bifacial hierarchical centripetal system defined at Peninj (de la Torre et al. 2003). However, although these cores display a system of bifacial and hierarchical exploitation, it is observed that the extractions are not distributed in a radial pattern (as in the case of the centripetal system) but in an anarchical way.

Type 7. Multi-facial irregular exploitation. This group is constituted by the cores that present several exploitation surfaces without a clear organization in the reduction process. In the ST Site Complex, cores of this category are always small-sized and with hardly any cortex. This suggests that they may be overexploited cores which could have been more systematically flaked in a previous stage of the reduction sequence.

This system of classifying cores is based on empirical criteria. That is, it has been developed on the basis of the collection recorded in Peninj. So it should not be considered a method of classification intended for general use. In fact, it is simply an analytical tool for understanding the strategies used for the reduction of cores and, for this purpose, it is a flexible and subjective methodological tool. So it can and should be modified as new finds are made and on the basis of methodological reflections aimed at understanding technological processes. In any case, we think it is a useful analytical tool for describing the record of the ST Site Complex, and it will be employed as such in the description given below.

The industry in the ST Complex sites

ST 30

A total of 86 stone tools have been retrieved from ST30 (Table 7.3). Most of the lithic material is well-preserved, with only 20% of the materials slightly rounded and a few pieces with post-depositional pseudo-retouching (Figure 7.2). Considering the well-preserved edges, and the fact that the few pieces with blunt edges could be explained by diagenetic processes affecting the lavas, it would seem reasonable to argue for the *in situ* position of the lithic collection. Nevertheless, it is evident that the industry has not preserved its original contextual integrity, as the scarcity of knapping debris indicates.

Table 7.3 Technological categories in ST30

	Basalt	Quartz	Nephelinite	Total	
	N	N	N	N	%
Cores	3	0	1	4	4.7
Retouched pieces	3	0	0	3	3.5
Flakes	17	0	2	19	22.1
Flake fragments	16	2	6	24	27.9
Chips	2	1	2	5	5.8
Angular fragments	10	3	2	15	17.4
Hammerstones	2	0	0	2	2.3
Anvils/ hammers.	1	2	0	3	3.5
Battered fragments	0	1	0	1	1.2
Unmodified material	9	1	0	10	11.6
Total	63	10	13	86	100

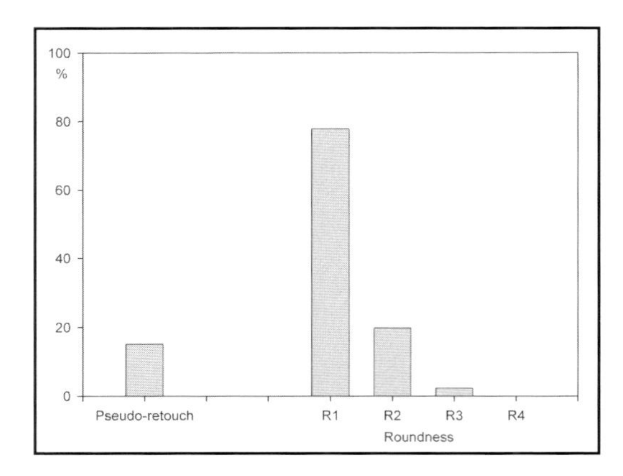

Figure 7.2 Roundness and pseudo-retouch in the lithics from ST30.

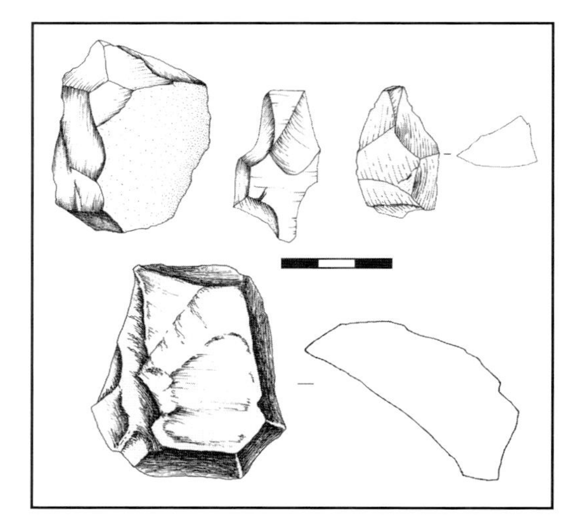

Figure 7.3 Edge-core flakes from ST30.

De la Torre et al. (2004) suggested that the unmodified lithic material could be the result of natural deposition and was not related to the archaeological collections at Peninj. In ST30, one of the assemblages of the Type Section with most unmodified pieces, the poor quality and irregularity of these objects is obvious, and they cannot be considered reserves of raw material for knapping. Thus these pieces are assumed to be natural and not anthropogenic.

Three pieces could possibly be classified as anvils, or passive bases for hammering, and one as an active hammerstone with fracture angles, since they display the classic features of objects of this kind, such as hinged scars, central depressions, ridges damaged by hammering, etc.

The knapping systems are not very clearly structured, there being two bifacial cores - one abrupt and the other simple (i.e., a bifacial chopper) - one irregular multifacial core and another partial abrupt unifacial core. The unidirectional and apparently not very well-prepared exploitation of the cores contrasts with the technical information provided by the knapping products, since some edge-core flakes (Figure 7.3) and complete flakes (Figure 7.4) seem to come from well-structured bifacial systems with successive rejuvenations. This, together with two Siret fragments that refit to form a possible handaxe flake (Figure 7.5) and the presence of some retouched pieces (Figure 7.6), suggest that certain lithic elements were brought to or taken from the site, although the collection is too small to justify such an assumption empirically.

ST6

In ST6 only 12 lithic pieces (5 flakes, 6 fragments of flake and a chip) have been recovered, all of lava and collected from the surface. The few complete flakes display the same technical features as the ST4 pieces (Figure 7.7) which, together with their topographical proximity (less than 20 m) and identical stratigraphic position, have led us to speculate that ST6 could be just a northwards dispersion of ST4.

ST15

Only 12 lithic artifacts have been documented. However, at least in ST15 there is a greater variety

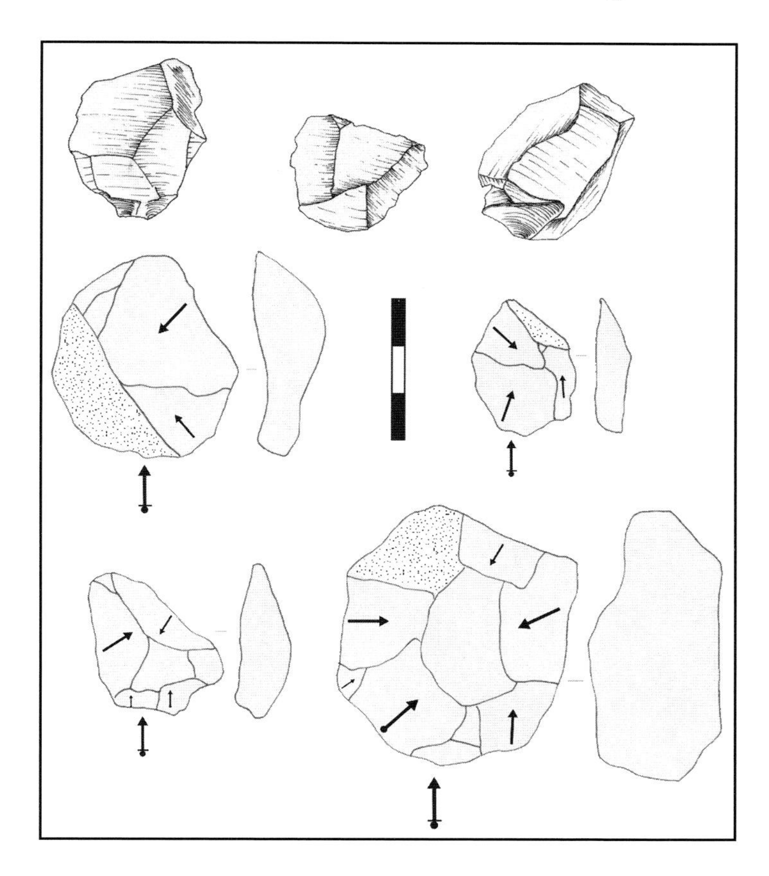

Figure 7.4 Flakes from ST30.

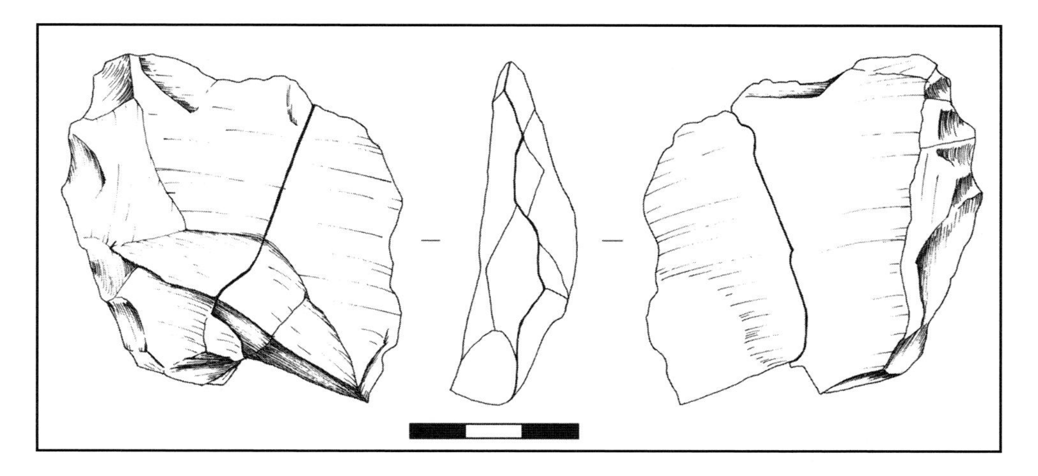

Figure 7.5 Refitting of two Siret fragments which form a possible handaxe flake, as suggested by the bifacial edge located in the section and the ventral face of the piece (drawn by N. Morán).

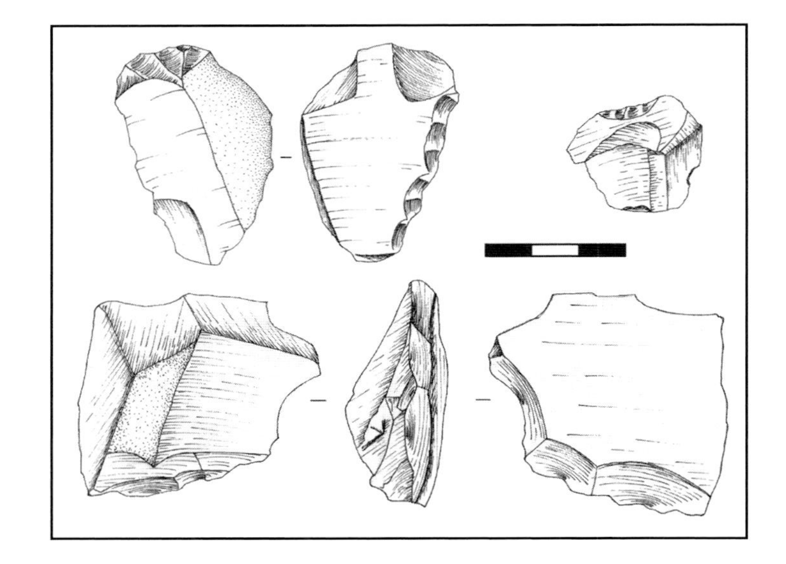

Figure 7.6 Some retouched flakes from ST30 (drawn by N. Morán).

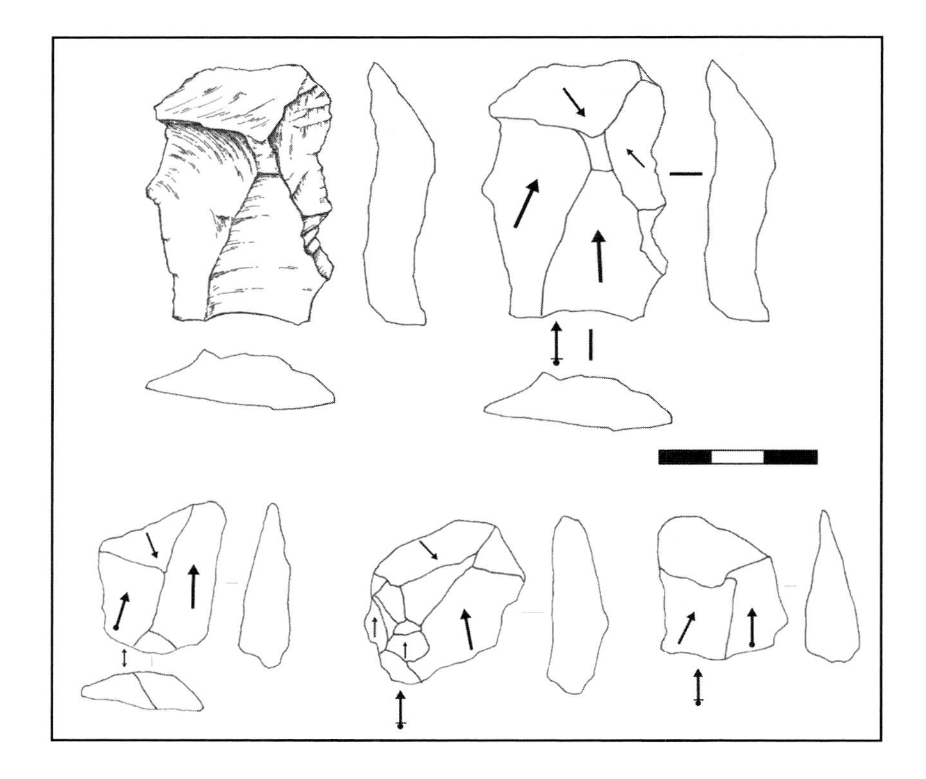

Figure 7.7 Flakes from ST6 (drawn by N. Morán).

Table 7.4 Technological categories in ST6 and ST15

	Basalt N		Quartz N		Nephelinite N		Total N	
Site	ST6	ST15	ST6	ST15	ST6	ST15	ST6	ST15
Cores	-	2	-	-	-	1	-	3
Retouched pieces	-	-	0	-	-	1	-	1
Flakes	5	2	-	-	-	2	5	4
Flake Fragments	6	3	-	-	-	-	6	3
Debris	-	-	-	-	1	-	1	-
Angular fragments	-	1	-	-	-	-	-	-
Hammerstones	-	-	-	-	-	-	-	-
Battered fragments	-	-	-	-	-	-	-	-
Unmodified material	-	-	-	-	-	-	-	-
Total	11	8	0	0	1	4	12	12

of categories than in ST6 (Table 7.4) and, in particular, there are some cores that enable the evaluation of the knapping systems applied by hominids. Together with cores that have not been much exploited (i.e., Figure 7.8), the small size of a core as clearly-structured as the one depicted in Figure 7.9:1 is surprising. It is a piece of lava that, despite its small size, was knapped using the bifacial hierarchical centripetal method. As will be seen in ST4, the craftsmen were applying conceptually rigid knapping schemes to pieces of different size, without being deterred by the technical and manual complications that might arise from using objects of unsuitable shape. Small flakes such as that shown in Figure 7.10:2, in which the whole of the core knapping surface has been detached, indicate that the reduction of small objects was not limited to isolated examples. A similar philosophy has also been adopted in the example shown in Figure 7.9:2. However, in this case the knapping surface has not been well-handled, and the scars with a simple angle rapidly exhausted the horizontal plane, producing a peripheral rather than a truly centripetal result.

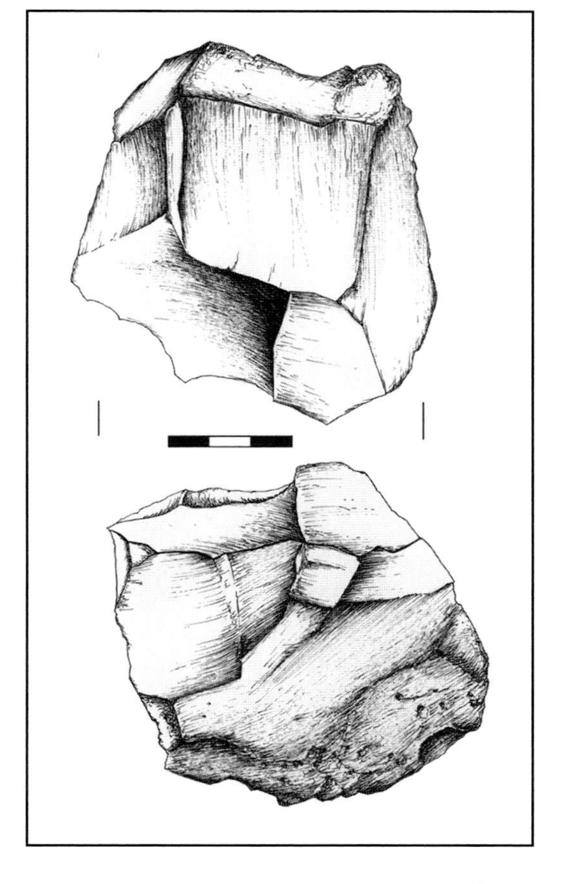

Figure 7.9 Cores from ST 15 (drawn by N. Morán).

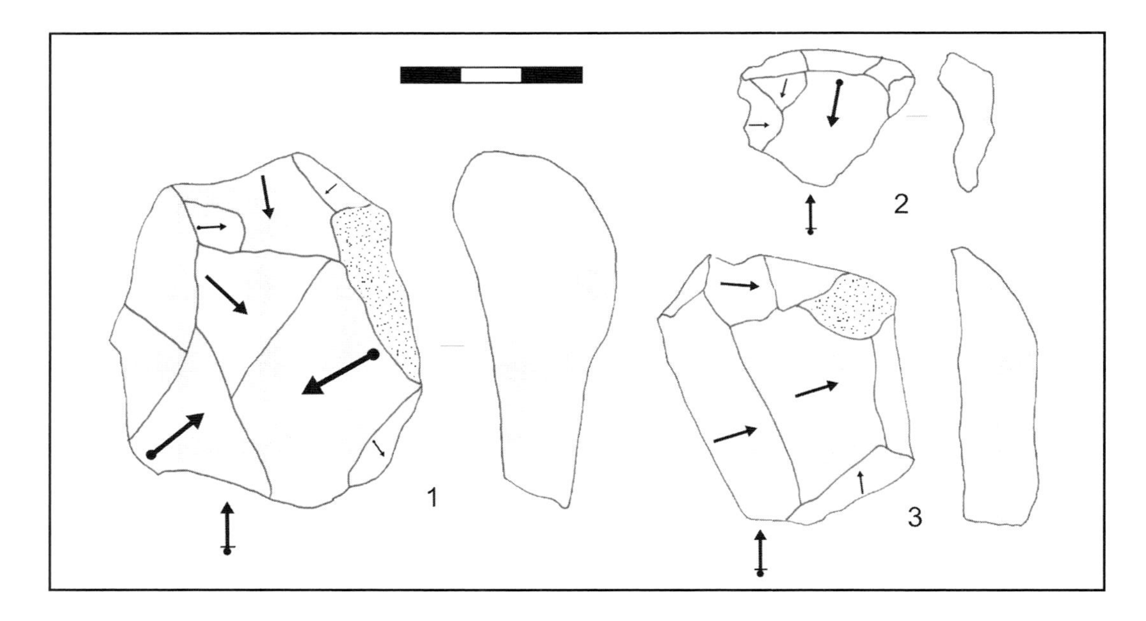

Figure 7.10 Diacritical schemes of flakes from ST15.

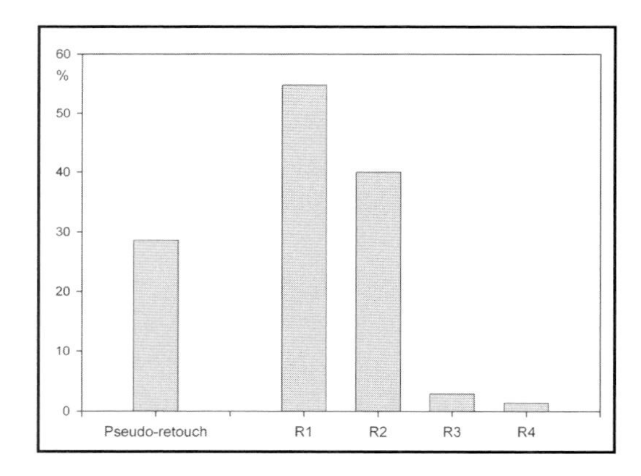

Figure 7.11 Roundness and pseudo-retouch in the lithics from ST31-32.

ST31-32

ST31-32 has 70 lithic items. The proportion of pieces with pseudo-retouch (28.6%) is greater than in other assemblages such as ST4 and ST30, as is the number of pieces that display signs of natural rolling and/or roundness (Figure 7.11). There are very few chips (Table 7.5), which indicate hydrological disturbance. The distribution of technical categories in ST31-32 is similar to that in other ST assemblages, with cores accounting for around 10%, and twice as many flakes and flake fragments together with some hammerstones.

Bifacial hierarchical centripetal cores predominate. In Figures 7.12-7.13 we show two examples that emphasize the importance of structured methods of knapping in the ST Site Complex, and the presence in the same site of cores worked in the same way but at different phases of reduction. As in ST4 and ST15, the bifacial hierarchical centripetal method is documented in different stages of exhaustion. This, together with the flakes shown in Figure 7.14, indicates that the raw material was worked in exactly the same way as that observed in other assemblages from the ST Site Complex.

Table 7.5 Technological categories in ST31-32

	Basalt	Quartz	Nephelinite	Total	
	N	N	N	N	%
Cores	6	-	-	6	8.6
Retouched pieces	-	-	1	1	1.4
Flakes	12	-	2	14	20
Flake Fragments	18	-	11	29	41.4
Debris	2	-	2	4	5.7
Angular fragments	7	-	1	8	11.4
Hammerstones	2	-	1	3	4.3
Battered fragments	-	-	-	-	-
Unmodified material	5	-	-	5	7.1
Total	52	-	18	70	100

ST3

Seventy-six lithic artifacts have been recovered. The material is particularly well-preserved: 5.7% of the pieces with pseudo-retouch, more than 75% of the artifacts with no signs of roundness, and only 12.9% that might be slightly rolled/rounded (Figure 7.15). This coincides with the distribution of the industry by sizes, and this time there is a predominance of smaller pieces, as would be expected in the case of assemblages in primary position. However, even though there is a higher proportion of small chips than in the case of ST4 or ST30 (Table 7.6), they do not reach the normal proportions of an assemblage in primary position. (Schick 1984).

The rest of the categories have a similar distribution to that of other assemblages with material in stratigraphy, such as ST4 or ST30. There are two aspects that stand out in ST3. The first concerns the methods of exploitation. Although some elements such as the edge-core flake in Figure 7.16:1 indicate that the bifacial hierarchical system is present in ST3, of the five cores documented, three are small irregular multifacial pieces and another two are abrupt unifacial. This is perhaps related to the generally small size of the industry, which appears to be the result of intensive exploitation of the raw material. An example is the core in Figure 7.16:2, which in another context would have been classified as a retouched piece, but considering the size of some of the products in ST3 (Figure 7.16:3-5) has been interpreted as a blank for the production of flakes.

The other idiosyncratic element of ST3 is the high percentage of retouched pieces (9.2%), which exceeds even the cores. Most of them are denticulate side scrapers and notches (Figure 7.17). The predominance of multifacial cores, which has been associated with the exhaustion of pieces previously worked using other knapping methods (de la Torre and Mora 2004; de la Torre et al. 2004), together with the small size of the pieces (Figure 7.16) and the large number of retouches, would appear to indicate that reduction was more intense in ST3.

ST2

No more than 15 artifacts were uncovered in the localities of ST2A, ST2D, and ST2G, which makes locality ST2 C-E the most important locality (n=150 lithic pieces). It is the highest absolute frequency of the whole of the ST Site Complex after ST4, but the dispersion of surface remains over more than 40 m^2 has to be taken into account, so the density cannot be compared to that of materials in stratigraphy such as ST4,

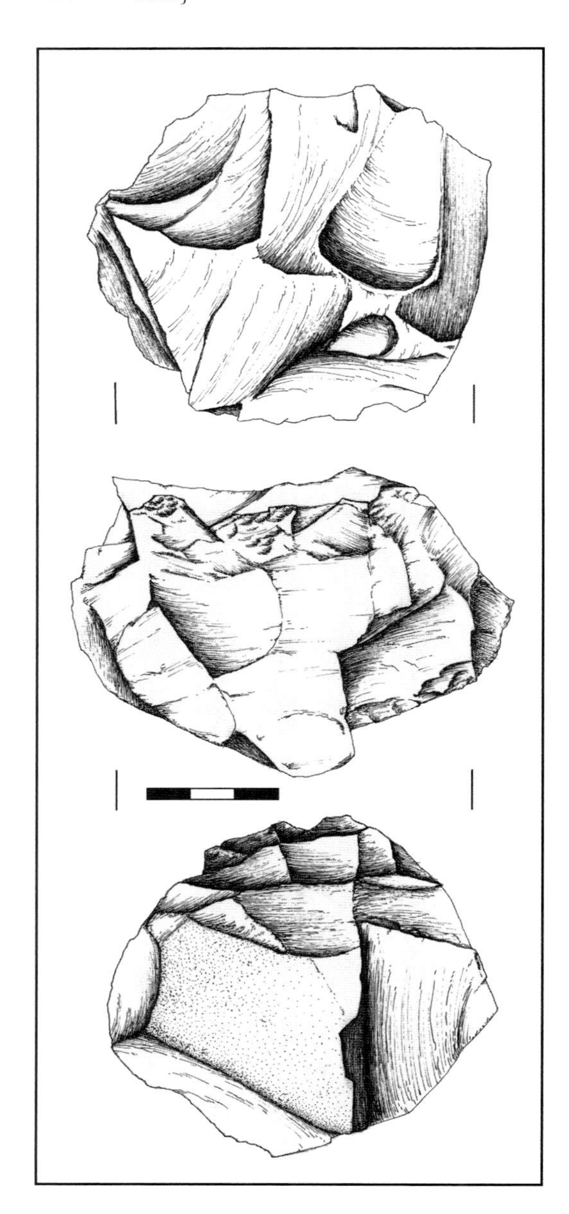

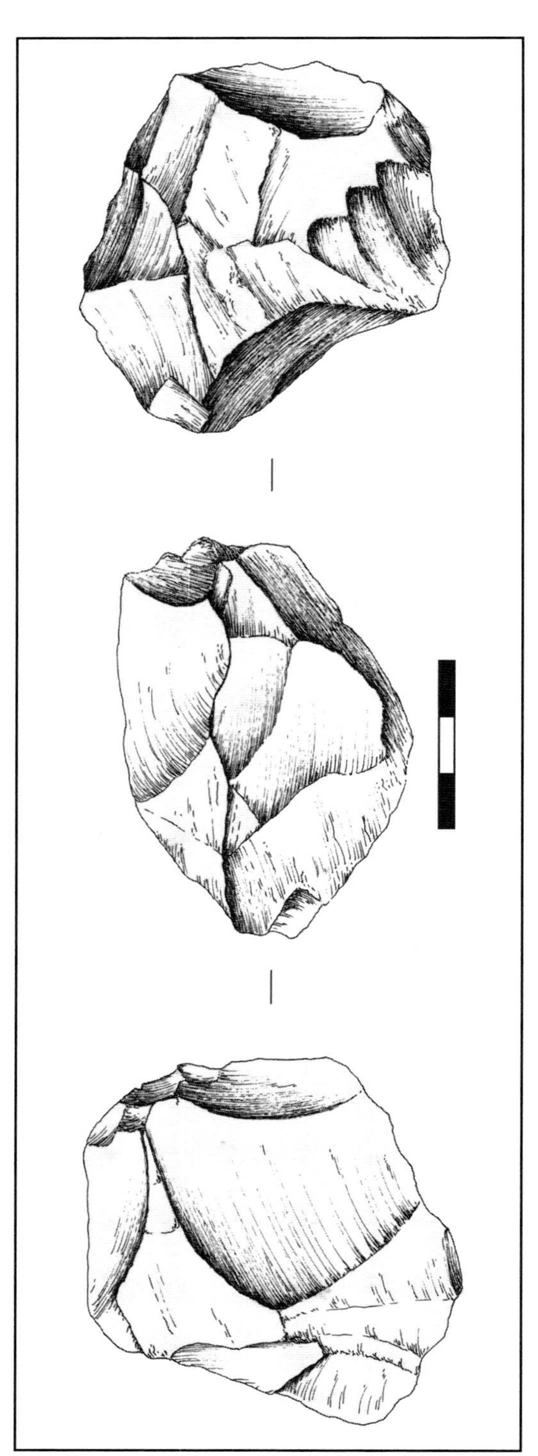

Figure 7.12 (above) Example of the hierarchical bifacial method from ST31-32.

Figure 7.13 (right) Example of the hierarchical bifacial method from ST31-32 in the last stage of exploitation.

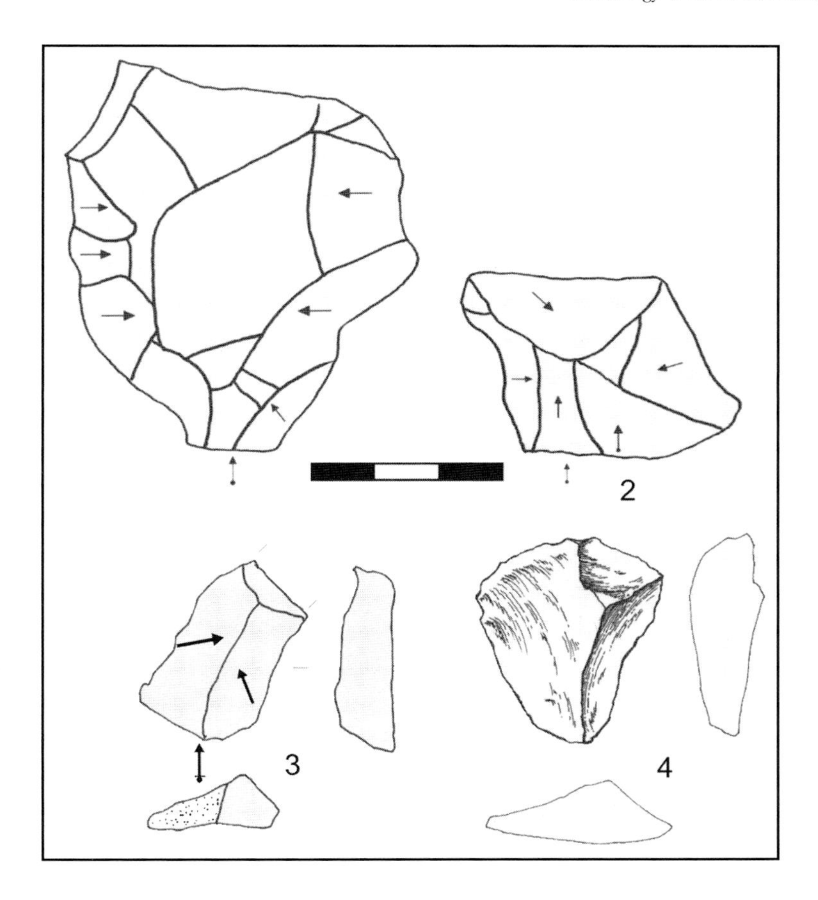

Figure 7.14 Flakes from ST31-32.

ST3 or ST30. The pieces usually have intact edges (69.2%), although they seem to have suffered more alteration than those of other assemblages such as ST3 (Figure 7.18).

There are a large number of complete flakes, which are also usually very well-made, with structured dorsal faces and well-produced sections (Figure 7.19). The proportion of retouched pieces is also relatively high (Table 7.7), practically identical to the 9.2% documented in ST3. As in the latter, in ST2C-E denticulate side scrapers on small flakes or flake fragments are predominant (Figure 7.20).

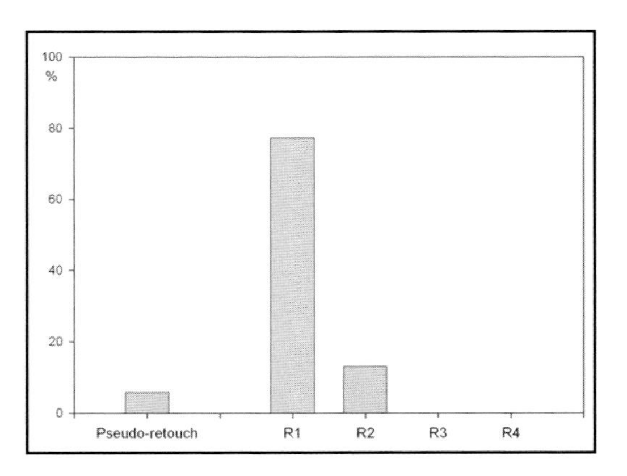

Figure 7.15 Roundness and pseudo-retouch in the lithics from ST3.

Table 7.6 Technological categories in ST3

	Basalt	Quartz	Nephelinite	Total	
	N	N	N	N	%
Cores	2	-	3	5	6.5
Retouched pieces	4	1	2	7	9.2
Flakes	14	1	2	17	22.3
Flake Fragments	16	3	5	24	31.5
Debris	10	0	2	12	15.7
Angular fragments	3	4	-	7	9.2
Hammerstones	-	2	-	2	2.6
Battered fragments	-	1	-	1	1.3
Unmodified material	1	-	-	1	1.3
Total	50	12	14	76	100

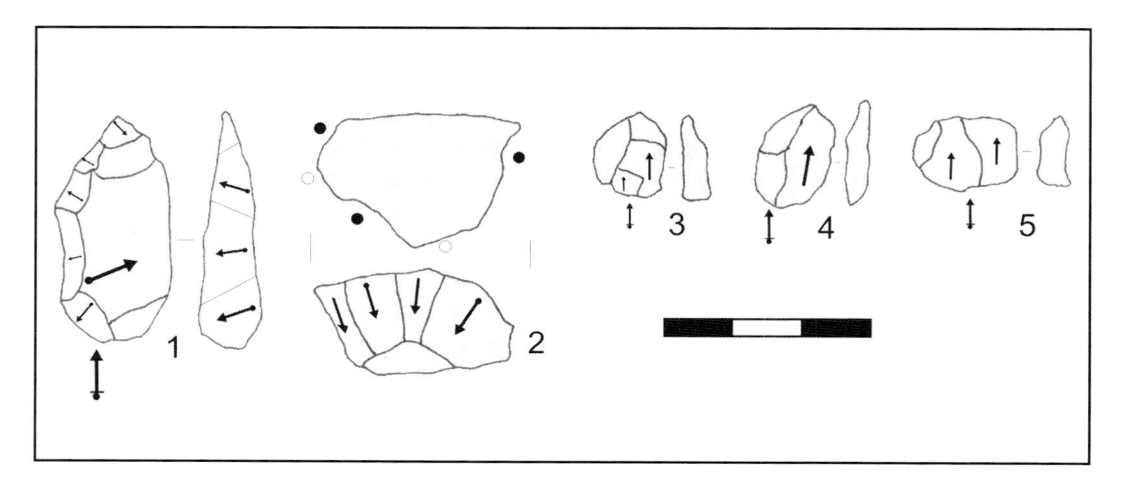

Figure 7.16 Diacritical schemes of lava artifacts from ST3: 1) edge-core flake; 2) abrupt unifacial core; 3-5) small whole flakes.

If the scarcity of small chips can be explained by the fact that the material was collected from the surface, the low percentage of cores (3.3%) is more difficult to interpret. Two cores were worked using multifacial methods, one with the unifacial centripetal system, and two with the bifacial hierarchical centripetal method. The latter are especially interesting: they represent, firstly, clear examples of the preparation of platforms and the hierarchization of surfaces. Moreover, and as occurs in other sites of the ST Site Complex, the cores are in various phases of reduction (the example shown in Figure 7.21 is very exhausted while that in Figure 7.22 seems to be in one of the early stages of exploitation), but the same knapping pattern has been used in both cases.

ST4

A total of 166 lithic pieces have been recovered. The material is generally intact, with only 10% of

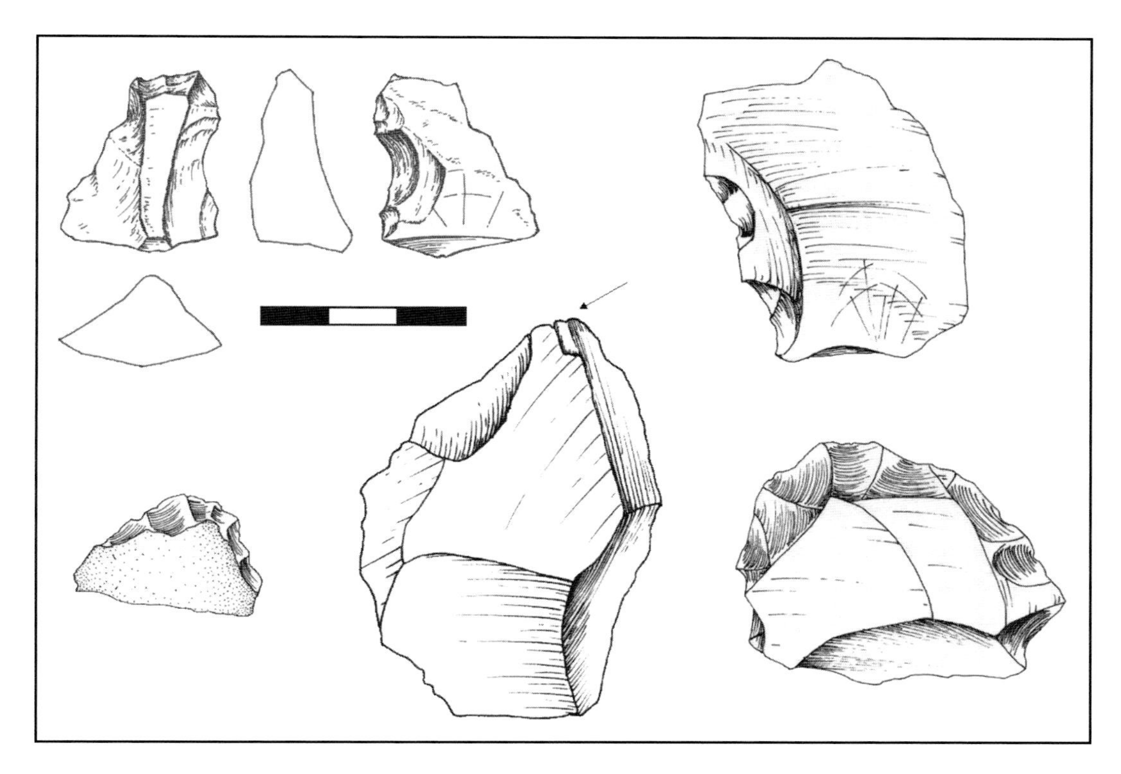

Figure 7.17 Retouched flakes and flake fragments from ST3 (drawn by N. Morán).

the artifacts displaying pseudo-retouch, and more than 78% of pieces show no trace of roundness (Figure 7.23). Despite this a certain hydrologic bias is obvious, since there are very few small chips (8.4%). In ST4, lava predominates, especially the various types of basalt (Table 7.8). There is little quartz and it cannot have been a primary element in the knapping processes.

The most important technological information is provided by cores. Four can be ascribed to the bifacial hierarchical centripetal system, three to the bifacial peripheral method and three to the bifacial abrupt method, two to other bifacial methods, six to the unifacial abrupt method (i.e., Figure 7.24) and one to the simple unifacial method (unifacial chopper). The presence of the

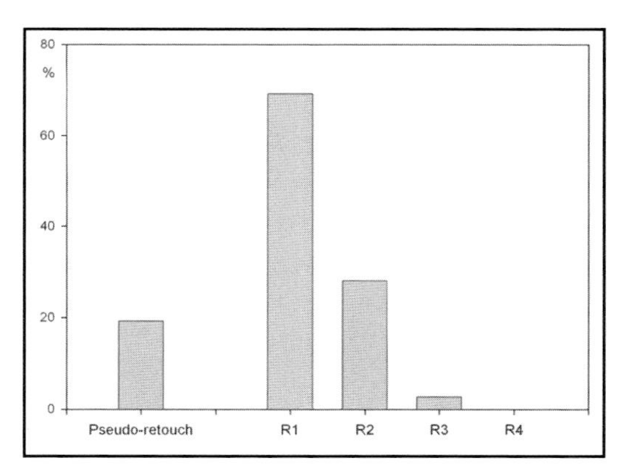

Figure 7.18 Roundness and pseudo-retouch in the lithics from ST2C-E.

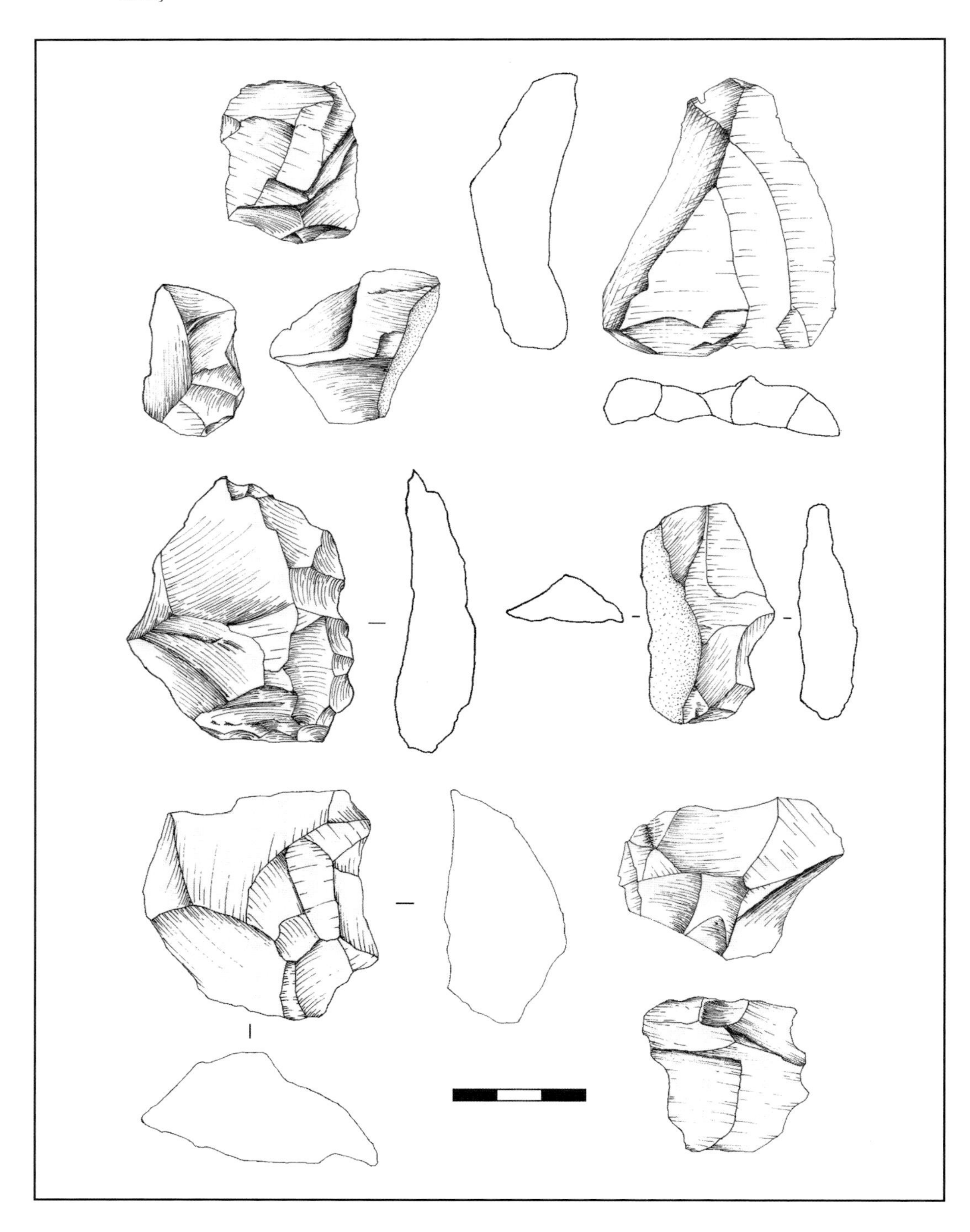

Figure 7.19 Flakes from ST2C-E (drawn by N. Morán).

Table 7.7 Technological categories in ST2C-E

	Basalt	Quartz	Nephelinite	Total	
	N	N	N	N	%
Cores	2	1	2	5	3.3
Retouched pieces	12	-	1	13	8.6
Flakes	36	1	5	43*	28.6
Flake Fragments	38	-	5	43	28.6
Debris	7	-	1	8	5.3
Angular fragments	25	6	7	38	25.3
Hammerstones	-	-	-	-	-
Battered fragments	-	-	-	-	-
Unmodified material	-	-	-	-	-
Total	120	8	21	150*	100

*A bone flake is included

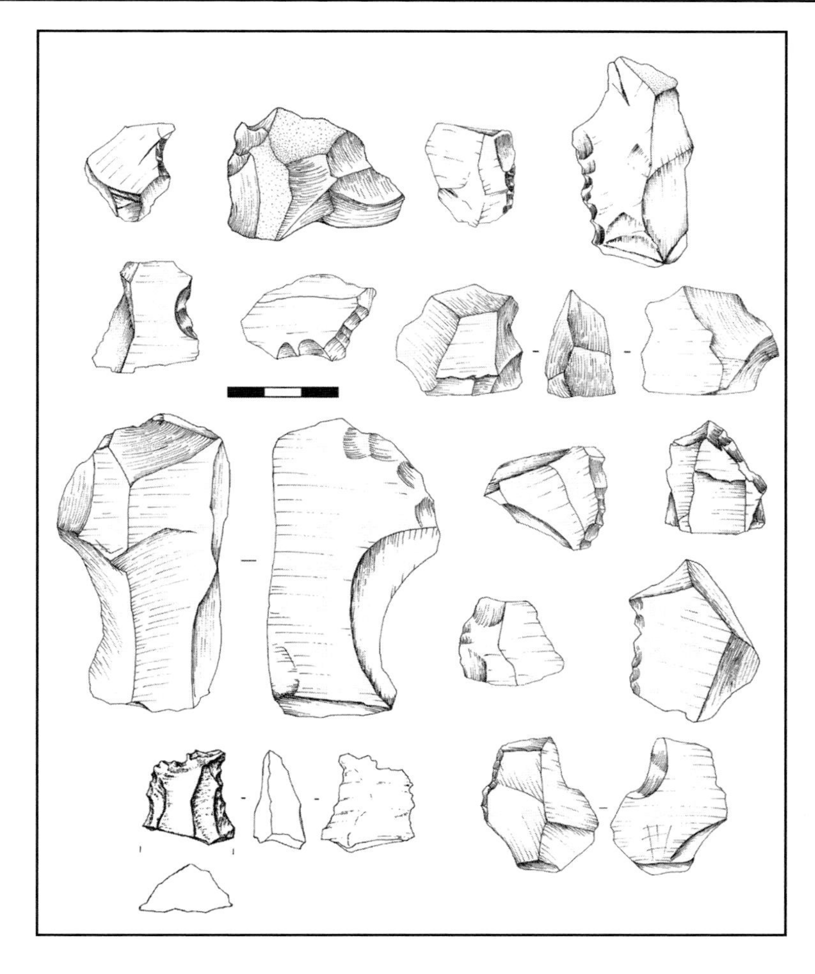

Figure 7.20 Retouched flakes and flake fragments from ST2C-E (drawn by N. Morán).

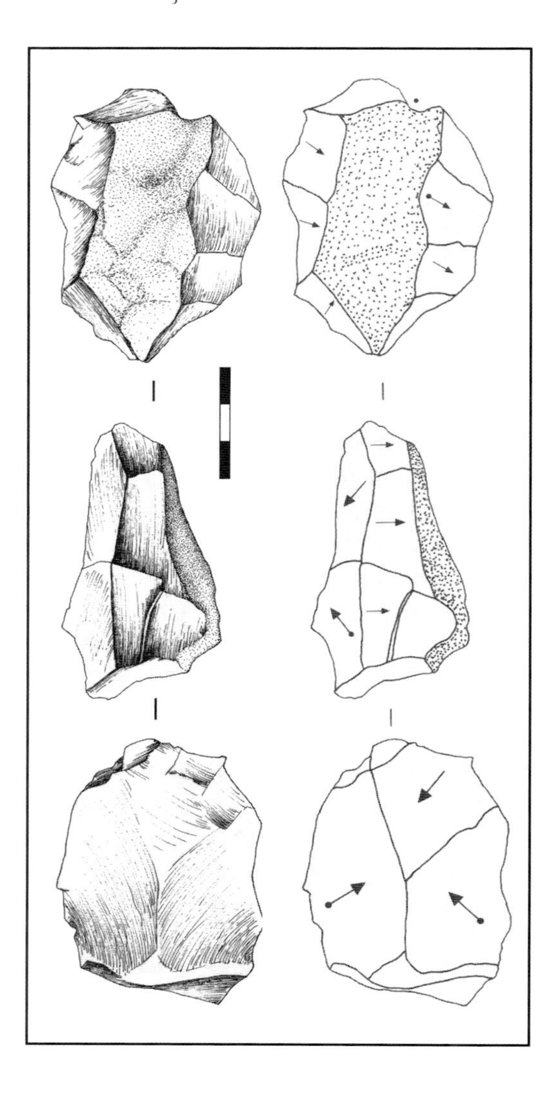

Figure 7.21 Hierarchical bifacial centripetal core from ST2C-E in a late stage of reduction (drawn by N. Morán).

bifacial hierarchical centripetal method in various phases of reduction is notable, from the initial knapping stages (Figure 7.25) and full exploitation (Figure 7.26) to those in which the core is completely exhausted (Figure 7.27). This is relevant when evaluating production strategies, since it indicates that blocks were brought to the site at different stages of reduction, and also suggests that intensive knapping took place in the settlement itself.

In Figure 7.28 we have an example of the abrupt unifacial method. After preparing the horizontal plane with a single detachment, this scar is used as a knapping platform on which to carry out recurrent exploitation of the transversal and sagittal planes using systematic abrupt unidirectional reduction.

In Figure 7.29:1, the scars on the horizontal plane serve as striking platforms on which flakes are extracted in the transversal plane, which leave under-developed and stepped scars, certainly due to the poor quality of the raw material. In Figure 7.29:2 a strategy which tends towards the bifacial hierarchical centripetal method can be observed. However, the exploitation surface in the horizontal plane is not well-worked, and displays scars with simple angles rather than flat angle detachments, ultimately exhausting the volumes and leaving the knapping surface useless. Despite these technical errors, it is important to note the small size of the core, which would have made knapping difficult. Even so, the exact same method of reduction was imposed on this diminutive nephelinite cobble as was used for larger blocks.

In general, the knapping products recovered in ST4 coincide with the characteristics observed in the cores (Figure 7.30), and there are examples obtained from centripetal knapping and others more typical of unidirectional production. They are well-made products, with fine, well-developed sections, although with striking platforms that are normally simple and unprepared.

Both the majority of flakes and cores seem to come from the system of débitage of small products, aimed at obtaining small flakes from more or less structured cores. There are only two examples that are inconsistent with this technological

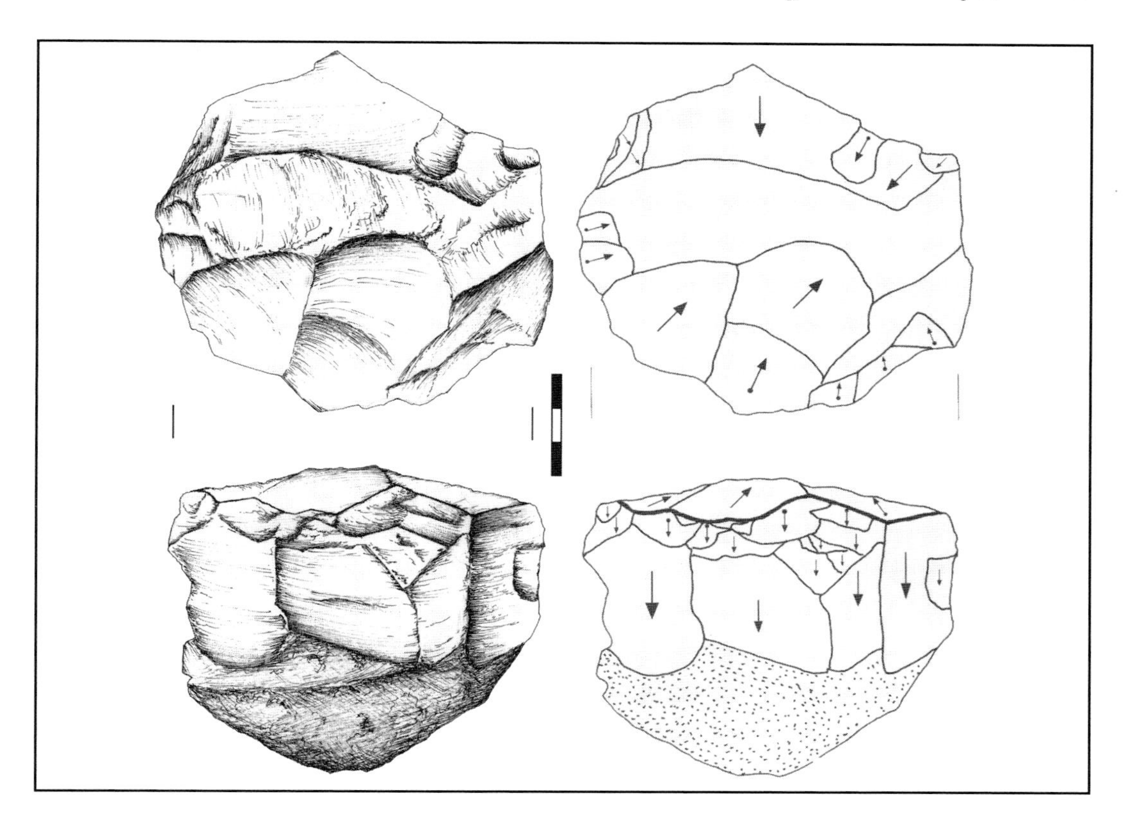

Figure 7.22 Hierarchical bifacial centripetal core from ST2C-E in an early stage of exploitation (drawn by N. Morán).

profile, and they are those shown in Figure 7.31. The first example, due to its large size (maximum length 6 cm), multifaceted striking platform, minimal thickness and the fine delineation of its sections, could lead to this flake being interpreted as a product of handaxe rejuvenation. The second example is difficult to classify as a retouched piece or a core, since the scars are too short to suggest that the aim was to obtain flakes, but at the same time they are rather too abrupt to create a functional edge. Irrespective of the correct classification, the fact is that the blank is a large flake, more typical of Acheulean than Oldowan assemblages.

The industry of the ST Site Complex

General Characteristics

Not all the sites furnish the same volume of information in terms of quantity or quality. With regard to the number of items (Figure 7.32), it can be observed that ST2C-E, ST3, ST4, ST30 and ST31-32 are the only ones in which we have a representative collection sufficient to evaluate the relationships between the categories. The total weight of raw materials brought to each site (Figure 7.33) underscores the importance of ST4. This is due to the abundance but small size of the cores in ST3, and the underrepresentation of cores and hammerstones in ST2C-E. In any

Table 7.8 Technological categories in ST4

	Basalt	Quartz	Nephelinite	Total	
	N	N	N	N	%
Cores	13	0	6	19	11.4
Retouched pieces	4	0	1	5	3
Flakes	31	2	4	37	22.3
Flake Fragments	25	15	12	52	31.3
Debris	6	6	2	14	8.4
Angular fragments	11	6	0	17	10.2
Hammerstones	3	2	0	5	3
Battered fragments	0	2	0	2	1.2
Unmodified material	15	0	0	15	9
Total	108	33	25	166	100

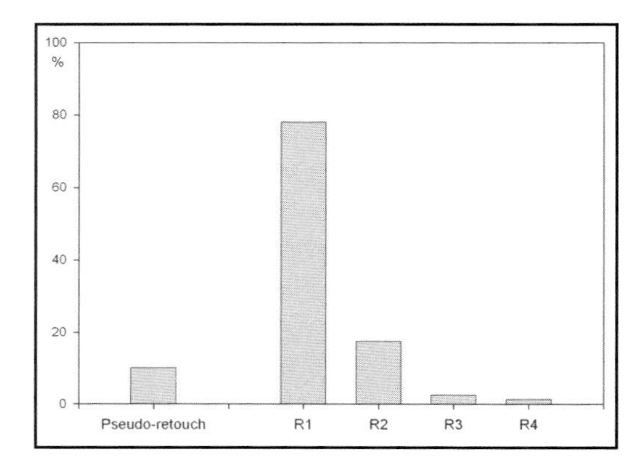

Figure 7.23 Roundness and pseudo-retouch in the lithics from ST4.

case, it is obvious that the 33,274 grams of lithic material worked are concentrated at specific localities such as ST4, ST31 and ST32.

With regard to the alterations caused by rounding, in Figure 7.34 it can be observed that the edges of most of the artifacts have been preserved intact (R1). Although in some of the sites, such as ST2C-E, there does seem to be a higher percentage of slightly rounded pieces (R2; perhaps rounded by diagenesis), signs of medium (R3) or severe rounding (R4) are virtually absent. Thus, the systematic absence of chips in any of the assemblages must be attributed to processes of hydraulic alteration contemporary with the formation of the sites or current processes of erosion (and no doubt both), assuming that the rest of the objects were not moved after their original deposition.

Table 7.9 shows the distribution by technological categories in the ST Site Complex. Figure 7.35, which synthesizes all the information about the Complex, shows the classification of technological groups typical of the débitage of small products. With the exception of small chips, which are underrepresented for taphonomic reasons, the débitage categories (flakes and flake fragments) are the dominant groups. The percentage of cores is more or less consistent with the proportions of flakes, but the retouched pieces appear with rather greater frequency than usual in the collections ascribed to Mode 1.

The raw materials

The X-ray diffraction analyses and the mineralogical analyses have resulted in the identification of several types of basalt (basanites, aphiric basalts, Hawaiian basalts and aphitic basaltic tuffs), and also pyroxenic nephelinites and quartz. However, the internal variability of the rocks cannot be macroscopically distinguished. For this reason,

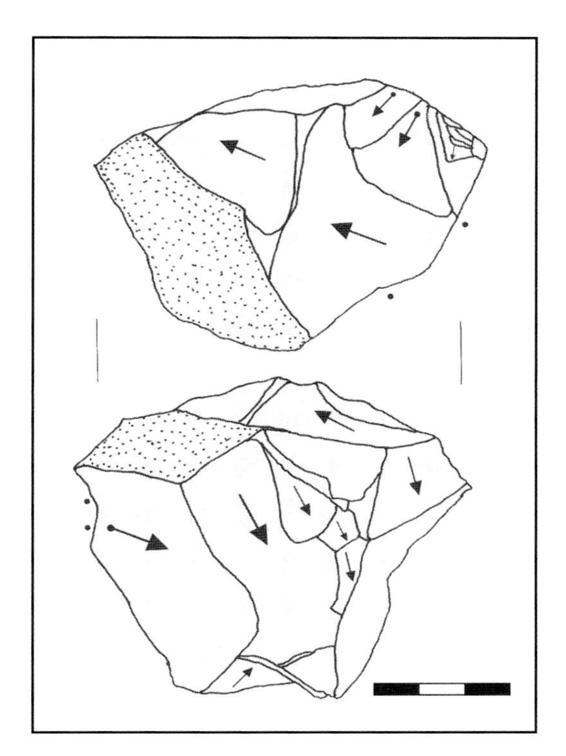

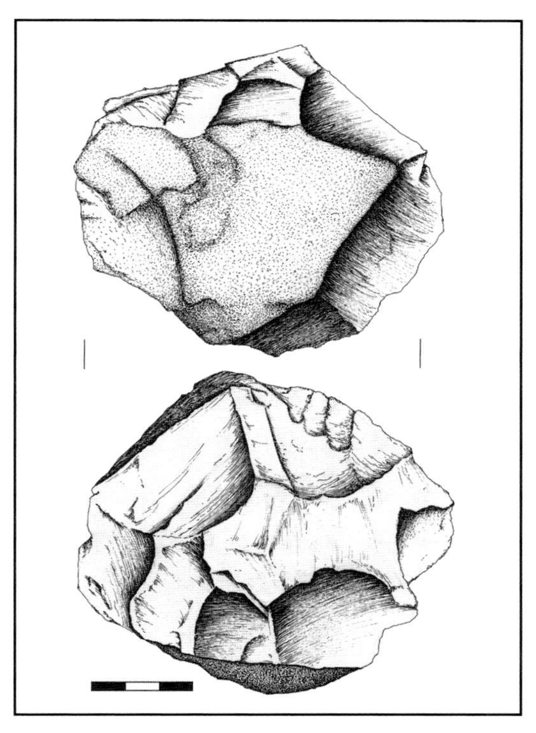

Figure 7.24 (above left) Basalt core from ST4 (drawn by N. Morán). The blank is a fragment of a large nodule that broke and which afterwards continued to be used as a core. It displays detachments on various planes, but independently on each platform. The part of the core that rested on the sediment has a different patina from that of the part that was exposed on the surface, which indicates that the piece took some time to be covered by sediment. Figure 7.25 (above right) Basalt core from ST4 in the first stages of reduction (drawn by N. Morán).

only three types of raw materials are clearly differentiated *de viso*: basalt, quartz and nephelinite. Quartz is the most difficult raw material to work with in an organized knapping process.

Considerable variation in the properties of the basalts that hominids used in the Type Section has been documented. Some artifacts were made from fine-grained basalts, which are very suitable for flaking. Other basalts are very porous, thick-grained and contain internal irregularities. Conchoidal fractures would have been hard to obtain from these types of basalts. Some nephelinites must have been highly prized by hominids, since they are fine-grained and produced sharp-edged tools without internal vesicular irregularities. Thus, nephelinites and some basalts were the best raw materials for making tools.

The source area of the quartzes are the metamorphic hills of Oldoinyo Ogol, to the west of the Peninj Group. The primary area for the basalts was certainly the Sambu volcano and the Hajaro lavas, and the majority of the nephelinites originally came from the Pliocene hills of Shirere and the Mozonik volcano. Despite having identified the source areas of the raw materials, it is difficult to determine the exact points in the Type Section

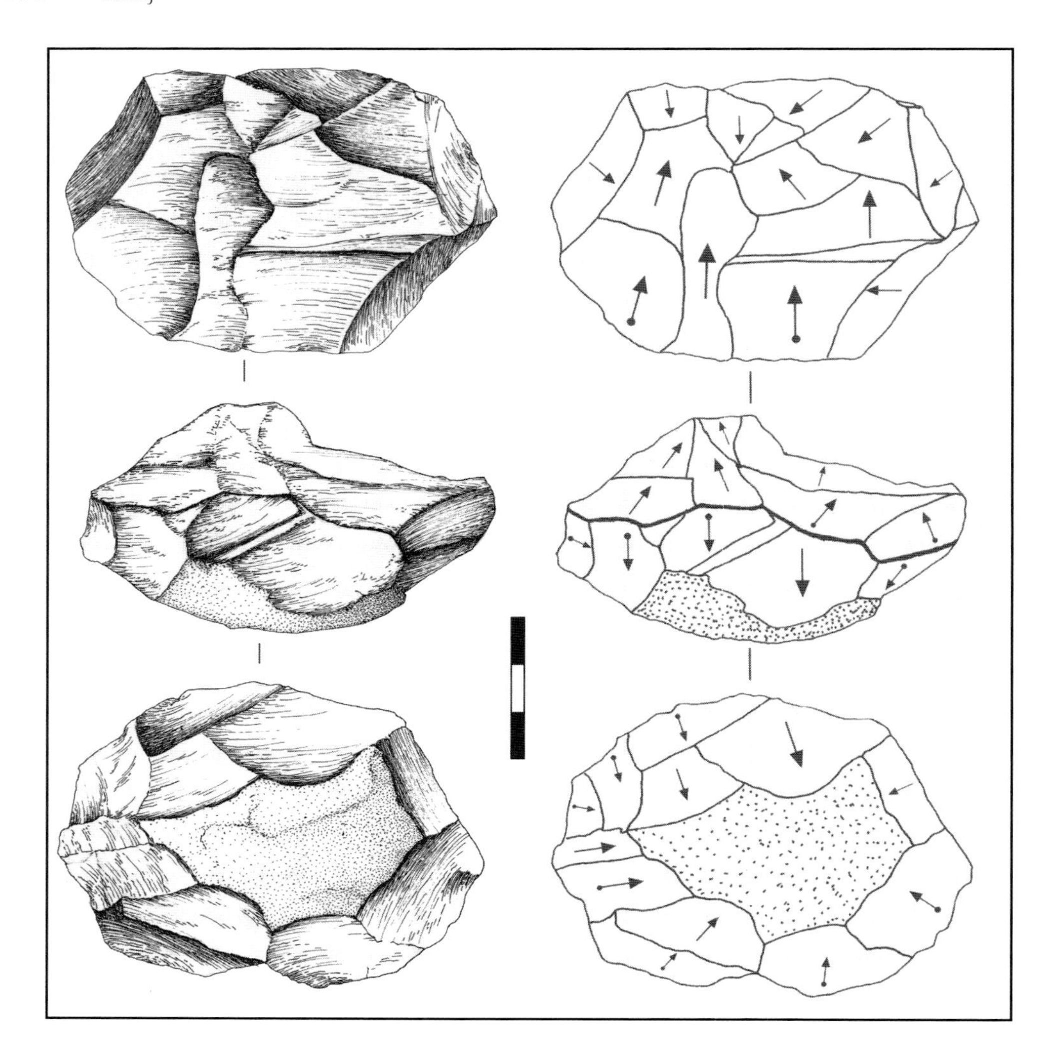

Figure 7.26 Hierarchical bifacial centripetal core from ST4 (drawn by N. Morán).

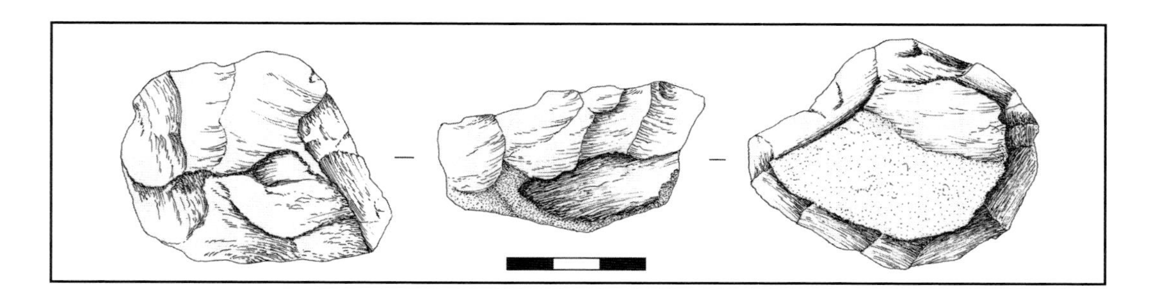

Figure 7.27 Hierarchical bifacial centripetal exhausted core from ST4 (drawn by N. Morán).

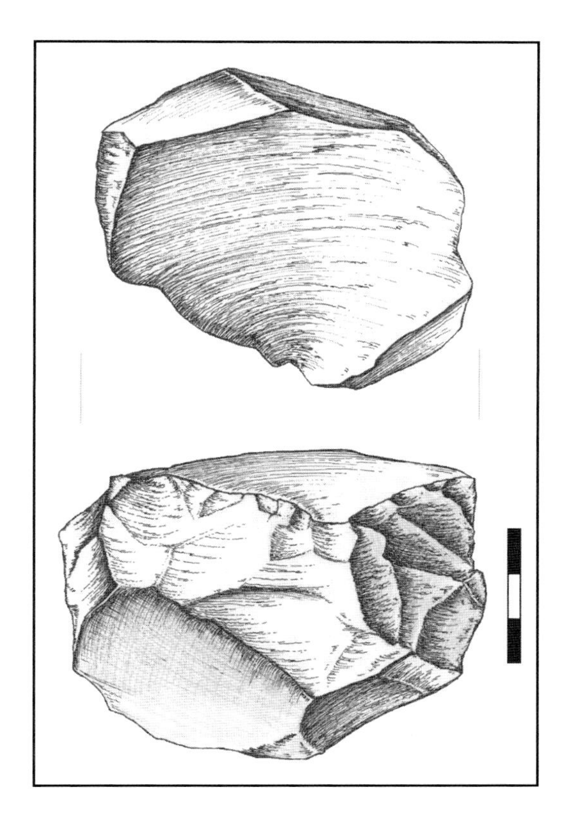

Figure 7.28 Unidirectional nephelinite abrupt unifacial core from ST4 (drawn by N. Morán). The large detachment on the horizontal plane has prepared a striking platform for working the other planes, although only a single part of the perimeter was worked, and not the whole. The craftsman must have realized that there are no good angles for knapping on the opposite side of the exploitation surface, since no attempt to reduce this part of the core can be seen. It can also be observed that there was no interest in rejuvenating the knapping platform once the necessary angles had been exhausted; after a couple of detachment sequences which left stepped scars, instead of rotating the core and rejuvenating the platform, it was simply abandoned.

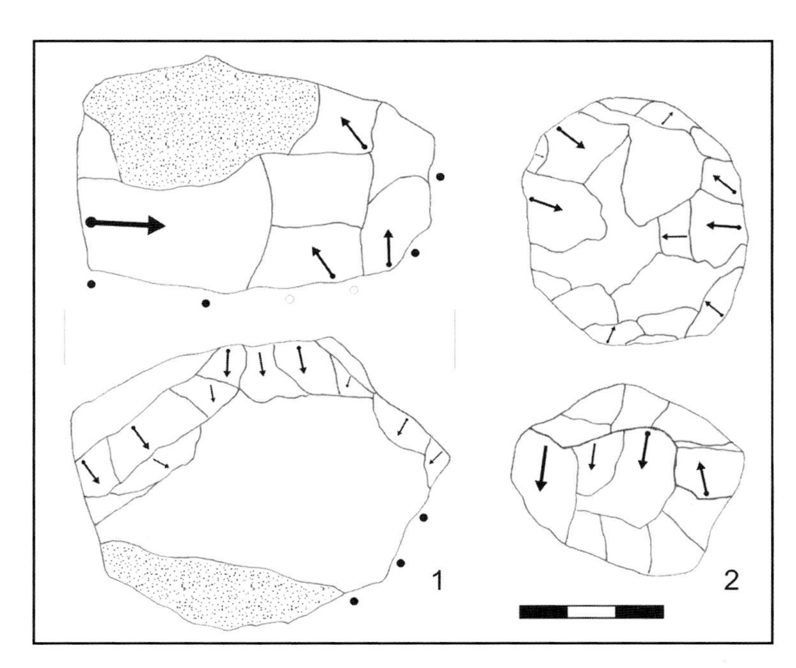

Figure 7.29 Diacritical schemes of cores from ST4: 1) partial abrupt bifacial core of basalt; 2) peripheral bifacial core of nephelinite.

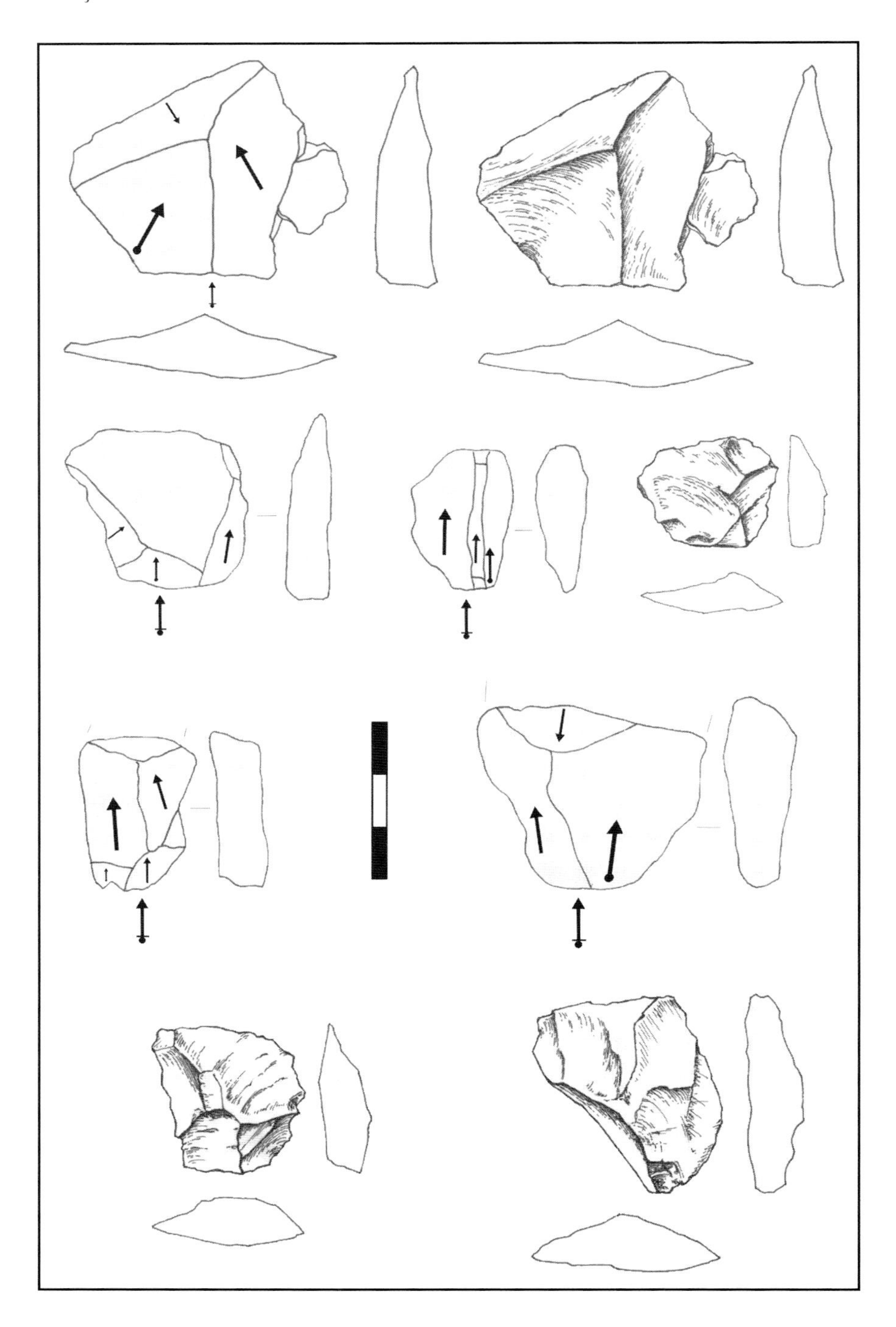

Figure 7.30 Lava flakes from ST4.

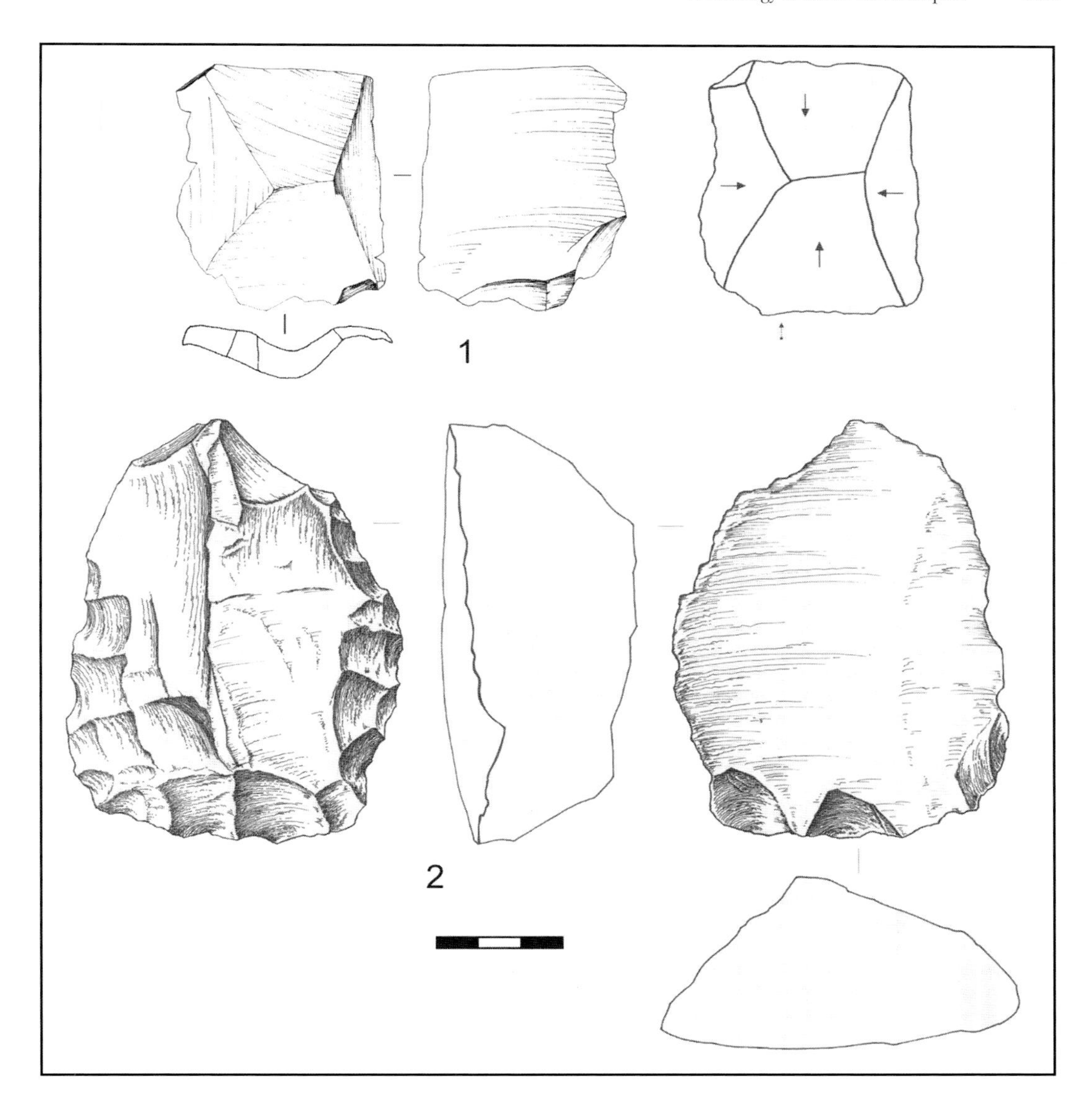

Figure 7.31 1) Possible handaxe basalt flake;
2) retouched piece/core on basalt core. Drawn by N. Morán.

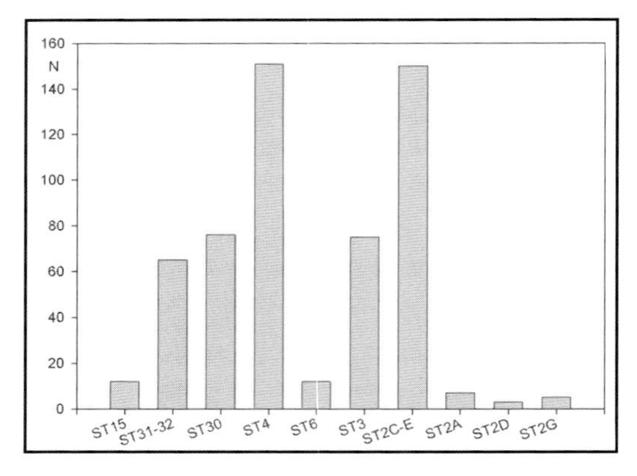

Figure 7.32 Number of lithic pieces in the ST Site Complex, excluding unmodified material.

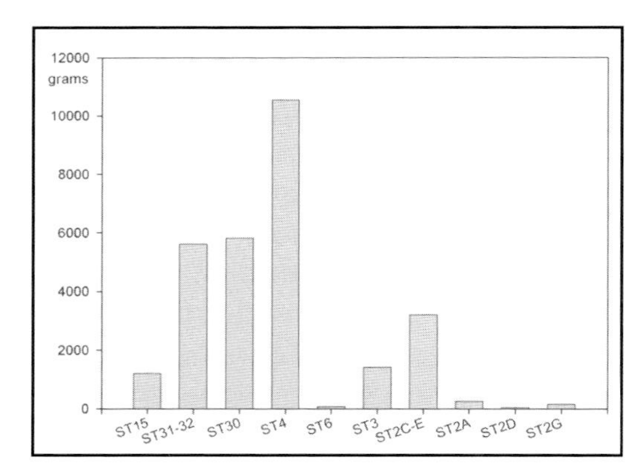

Figure 7.33 Total weight of worked lithic material at the ST Site Complex, excluding unmodified pieces.

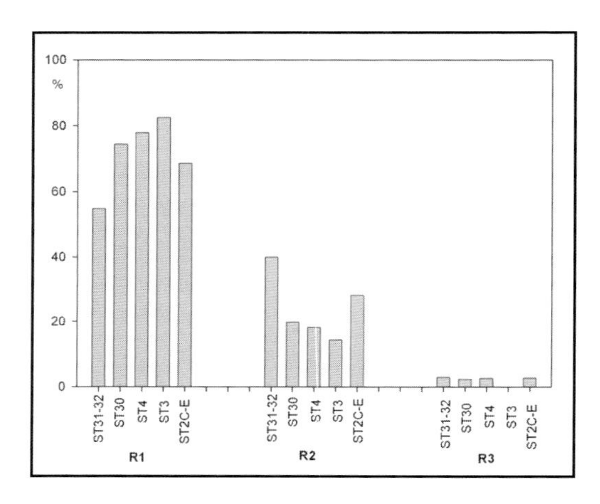

Figure 7.34 Rolling /roundness index in the most relevant sites from the ST Site Complex.

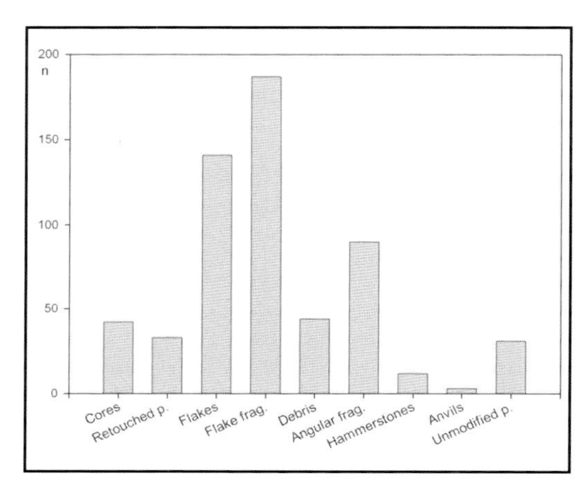

Figure 7.35 Technological categories in the ST Site Complex.

that the hominids used to obtain supplies. In the first place, this is because of the configuration of the landscape in period when the sites were formed: Paleotopographic reconstructions of the Type Section indicate a deltaic environment with little capacity for traction. It is therefore unlikely that there were any stream bars contemporary with the sites from which the hominids might have been able to obtain cobbles. The most plausible hypothesis is that a riverbed in the lower part of the sequence would have been exposed during the period when the sites where being formed, although this would also appear to have been unlikely, since the Upper Sandy Clays do not constitute an erosion stage of the underlying sediments.

The other major problem for identifying the sources of raw material procurement is the

Table 7.9 Technological categories in the ST Site Complex

	ST15	ST31-32	ST30	ST4	ST6	ST3	ST2C-E	ST2A	ST2D	ST2G	TOTAL	
	N	N	N	N	N	N	N	N	N	N	N	%
Cores	3	6	4	19	0	5	5	0	0	0	42	7.1
Retouched pieces	1	1	3	5	0	7	13	1	1	1	33	5.6
Flakes	4	14	19	37	5	17	43	0	0	2	141	24
Flake fragments	3	29	24	52	6	24	43	4	1	1	187	31.8
Chips	0	4	5	14	1	12	8	0	0	0	44	7.4
Angular fragments	1	8	15	17	0	7	38	2	1	1	90	15.3
Hammerstones	0	3	2	5	0	2	0	0	0	0	12	2
Anvils	0	0	3	0	0	0	0	0	0	0	3	0.5
Battered fragments	0	0	1	2	0	1	0	0	0	0	4	0.6
Unmodified material	0	5	10	15	0	1	0	0	0	0	31	5.2
Total	12	70	86	166	12	76	150	7	3	5	587	100

current configuration of the sedimentary outcrops themselves; the Type Section constitutes a small sedimentary area, in which the Plio-Pleistocene deposits exposed on the surface are not extensive. In fact, areas such as the ST Site Complex are on the edge of one of the outcrops. For this reason, the streams where the hominids obtained the cobbles may have been a few hundred meters from the sites, but we are unable to identify them due to the absence of Plio-Pleistocene outcrops.

In the case of the nephelinites, it had been assumed that the rounded character of some of the cores indicated their fluvial origin, a river carrying the blocks down from the Shirere hills and transforming them into cobbles on their way to the Type Section. However, in the Shirere hills, located some 5–7 km from the delta of the river Peninj, blocks that have fallen down the hillside have been documented that look very similar to stream cobbles, with rounded edges caused by diagenesis. So it is not possible to claim categorically that the archaeological examples came from stream bars, since their appearance suggests that they could also have been collected directly by the hominids in the source areas.

With regard to basalts, hominids certainly did not go to the Sambu hillsides and these lavas were available in some part of the landscape of the Type Section. Since in the aggradation phase of the Upper Sandy Clays the typically deltaic sedimentation did not have the capacity to drag cobbles, these must have been obtained from the gravel bars deposited in the previous phase. The general features of the bars of Lower Sandy Clays at least 1 km away from the ST Site Complex have been compared; there the absence of nephelinites and quartzes is notable, and only basalts have been documented. A random selection of cobbles from the stream bar has been compared with the dimensions of the cores from the ST Site Complex (Figure 7.36), and it can be seen that the natural cobbles are systematically larger than the cores. A Student's t-test was carried out, and this showed that the mean maximum lengths of the two samples were not the same. It can be assumed that, by definition, the reserves of raw material must have been larger than the cores that came from them, so we could infer that cobbles of this type would be the typical blanks used to produce cores. In the analysis of the stream bar it was also observed that a good part of the cobbles

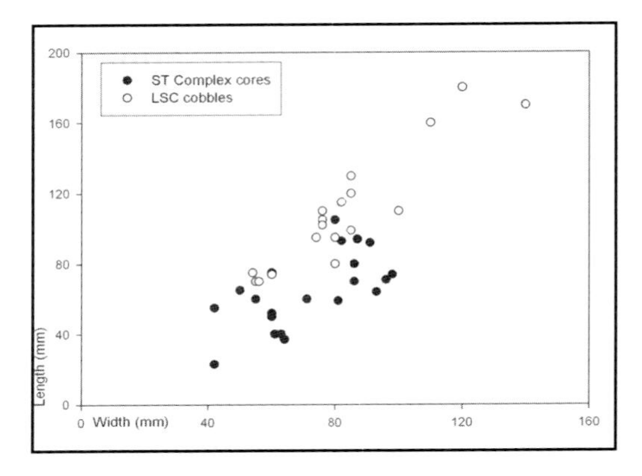

Figure 7.36 Dimensions of the cores from the ST Site Complex and a cobble sample from the Lower Sandy Clays.

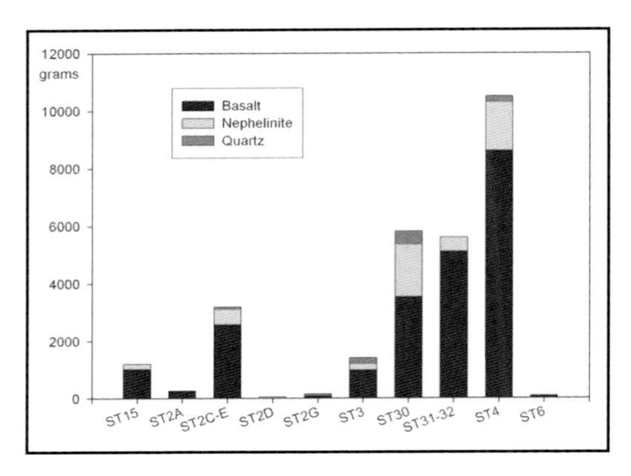

Figure 7.37 Weight of each raw material in the ST Site Complex.

were of poor quality and covered with vacuoles, which shows how difficult it must have been to find blanks suitable for knapping.

On the North Escarpment, situated in an area in which large blocks of basalt must have been abundant, the technological processes were centered mainly on obtaining enormous blanks. The paleographic environment of the Type Section, some eight kilometers to the southeast of the North Escarpment, was different: in a deltaic medium with low-energy detritic material, the availability of raw material must have been very limited. One possibility would have been that pieces were brought from the area of the escarpments. However, at a petrological level, the artifacts documented in the ST Site Complex bear no similarity to those from the escarpments, and we do not find large objects as we do in those areas either. The hominids who roamed the ST Site Complex concentrated on working small cobbles. In fact, only small cobbles would have been available in a deltaic riverbed such as the Peninj in that area, and moreover they could not have been abundant. The hominids seem to have been well aware of this shortage; in many cases they worked the cores until they were exhausted and quite frequently used a structured method of reduction, which would maximize the yield of a resource that was scarce in this area.

In Table 7.10 it can be observed that the basalts, with more than 22 kg, were the raw materials most frequently used to obtain artifacts in the ST Site Complex. The quartz was only used sporadically (barely 1 kg in total), while the nephelinites (with around 5 kg) were in between. This general pattern can obviously also be observed in individual cases (Figure 7.37), although we see that most of the nephelinites are concentrated in sites such as ST4 and ST30, while ST4 and ST3 have the largest proportions of quartz.

It is also interesting to explore the possibility that different raw materials were used for different categories of artifacts. If we look at all the material from all the sites (Figure 7.38), we see that the basalts dominate all the categories, and are the most important both in general in each site and also in their distribution by categories. A Lien test was carried out (Figure 7.39) in order to see if, in terms of frequencies, any less obvious pattern

Table 7.10 Weight (grams) of the knapped material in the ST Site Complex

	Basalt	Nephelinite	Quartz	Total
ST15	1,028	180	0	1,208
ST2A	248	10	0	258
ST2C-E	2,588	509	94	3,191
ST2D	14	32	0	46
ST2G	94	0	58	152
ST3	997	213	206	1,416
ST30	3,536	1,822	455	5,813
ST31-32	5,115	492	0	5,607
ST4	8,625	1,683	198	10,506
ST6	77	1	0	78
Total	22,322	4,942	1,011	28,275

could be observed. This documented a clear preference for quartz for hammerstones, which corroborates earlier studies in which the same selection was observed (de la Torre et al. 2003).

The technological categories of the ST Site Complex

With respect to the unmodified material, although the unmodified blocks of lava were originally considered manuports (de la Torre et al. 2003), they were subsequently shown to have little connection with the actual lithic industry (de la Torre et al. 2004; de la Torre and Mora 2004). The question is of some importance, because if we consider the unmodified objects to be manuports (overall weight of the unmodified material = 11,974 grams) we would increase the little more than 33 kg of stone that had been worked to 45,248 grams for the total weight of raw material brought to the ST Site Complex.

The hammerstones account for 1.9% of all the artifacts, there being a certain preference in the selection of quartz cobbles. In ST30 we cannot exclude the possibility that some of the quartz objects were also passive percussion pieces. Although the features which are systematically repeated in the tabular blocks of Olduvai cannot be identified so clearly (Mora and de la Torre 2005), in the three examples from ST30 which could be anvils we find scars and percussion steps associated bidirectionally, and these are perhaps linked to their use as passive elements. However, given that there are no other examples in the ST Site Complex, they have been included in the general recounts as percussion elements, without specifying the technical gestures that produced them. In ST31-32, ST4, and once again in ST30, other hammerstones display irregular planes that could lead to them being classified as hammerstones with fracture angles and not as simple knapping hammerstones. However, in the absence of a systematic repetition of the features described in Olduvai (Mora and de la Torre 2005), in this case too they have been included in the general category of hammerstones.

In the ST Site Complex there are a total of 141 complete flakes (Table 7.9). Out of this sample, at least 16 flakes are also edge-core flakes, which indicates some systematic rejuvenation and reactivation of the knapping surfaces. The dimensions of these flakes, with an average length ranging from 3 to 4 cm (Table 7.11) and similar width, mean they are of a homogeneous quadrangular size and shape (Figure 7.40).

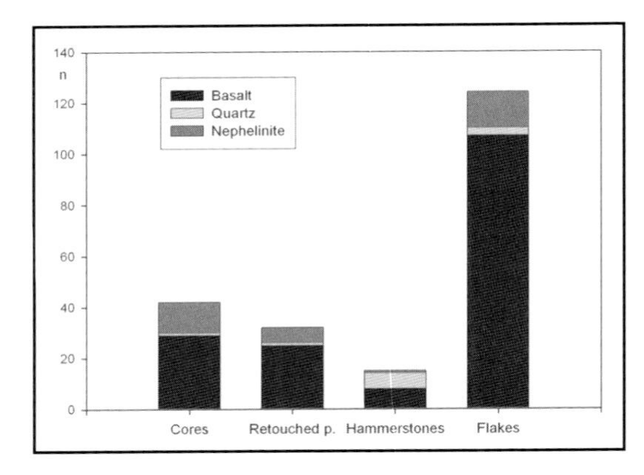

Figure 7.38 Raw materials by main technological categories in the ST Site Complex.

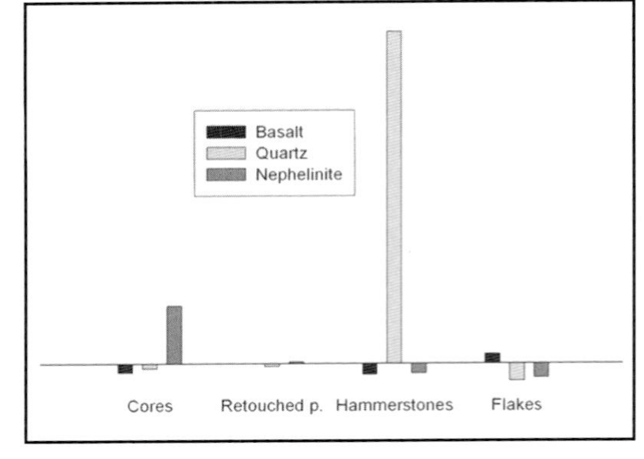

Figure 7.39 Lien Test comparing raw materials and technological categories.

At Peninj it is difficult to discern which part of the flakes may be considered cortical since, in the unrounded blocks, the original external surface is usually distinguished from the inside by a very slight and often inconspicuous weathering layer. This dilemma leads to an underrepresentation of knapping products with remains of cortex. In any case, our estimate (Table 7.12) indicates that cortical flakes are very scarce throughout the ST Site Complex. In fact, by applying Toth's types (1982) it can be observed that the first generation flakes (types I and IV) are absent in all the sites (Figure 7.41). In the ST Site Complex we have recurrent and structured knapping systems. This is consistent with the predominance of flakes in which the cortex has already disappeared. Furthermore, this dynamic of cortical representation may also indicate the intensity of the knapping processes, and the possible separation in time and space between the initial flaking and the input of previously-decorticated cores to the sites.

In Figure 7.42, it is observed that most of the knapping platforms of the flakes were not prepared before their detachment, although up to 11.8% of these striking platforms are dihedral or multifaceted, which is a higher percentage than in many of the Olduvai assemblages (de la Torre 2005). The number of detachments on the dorsal faces of the flakes (Figure 7.43) and the direction of those preliminary scars (Figure 7.44) also provide information about knapping methods. 43.5% of the complete flakes have a minimum of three to four preliminary detachments, and up to 23.1% display an even greater number of previous scars. Like the small percentages of cortex and the structured morphology of the sections, the ranges of preliminary extractions from the flakes suggest that the Peninj cores were exploited systematically.

Furthermore, the flakes in which it has been possible to reconstruct the diacritical pattern display knapping patterns typical of different ways of rotating the cores (Figure 7.44), which include centripetal methods indicative of recurrent working of the same débitage surfaces (Figure 7.45). In short, the flakes display a set of homogeneous characteristics: they measure between 3-4 cm, and are basically quadrangular in shape. Technically, it can be seen that the striking platforms are not generally prepared, although they

Table 7.11 Dimensions (in mm) and weight (in grams) of the whole flakes in the ST Site Complex

	Minimum	Maximum	Mean	Std. Deviation
Length	12	98	33.4	13.79
Width	3	38	10.9	5.8
Thickness	12	77	36.1	14.57
Weight	1	107	20.3	24.6

Table 7.12 Percentages of cortex in striking platforms and dorsal faces of the whole flakes

	Striking platform					
	Cortical		Non-cortical		Total	
Dorsal face	N	%	N	%	N	%
Full cortex	0	0	0	0	0	0
Cortex > 50%	0	0	5	3.7	5	3.7
Cortex < 50%	5	3.7	19	14	24	17.6
Non-cortical	6	4.4	101	74.3	107	78.7
Total	11	8.1	125	92	136	100

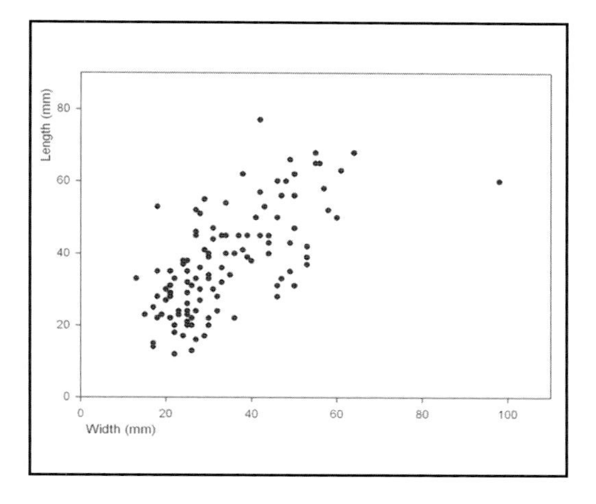

Figure 7.40 Dimensions of the whole flakes from the ST Site Complex.

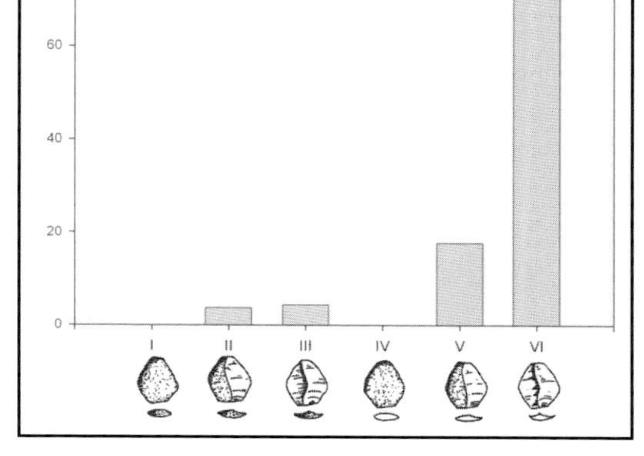

Figure 7.41 Toth's types in the whole flakes from the ST Site Complex.

have a relatively high index of faceted platforms. Moreover, the fine sections and the dorsal faces with little cortex and a good number of earlier detachments indicate that the knapping products were produced by effective, recurrent and well organized methods of exploitation.

In the whole of the ST Site Complex there are 42 cores, which are mainly concentrated in ST4, ST31-32 and ST30. Although they are very varied in size (Table 7.13), in general the cores are fairly large, with a mean weight of 340 grams per piece. To the naked eye there seems to be a significant

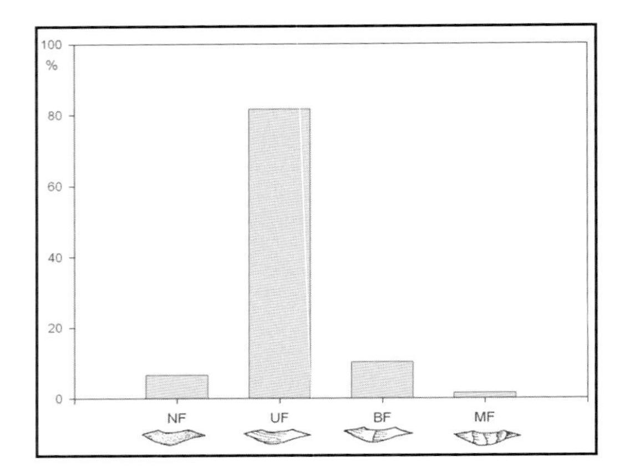

Figure 7.42 Types of striking platforms in whole flakes.

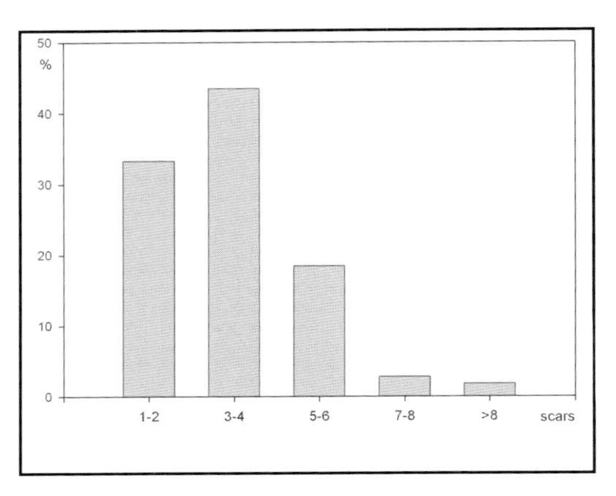

Figure 7.43 Number of scars on the whole flakes.

difference in the sizes of the basalt and nephelinite cores (Table 7.14), and this is confirmed by the Student's t-test, which shows that their mean maximum length (p=0.08) and weight (p=0.08) is not the same. This difference in size of the cores (Figure 7.46) is interesting, although it is difficult to determine whether it is due to the fact that the natural blocks of nephelinite available were smaller, that they were subject to more reduction than the basalt, or a combination of both factors.

In general, cores from the ST Site Complex were intensively exploited, as we documented a mean of 9.59 flakes per core, which is the minimum calculated on the basis of the last reduction sequence preserved in each piece (Figure 7.47). By using the type of calculations that McNabb (1998) carried out in the Olduvai sites, and given that we have documented a minimum of two detachments and a maximum of 19 in the cores from the ST Site Complex, a minimum of 84 and a maximum of 798 flakes must have been produced. Counting both flakes and fragments of flakes, we have around 300 knapping products (Table 7.9), which gives an average compared with the preliminary calculations and is consistent with the proportions of cores and débitage.

With regard to the knapping methods, there are two main systems (Figure 7.48). One is the bifacial hierarchical centripetal strategy, with 11 examples (26.2% of the total). This method is the most important of those documented in the ST Site Complex in terms of both quantity and quality. As has already be argued in other studies (de la Torre and Mora 2004; de la Torre et al. 2003, 2004), the presence of a method that implies the division and hierarchization of the knapping surfaces and the maintenance of this structure throughout the entire débitage process in sites as old as Peninj implies considerable technical knowledge and skill. Examples (such as those shown in Figures 7.9, 7.21, and 7.25-7.27, etc.) varying in size but all knapped using the same technological strategy, prove that the craftsmen were imposing the basic design for the knapping method on the blocks, following the same technical pattern throughout the entire sequence of reduction, and exploiting the cores in an orderly, rational and effective way.

Together with the hierarchical centripetal method, abrupt unifacial strategies predominate, including with 26.2% of all the cores.

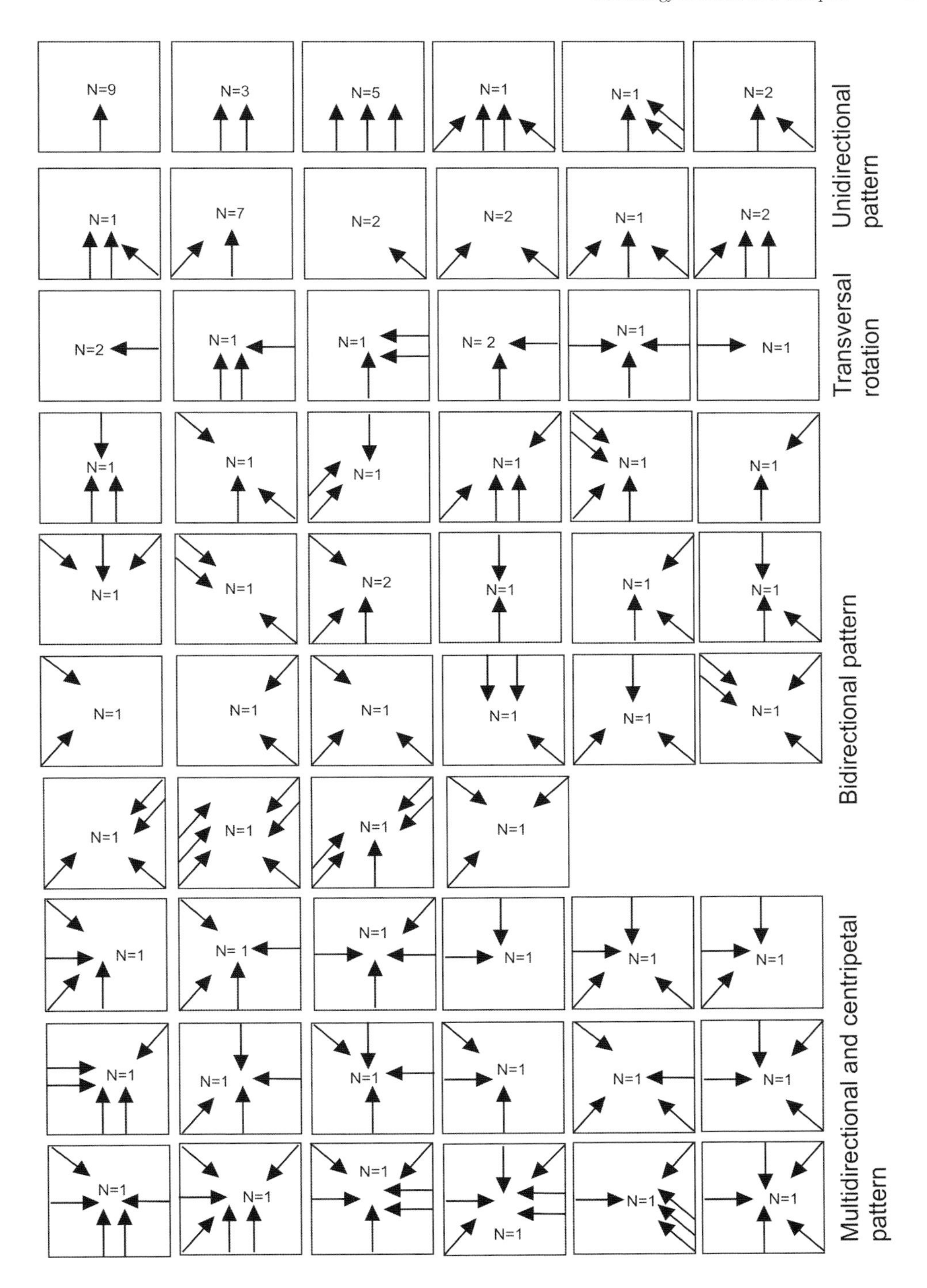

Figure 7.44 Direction of previous scars in the dorsal faces of the whole flakes from the ST Site Complex.

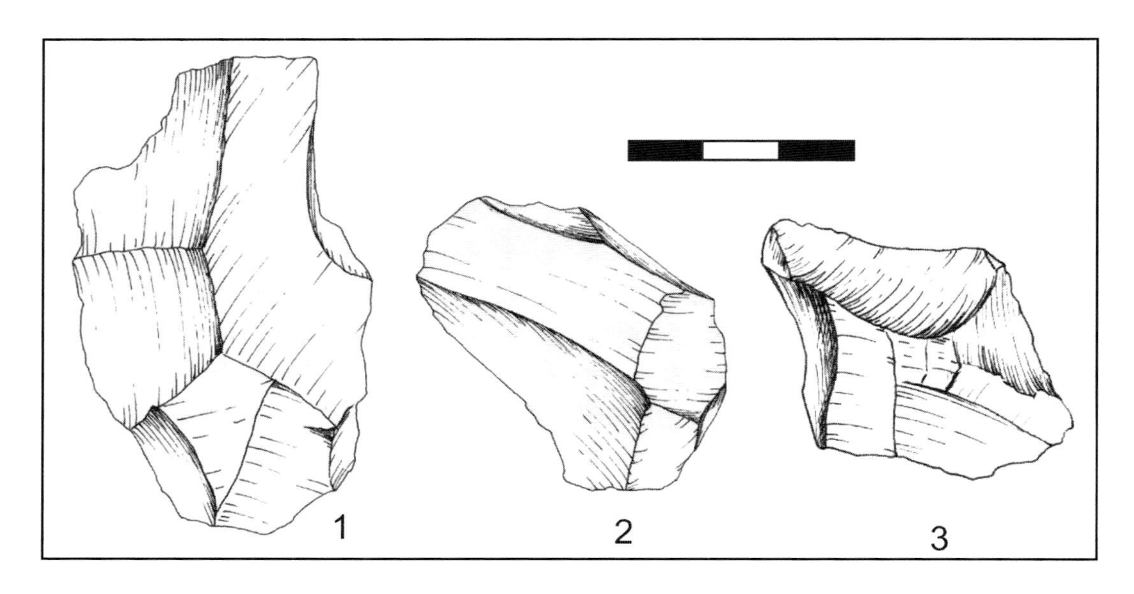

Figure 7.45 Centripetal flakes from the ST Site Complex (drawn by N. Morán): 1) Flake from ST2A; 2) flake from ST2G; 3) flake from ST31-32.

Table 7.13 Dimensions (in mm) of the cores in the ST Site Complex

	Minimum	Maximum	Mean	Std. Deviation
Length	23	106	62.2	22.12
Width	22	98	63.1	19.54
Thickness	15	110	53.5	22.86
Weight	11	1,175	343.4	302.47

Table 7.14 Dimensions (in mm) and weight (in grams) of the cores, distinguishing between nephelinite and basalt cores

Material	Minimum	Maximum	Mean	Std. Deviation
BASALT				
Length	23	106	69	21.32
Width	40	98	70.5	17.58
Thickness	26	110	60.32	23.53
Weight	54	1,175	434.54	319.61
NEPHELINITE				
Length	25	94	50.23	17.14
Width	22	77	49.15	15.03
Thickness	15	61	40.77	14.31
Weight	11	584	170.38	152.99

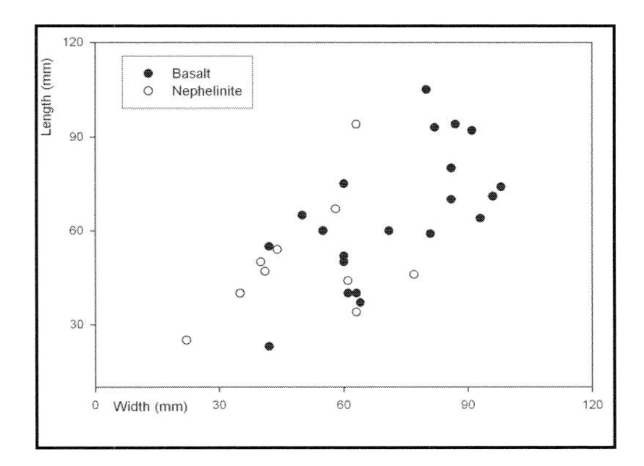

Figure 7.46 Dimensions of the cores from the ST Site Complex.

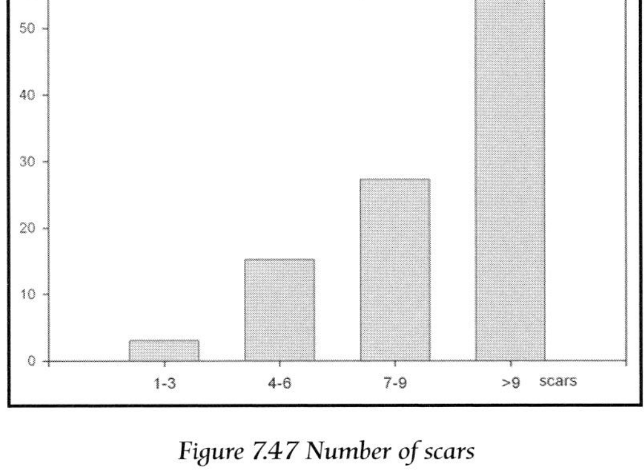

Figure 7.47 Number of scars on the cores from the ST Site Complex.

Although the knapping philosophy is the same as that previously described at Olduvai (de la Torre 2005), taking advantage of the natural platforms in order to work the transversal and sagittal planes, the cores of the ST Site Complex display more recurrent exploitation. While in Olduvai cortical platforms were used to detach short sequences of flakes in a single plane, leaving the rest of the core unworked, in Peninj it is not unusual to find cores in which, while the platforms were still not prepared, the whole of the perimeter of the piece was knapped repeatedly, and reduction of the same areas of débitage continued until they were exhausted.

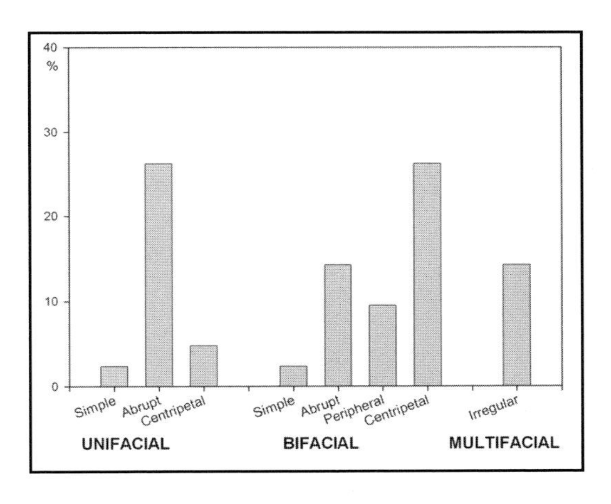

Figure 7.48 Types of exploitation on the cores from the ST Site Complex.

The rest of the cores, exploited using complementary strategies to those already described, also tend to display systematic knapping. In fact, the abundance of irregular multifacial cores (Figure 7.48) was interpreted as the final phase of the reduction of cores that to begin with would have been worked using orderly knapping methods. When they were too small to continue in this way were they worked taking advantage of all the angles available without imposing an orderly structure of débitage (de la Torre and Mora 2004). The greater extent of recurrent exploitation that we observe in the ST Site Complex compared with Olduvai (de la Torre 2005) could be due to the more intensive exploitation of the resources available, or simply to a better knowledge of the possibility of reactivating cores and the technical methods necessary to do so. It is difficult to determine which of the two options guided the technical strategies adopted in the ST Site Complex, but what is certain is that these craftsmen very obviously displayed considerable mastery over the way in which knapping products could be obtained.

There is a relatively high proportion (5.6%) of retouched pieces in the ST Site Complex, particularly compared with some of the sites of Olduvai Bed I, where they are practically non-existent (de la Torre 2005). In the ST Site Complex simple and denticulate side scrapers predominate. These, together with the notches, account for the majority of the 33 retouched artifacts (see again Figures 7.6, 7.15, 7.20 and also 7.49). Together with these pieces we have two possible burins (one of them also a sidescraper) and two that are probably endscrapers. In any case, there is no typological standardization, and usually flakes or fragments of flake were chosen in which the edges were modified with a denticulate retouch restricted to a single edge of the piece.

Basalt is the predominant raw material among the retouched pieces and follows the general trend of the other categories. It cannot be said that different criteria were used in its selection. However, the dimensions of the retouches, with an average length of about 4 cm and despite being in the range of variation of the complete flakes (Figure 7.50), are generally larger than the normal products of knapping (Table 7.15). This fact is more important if we bear in mind that up to 76% of the retouched pieces are obtained from blanks that are fragments of flakes rather than complete flakes, and even so they are still bigger and heavier than the latter. For these reasons it would appear that larger pieces were selected.

Technological evaluation

Having described the collections, we now discuss the nature of the systems of exploitation represented in the ST Site Complex. The abrupt unifacial knapping method is one of the most important. This consists of the detachment of longitudinal and parallel flakes from the striking platform, which may or may not have been prepared, and which forms an approximate right angle with the knapping surface. In the Peninj cores we find both the unidirectional (Figure 7.24) and bidirectional (Figure 7.51) versions of this system of reduction represented.

Together with the abrupt unifacial system, centripetal knapping predominates, both uni- and bifacial. As has been said, the unifacial centripetal system is characterized by the detachment of radial flakes from the knapping surface from striking platforms that have not been prepared. The bifacial hierarchical centripetal system, on the other hand, uses two knapping surfaces related to each other by a plane of intersection that divides the core into two parts. In this strategy, the principal knapping surface is used for the radial detachment of short, wide flakes. These products are obtained from the secondary surface, used as a striking platform prepared by longitudinal detachments that are parallel with each other (e.g. Figure 7.22).

Both uni- and bifacial centripetal strategies can be the result of the same sequence of reduction, and could also have involved irregular multifacial cores. In this hypothesis, formulated elsewhere (de la Torre et al. 2004), various phases can be distinguished (Figure 7.52);

Phase 1. The blocks would start being exploited in a single plane, i.e. the horizontal plane, following a radial pattern of knapping. In this phase the striking platforms would be natural, that is, the cortical surfaces would be used as planes for exploiting the cores in a unifacial and centripetal way.

Phase 2. As the process of reduction advanced, the removal of flakes from the core would result in the disappearance of the convexities necessary for continuing to exploit the knapping surface.

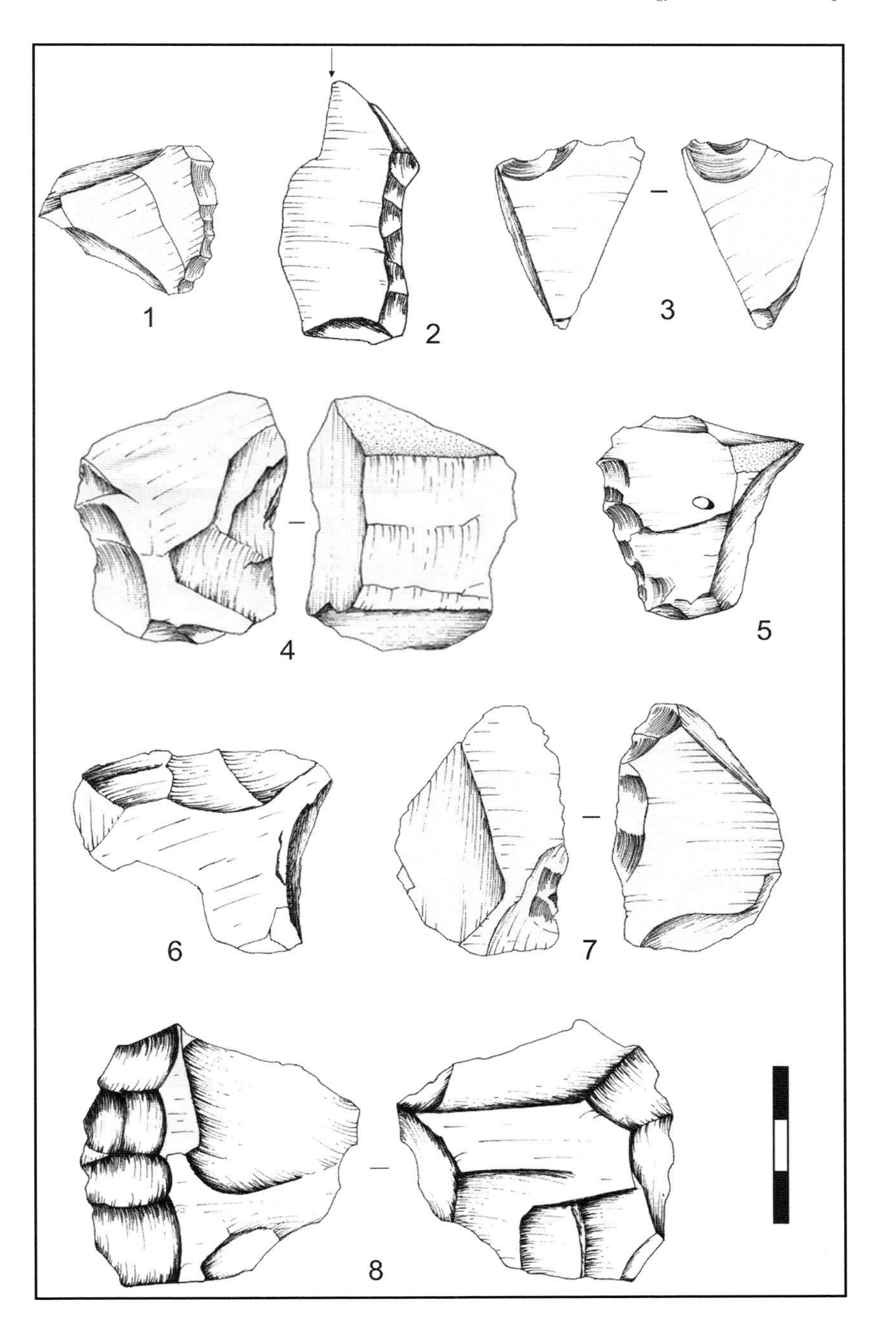

Figure 7.49 Retouched flakes and fragments from the ST Site Complex (drawn by N. Morán): 1) Sidescraper from ST2D; 2) burin and sidescraper from ST31-32; 3) alternating notch from ST15; 4) sidescraper from ST2G; 5) sidescraper from ST2A; 6-8) sidescrapers from ST4.

Phase 3. This would make it necessary to adopt a different approach to the structure of the core, and the natural transversal and sagittal planes would then be worked in order to reactivate the volume of the knapping surface.

Phase 4. The preparation of the natural planes would have produced a bifacial shape, in which a plane of intersection separated the main knapping surface, which continues to be centripetal in character, from a subordinate or preparation surface with parallel detachments around the whole of the core's perimeter.

Phase 5. Once the bifacial hierarchical centripetal system had been adopted, the knapping of the core would continue on this model throughout a long process of reduction. A stage of full exploitation would be reached, in which the cores would display a structure produced entirely by the technical strategy employed.

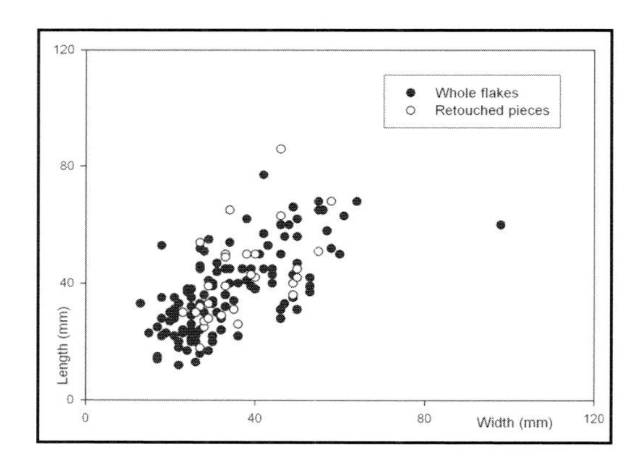

Figure 7.50 Dimensions of the whole flakes and retouched pieces from the ST Site Complex.

Phase 6. Some cores would continue to be worked until a point was reached when it would be impossible (either through lack of technical skill or due to the limitations of the raw material) to maintain the bifacial structure. In this case, the inability to continue creating convexities would lead to taking advantage of all the angles available without concern for the general configuration of the piece. In this way, irregular multifacial exploitation would be reached that would reflect the clear exhaustion of the cores.

This reduction hypothesis is supported by various arguments. The unifacial centripetal cores are always larger, followed by the bifacial centripetal and multifacial cores. The main characteristic of a core as the process of reduction advances is its loss of material, so that the obvious differences in weight found in the three exploitation strategies provides further evidence to defend our model. Moreover, the reduction hypothesis proposed is not based on simple inferences, as the collection contains real examples of each of the phases (Figure 5.53). To sum up, we consider the cores of the ST Site Complex to represent the continuity of a single technological sequence, which would commence with the exploitation of a single surface of the block (unifacial centripetal method), would continue with systematic reduction, leading to the bifacial centripetal strategy, in which the cores display a

Table 7.15 Dimensions (in mm) and weight (in grams) of the retouched pieces in the ST Site Complex

	Minimum	Maximum	Mean	Std. Deviation
Length	18	86	41.21	14.7
Width	21	58	36.94	10.44
Thickness	7	40	15.7	6.18
Weight	4	187	32.82	34.9

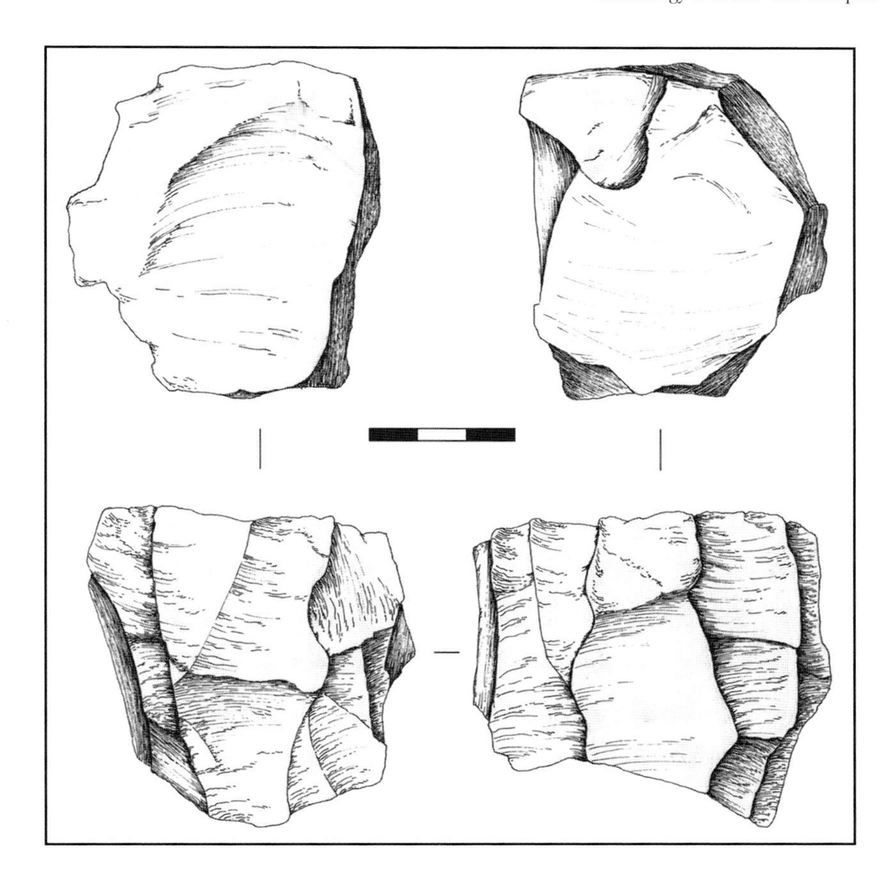

Figure 7.51 Bidirectional abrupt unifacial core from ST4 (drawn by N. Morán).

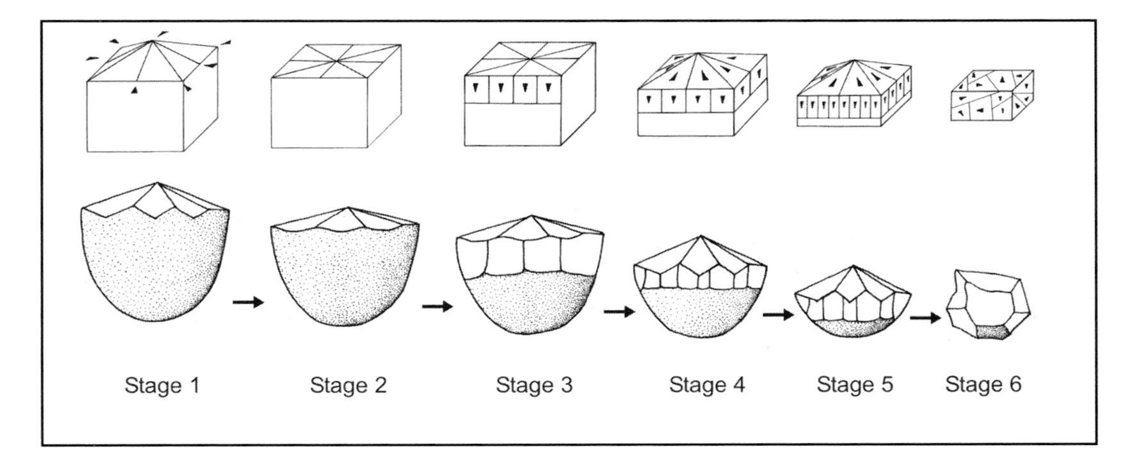

Figure 7.52 Ideal schemes of our reduction hypothesis of the centripetal cores from the ST Site Complex.

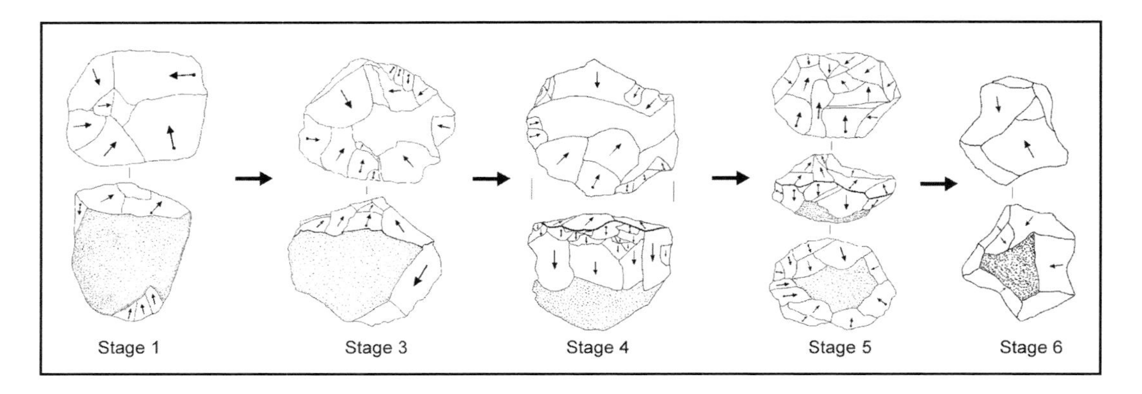

Figure 7.53 Our reduction hypothesis through archaeological examples from the ST Site Complex.

complex structure, and finally to the multifacial, with its irregular morphology. In this way, it can be proposed that the only method of organized knapping would be that represented by the bifacial centripetal stage, the point when the core is in full exploitation phase. This is why it is essential to look in depth at the nature of this system of knapping.

It is relevant to emphasize what we understand by the bifacial hierarchical centripetal system. These cores are characterized by displaying two opposite surfaces, separated by a plane of intersection which bisects the core through its perimeter.

The main surface is worked using a radial pattern of exploitation, which produces short, wide flakes with centripetal dorsal faces. Their scars form an angle parallel or sub-parallel to the core's plane of intersection, which makes it possible to maintain sufficient convexities on the knapping surface to continue the reduction.

The knapping platform or subordinate surface is characterized by its parallel detachments, which form a secant angle to the plane of intersection. The scars produced by these detachments appear to act as prepared striking platforms for obtaining flakes from the main surface.

This method of knapping denotes a certain complexity, since it does not just mean achieving a bifacial form of exploitation through an artificial plane of intersection, but maintaining this structure throughout the whole of the reduction process. The bifacial structure is not maintained simply by alternating blows, but by configuring one of the surfaces as a subordinate plane which serves to exploit the main surface. The concern for maintaining sufficient convexities in the main surface is constant, hence the recurrence of the products of preparation (Figure 7.54), which serve to re-establish the necessary angles. Thus, this process would continue until the pieces were exhausted, like the one in Figure 7.21. In this the recurrence of the same pattern of knapping is evident, and the morphology of the parallel scars on the subordinate surface indicate that these detachments were previous to the state of reduction in which the core was found. However, the core maintained the same pattern of reduction until it was abandoned, probably due to the absence of the necessary convexities on the knapping surface.

What we are trying to emphasize is that the craftsmen continued to use the same method of knapping throughout a long process of reduction, which would include not just the unifacial stages but also various stages of production within the bifacial hierarchical system (see Figures 7.12-7.13, 7.21-7.22, 7.26, etc.). This whole process is depicted in Figure 7.55, which shows artifacts from the ST Site Complex ascribed to different phases of knapping of the bifacial hierarchical centripetal method. This is an important fact, since it reveals the hominids' ability both to obtain complex reduction structures from a bifacial plane of intersection, and to maintain them throughout the whole of the knapping sequence. The technical, conceptual and abstract skills involved are thus more than evident.

It is worth asking ourselves what the bifacial hierarchical centripetal method we have defined here consists of, or what it corresponds to. The first impression is that this is a technology typical of the Middle Paleolithic. In fact, de la Torre et al. (2003) have compared the Peninj technology with the Levallois method since, according to Böeda´s (1994-1995) criteria for the recurrent centripetal Levallois method, the two techniques can be similar. However, recent trends in defining the discoid method (see papers in Peresani ed. 2003; Lenoir and Turq 1995; Slimak 1998-1999 etc.), tend to restrict the definition of Levallois and to amplify the concept of the discoid method. According to these proposals, the strategy identified at Peninj could never be considered within the definition of the Levallois method, since there is no preparation of convexities on the main surface for the detachment of preferential flakes, and no preferential products of this kind documented, of course. Therefore, if we accept the strictest definition of the Levallois technique, the bifacial hierarchical centripetal method found in Peninj would be closer to the various discoid methods. Even so, the ancient chronology in which the method is documented continues to make the technology of the Peninj ST Site Complex a basic point of reference for arguing the technical skills of the first hominids.

Conclusions: The operational sequence

According to Geneste's models (1985, 1991), a system of lithic production can be organized chronologically. Thus, the sequence begins with the phase of obtaining the raw material. It then continues with the initial transformation and modelling, or what amounts to the same thing, preparation of the core. The next stage is that of débitage, in which the products and blanks are obtained. It continues with the transformation phase, in which the blanks are made into tools by retouching, and the consumption stage, in which the retouched pieces or basic products are used as instruments for specific actions. Finally, the stage of abandonment of the lithic resources follows. Geneste's model (1985:250) was designed for operational sequences much more complex than that of Peninj, and cannot therefore be applied directly to the record of the ST Site Complex. However, the philosophy implicit in his model is valid for studying the assemblages of this chronology, so we can use it to investigate the nature of the Peninj operational sequence.

Although we cannot exclude other possible functions, examples such as those of Figure 7.56 suggest that at least in the ST Site Complex the unmodified lithic material recovered to date was not a potential reserve for subsequent use as cores. In this way, Phase 0, or the procurement stage according to Geneste (1985), would not be represented in the ST Site Complex. In this case, we find ourselves with an interesting pattern, one which implies that the hominids were intensively exploiting a highly-appreciated resource (that was perhaps scarce in the immediate environment), and in which the accumulation of reserves of raw material at specific points in the landscape was not necessary (or possible).

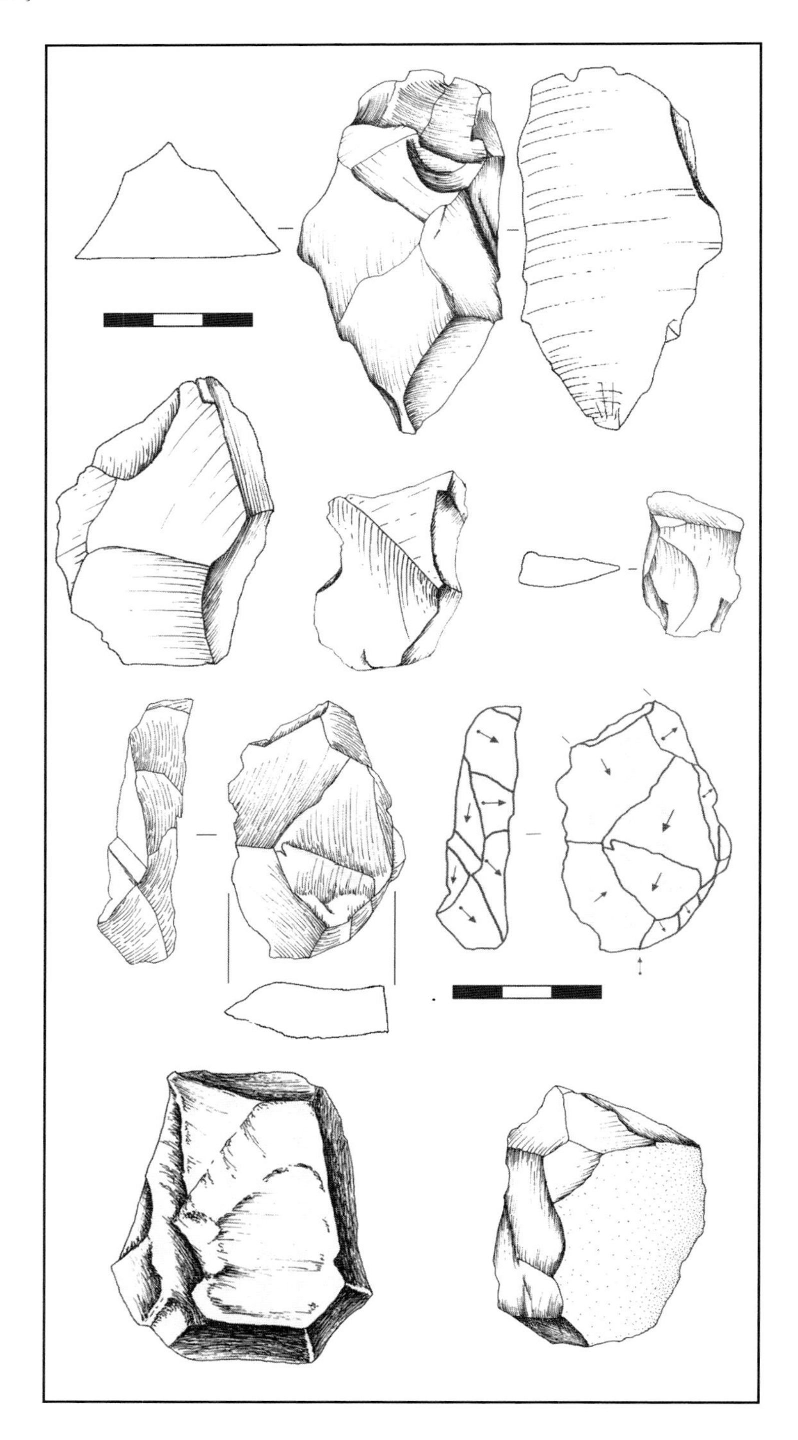

Figure 7.54 Edge-core flakes from the ST Site Complex.

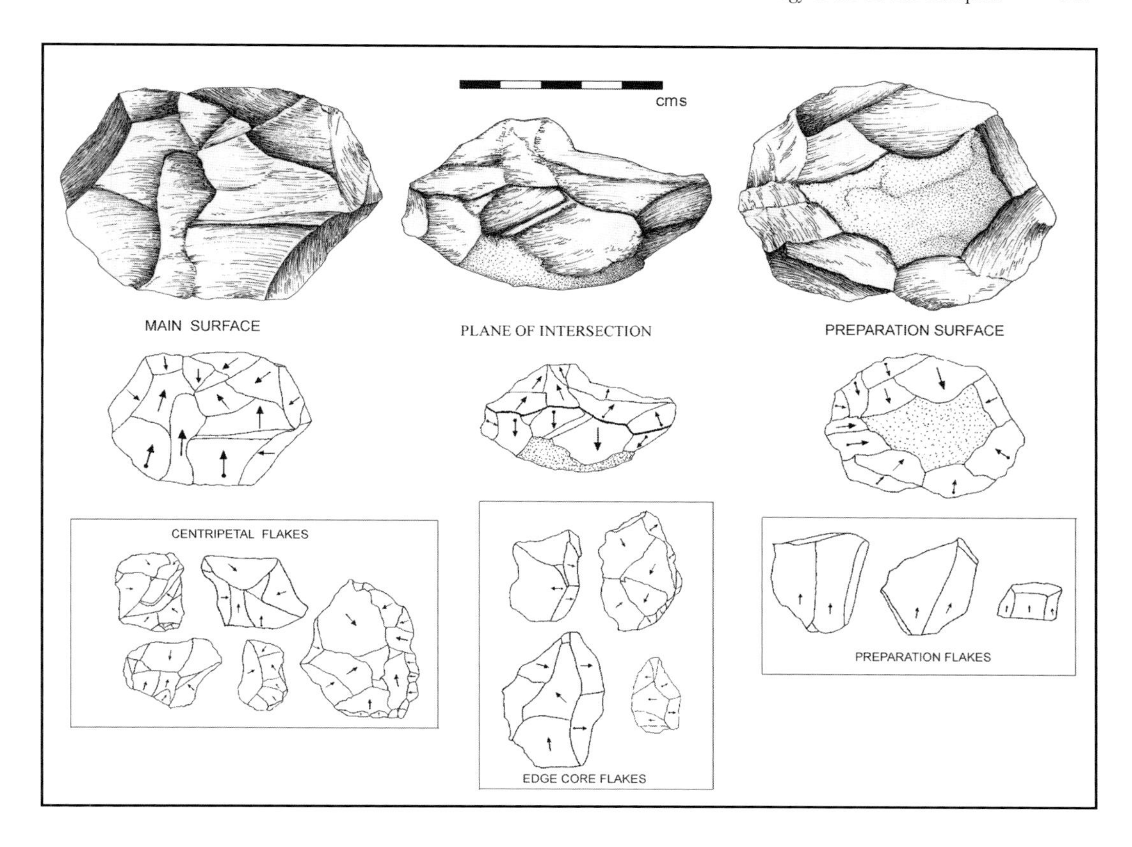

Figure 7.55 Example of the hierarchical centripetal exploitation method at Peninj and a diacritical scheme showing the products corresponding to each configuration stage (after de la Torre et al. 2003). The centripetal flakes are obtained from the flaking surface. The maintenance of the convexities is obtained through the rejuvenation products (edge-core flakes), which reactivate the adequate angles on the plane created by the intersection of both surfaces. The longitudinal and parallel flakes around all the perimeter of the core enable the preparation of the flaking surface to continue with the exploitation of the main surface.

This would be consistent with the low indices of pieces that can be ascribed to Phase 1 or the roughing out/de-cortication phase *sensu* Geneste (1985). Thus, as can be observed in Figure 7.57 and in our description of the flakes, only 3.7% could be included in the processes of initial roughing out. It could be proposed, then, that the initial stages of reduction of the cores would have been done at the places where they were obtained, which would also explain why cores that had been previously decorticated, rather than natural blocks, were brought to the sites.

What is obvious from looking at Figure 7.57 and the patterns described in this chapter in general, is that the sites of the ST Site Complex were basically places where flakes were produced, that were in all probability related to processing the faunal remains with which they were associated. In a previous work (de la Torre and Mora 2004) the choppers and polyhedrons of Peninj were included in the façonnage stage, as suggested by

Figure 7.56 Unmodified material from the ST Site Complex. Note the irregular morphology of the blocks, as well as their vesicular texture, that make difficult their use for knapping.

Inizan et al. (1995), Texier and Roche (1995), etc. After analyzing the Olduvai collections (de la Torre 2005) it has been observed, however, that it makes more sense to consider these objects simply as cores, so now we only include retouches on flakes or fragments in Geneste's Phase 3 (1985). Despite their percentage being irrelevant if we compare it with the production categories (cores, flakes, flakes fragments, etc.), 5.6% of retouched pieces is a relatively high proportion, particularly if we compare it with the flakes as a whole.

We could thus cite some of the arguments that link the frequencies of retouched pieces to the intensity of reduction (e.g., Dibble 1988), and consider that the relative abundance of small retouches in the ST Site Complex could be related to the recurrent exploitation that we see in other categories, such as the flakes that generally contain no cortex and the cores with multiple scars. This, however, is an hypothesis that is difficult to prove, as difficult as it is at present to link artifacts analytically with a specific function (i.e., through use-wear analysis). Contextually, and given the clear association between industry and fauna that we find in all the sites, it is assumed that the use and consumption phase of the artifacts is related with the processing carcasses. Even so, it would be necessary to have systematic functional analyses to create a clearly verified contextual framework.

In any case, we propose that a fragmented operational sequence existed in the ST Site Complex. Many of the gaps can be attributed to post-depositional causes, both contemporary with the formation of the sites and occasioned by current processes of erosion. Without doubt, these agents have contributed to categories such as small chips being very under-represented in all the assemblages. Even so, there also seem to be gaps in the operational sequence linked to the type of technological strategies that they generated: the cores seem to have been transported to the sites after being roughed out, since flakes from the initial de-cortication phase are not documented. This implies deliberate management of the landscape, in which the phases of procurement, roughing out and producing the blanks would have been separated geographically and chronologically. Moreover, these products were obtained using knapping methods that we can consider planned: the predominance of cores with hierarchization of surfaces, rejuvenation of edges and structuring of the débitage surfaces indicates a well-planned strategy of reduction that is applied to blocks of raw material of different sizes and quality. In short, the hominids of the ST Site Complex pursued planned technological strategies, which must imply yet another way in which the Peninj landscape was subjected to structured management.

In the opinion of Pelegrin (1990), complex operational sequences are those which consist of several stages, marked by changes in the operations (preparation, débitage, rejuvenation, etc.) and/or in the reduction techniques used. According to this author, these sequences will lead to obtaining standardized products that are

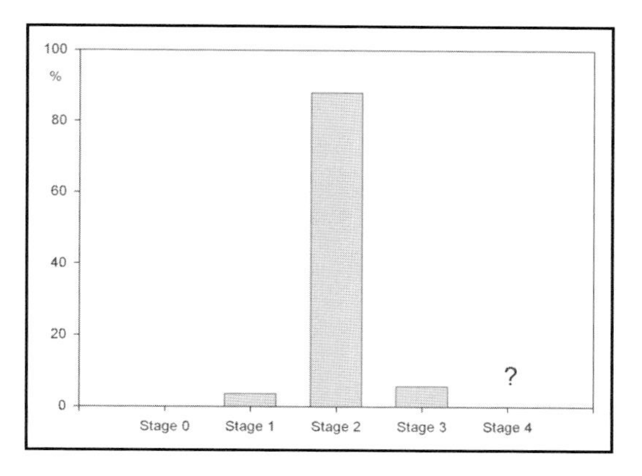

Figure 7.57 Application of Geneste´s (1985) stages to the collection from the ST Site Complex.

defined independently of the raw material used, and that at the same time require a complex organization of predetermined and predetermining detachments (Pelegrin 1990). We can see all this in the more than 28 kg of lithic material knapped in Peninj: the cores were brought to the site pre-shaped, they were prepared for exploitation using specific knapping methods, they were rejuvenated when they lost their convexities, they maintained the débitage method throughout the whole process of reduction, centripetal products were obtained that in turn conditioned the detachment of subsequent flakes. In short, we propose that the technological processes involved required precise technical knowledge and great capacity for abstract thinking and planning, and indicate the existence of technological strategies that had been well-structured by the hominids that occupied the ST Site Complex.

Acknowledgements

All the drawings of the lithics have been done by Noemi Morán.

8

ST-69: An Acheulean Assemblage in the Moinik Formation of Type Section

Fernando Diez-Martín, Luis Luque, and Manuel Domínguez-Rodrigo

Geology and paleoenvironment

Geological and chronological context

The ST69 site is located over one of the hills of the Type Section area, just outside the central area known as Maritanane. Archaeological remains are found on fine laminated tephra in the Upper Tuff member of Moinik Formation. No lithics have been found in *situ*, but their position on the relief is clearly related to erionitic tephra layers (Figures 8.1-8.2). In spite of the homogeneous appearance of the Upper Tuff member, a number of sedimentological features suggest intense depositional changes during its accumulation. Almost all tephra beds (totalling up to 13 m thick) have been deposited under a lacustrine environment. Geochemical analysis indicates that high alkaline waters reacted with trachytic tuffs to give rise to erionite (Isaac 1965;

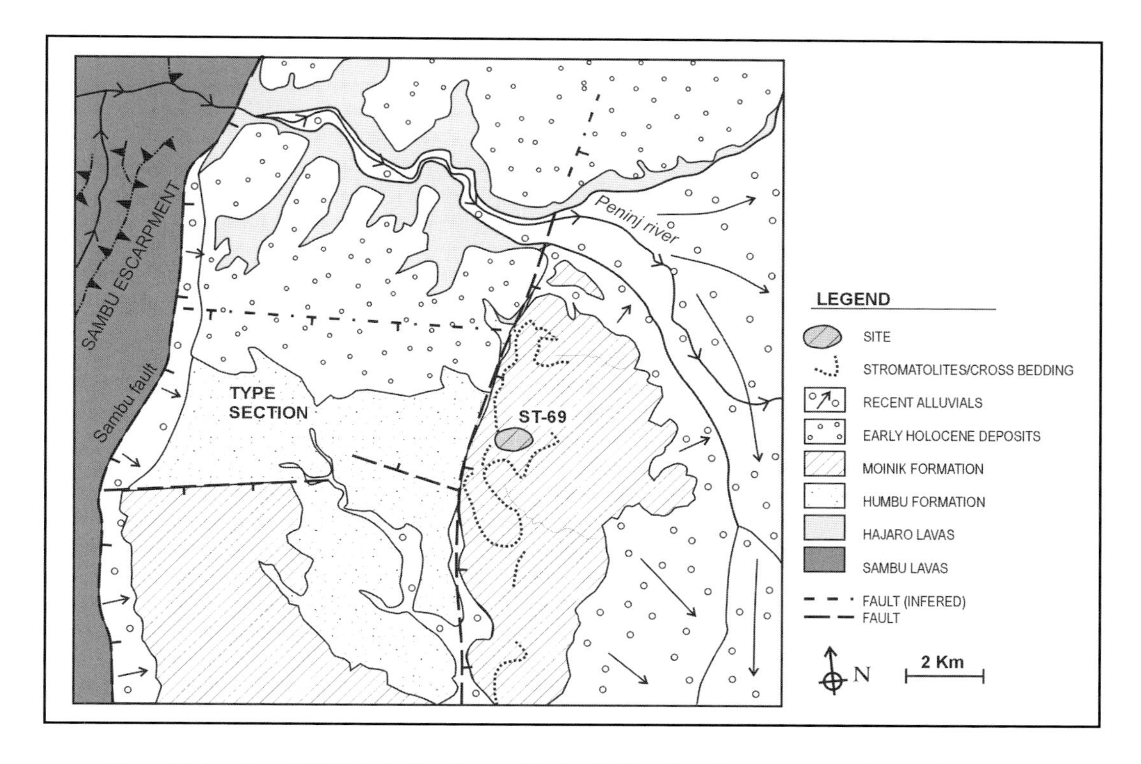

Figure 8.1 Location of ST-69 in the Type Section. The hatched line shows linear outcrops where sandy sediments (interbedded between the Upper Tuff and a cross-bedding layer) have been associated with the lithic industry.

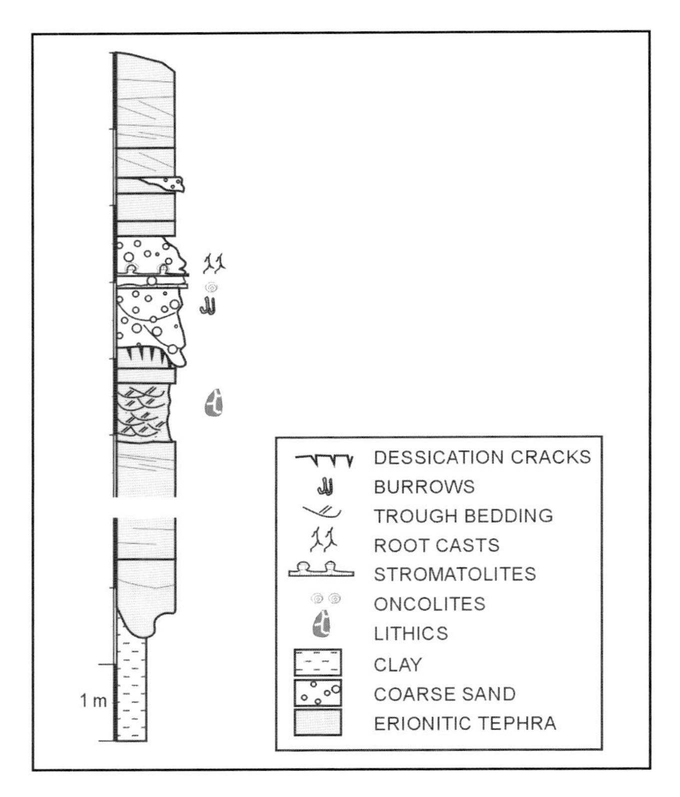

Figure 8.2 Upper Tuff stratigraphic section in the eastern area of the Type Section. The main stratigraphic features of inter-bedded sands and the cross-bedded layer are shown.

Icole et al. 1987). Detailed geological studies show that some non-eruptive events occurred at the time of lake level change. These events hardly influenced local paleoenvironments during this time. During low lake level times, erosion affected the area creating a paleorelief that showed a network of gullies near the shoreline. These gullies, up to 2 m deep, were filled by sandy deposits carried by flooding. An abrupt lake level rise led to an alluvial retreat and, under optimum pH and CO^3, concentration of cyanobacteria colonies gave rise to stromatolite growth. In other cases, smaller lake-level oscillations produced tephra and sand reworking and remobilization near the shore. Waves and flows gave rise to sand waves and ripples that today can be seen as crossed-lamination bedding.

Near ST69, an outcrop showing a regressive sequence has been recorded (Figure 8.2). Erosion has created a gully 2 m deep in the tephra layers. The highest layer of the sequence is a strongly weathered white tephra showing mudcracks that have been filled by well-cemented sandy sediments. The resulting tephra polygons show a characteristic honeycomb structure. Alluvial fill consists of gravel and sands, rich in quartz and pyroxenes, poorly sorted, and grading upward, showing oncolites, rootcasts and burrowing on the top. At least three sequences have been described in this fill, separated by two stromatolitic crusts. The

Table 8.1 Lithic assemblage sorted by raw material and tool type (percentage representation by tool type category is provided in italics)

	Basalt (%)	Nephelinite (%)	Tool Type
Bifaces			
flake handaxes	14 (*48.27*)	15 (*51.72*)	29 (*54.71*)
nodule handaxes	9 (*75*)	3 (*25*)	12 (*22.64*)
cleavers	3 (*50*)	3 (*50*)	6 (*11.32*)
flakes	3 (*75*)	1 (*25*)	4 (*7.54*)
cores	2 (*100*)	-	2 (*3.77*)
raw material	31 (*58.49*)	22 (*41.5*)	53

lower crust is thin (3 cm) and discontinuous, but the youngest can reach 15 cm in thickness and is arborescent in shape. A few hundred meters away, the same sandy layers can be found, although they are less developed. Some other sandy tephra layers have been observed higher than the main one of the sequence, suggesting new low lake levels.

No lithics have been found in the sand layers, but the remains most closely associated with this context are found in a stratum less than 1 m below the main erosive one. A thick layer of sandy tephra that shows intense crossed-lamination bedding is widely distributed in the area. This layer is 0.7 to 1.8 m thick and is found 0.5 to 1 m below the white tephra layer. Burrowing is intense in this bed. A number of large lithics have been found associated with this layer.

ST69 has not been found *in situ* and therefore its chronology can not be assured. Lithic artifacts are associated with the middle-upper levels of the Upper Tuff Member of Moinik Formation. Following Manega (1993), Isaac and Curtis (1964) and Foster et al. (1997), the boundary between the Humbu and Moinik Formations has been dated to around 1.35-1.2 Ma. Isaac (1965, 1967) suggests that the sedimentation of the Peninj Group terminated at the beginning of intense tectonic activity that gave rise to the modern narrow and deep basin. Taking into account regional tectonic data, MacIntyre et al. (1974) assumed that the main displacements took place between 1.15 and 1.2 Ma. Volcanic ashes in the Upper Tuff Member of the Moinik Formation constitute the final sedimentary unit of the Peninj Group and therefore, an age near 1.2 Ma can be inferred for this site.

Paleoenvironment

ST69 was formed in a lacustrine plain during a time of low lake levels. The loamy surface was locally eroded wherever it was exposed. Water flowed through narrow gullies to reach the alkaline lake waters. Other shallow underwater areas were under the influence of water currents that gave rise to sand waves and ripples on reworked ashes and sand. These sub-environments could be low-energy distal deltaic areas. This atypical site-formation environment did not allow bone preservation, which rarely occurs in the area.

The lithic assemblage

A survey was conducted in the 2004 season at the locality of ST69 over an area in excess of 300 m^2 (Figure 8.1). Here an assemblage of 54 lithic objects was recovered. About 88.8% of these artifacts are handaxes and cleavers (n=48), the remaining being four large flakes (7.4%) and two cores (3.7%). Three types of local or semi-local raw materials (mainly igneous) were used for the

Figure 8.3 Basalt with trachytic texture. Basic mineralogy: plagioclase, pyroxene (clinopyroxene) and opaque (probably iron oxides). Accessories: sphene. Microscope photograph: polarized light, photograph width 10 mm.

production of these artifacts (Table 8.1): basalt (57%); nephelinite (41%); and residually represented, lacustrine flint (2%; Figures 8.3-8.4). Despite the range of quality already reported in the basalts used by humans in other collections from the Type Section (de la Torre et al. 2003), at ST69 grayish and grayish-black basalt cobbles are generally fine-grained and have a homogeneous texture. Slightly porous surfaces, fissures and weakness planes are sometimes observed in blanks, but as a general rule, most specimens are optimal for conchoidal fracture. It appears that the hominids who created ST-69 were selecting the most suitable raw materials available for the production of large flakes and their bifacial configuration activities.

According to the external characteristics observed in the lithic objects, three weathering groups have been established: W1 includes 20 fresh artifacts (37%) that show no indication of fluvial traction or weathering; W2 includes 26 artifacts (48%) that tend to present slight to medium signs of abrasion on their surfaces and edges; W3 includes eight clearly eroded bifaces (15%), one of which is a heavily patinated core handaxe on basalt. Although paleoenvironmental information

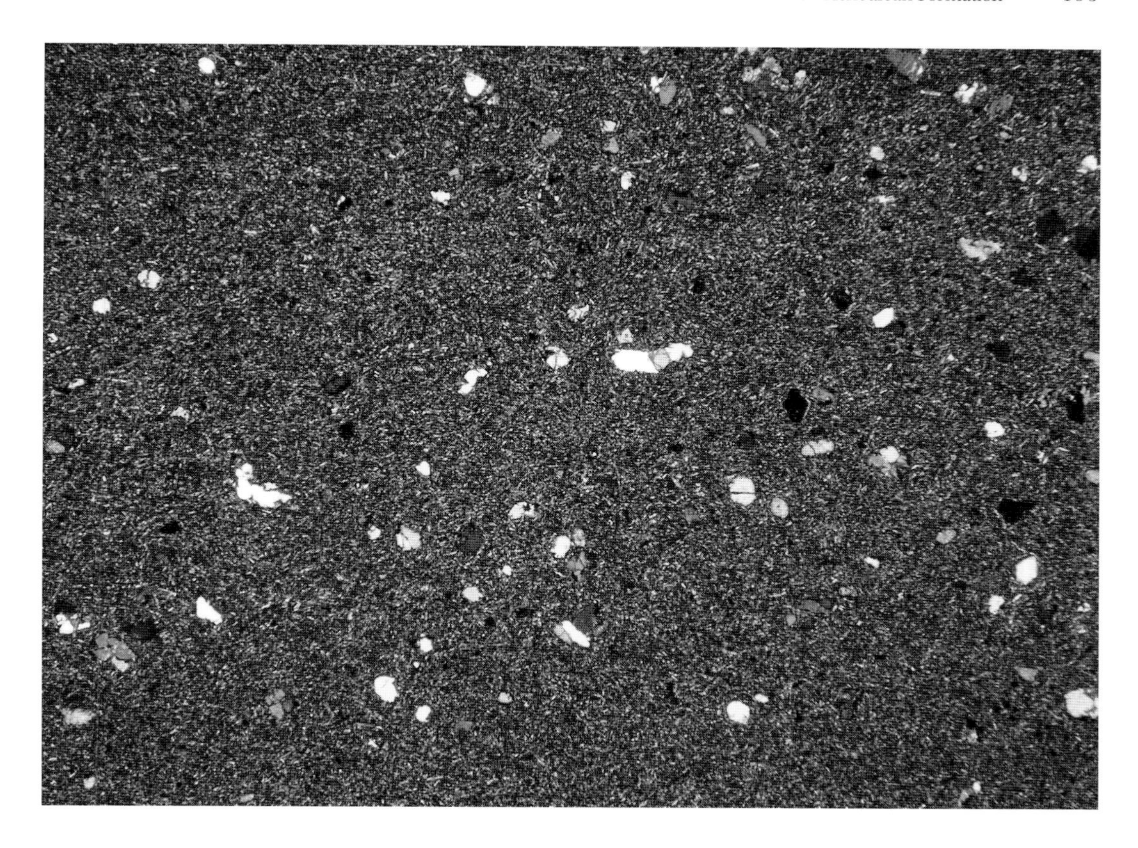

Figure 8.4 Fine-grained olivinic basalt (nephelinite) with trachytic texture (fluid/ microlithic). Basic mineralogy: plagioclase, olivine and opaque (probably iron oxides). Accessories: sphene. Microscope photograph: polarized light, photograph width 10 mm.

(above) suggests a low-energy environment for this site, weathering data suggest a probable heterogeneous origin for the accumulation of the archaeological remains, where reworking produced by fluvial forces might have been relevant in certain cases, or alternatively, prolonged exposure resulted in subaerial weathering that could be easily mistaken for abrasion produced by fluvial transport.

Bifaces

As stated above, a total number of 48 bifaces have been retrieved at ST69, of which 29 are handaxes made on flake blanks (here we include one object whose support has not been securely identified), 13 are handaxes made on nodules and six are cleavers. Of the total biface sample, 55% were made on basalt, 43% on nephelinite and 2% on flint. As can be seen in Tables 8.2-8.3, mean measurement and weight data show that the largest and heaviest artifacts were made on flake supports. Taking into account the scarce number of artifacts made on nodule blanks, no significant differences can be observed between the sub-assemblages. In fact, the relationship between length and width in the entire collection

Table 8.2 Dimensions and weight of bifaces performed on flake support

	minimum	maximum	mean	std. deviation
length	108	260	171.37	34.52
width	65	130	92.91	14.53
thickness	25	57	41.34	7.63
weight	218	1508	759.8	338.96

Table 8.3 Dimensions and weight of bifaces performed on nodule support

	minimum	maximum	mean	std. deviation
length	111	220	146.38	26.91
width	72	105	81.76	9.8
thickness	38	60	46.07	5.88
weight	269	1183	605.38	233.34

tends to be arranged within a fairly homogeneous pattern (Figure 8.3). When measures are broken down by raw material type, basalt artifacts show the longest mean values, while nephelinite bifaces show the greatest breadth and thickness.

Among the 29 flake handaxes, 28% have preserved their striking platform. No cortex has been observed in any of these butts and most of them tend to be uni-faceted (75%). In one case a bi-faceted striking platform has been observed, while in other, the platform has been altered by retouch. In relation to the technological axis of these artifacts, when observed, the knapping direction is in most cases lateral (preferentially right lateral), although proximal directionalities have also been recognized. However, in the majority of handaxes made on flakes (72%), traces of striking platforms have been removed by negative scars due to later reduction or retouch. In fact, at ST-69 most of the handaxes performed on flakes have been subject to some sort of knapping on both surfaces. The mean number of negative scars observed on ventral surfaces is 5 (the maximum being 13). In many cases these knapping processes have been undertaken in order to reduce thickness produced by the striking platforms.

Ventral surfaces of the flake handaxe collection show no signs of bulbs. These surfaces are convex (52%), flat (45%), and concave (3%). Ventral convexities have always been obtained through simple, mostly tending to centripetal, flaking or retouch around the perimeter of the piece. This seems to be the most efficient way to reduce ventral flatness and to improve cutting edges on the perimeter and bilateral convexities. Previous negative scars on dorsal surfaces and dorsal ridges can help reconstruct the way in which cores were managed to produce these large supports. A total of 85.7% of dorsal surfaces shows no cortex at all, with a mean number of eight negative scars per artifact and a mean largest dimension of 42 x 57 mm. When cortical areas are preserved, they tend to be one-quarter of the dorsal face. One case has been observed where the dorsal face is almost completely cortical. A clear centripetal exploitation strategy has been observed in 75.7% of these artifacts. However, a large flake has been obtained in which the position of the dorsal ridges suggest a unipolar reduction model. Most of the handaxes performed on flakes retain no cortex on their base. The majority of these objects (69%) are pieces with a rather sharp base (mostly convex and residually straight), although some thick base areas have been recorded as well.

Most of the natural edges (58.6%) have been retouched to some extent using hard-hammer percussion technique. Fifty-three percent of these artifacts have been retouched in both faces, while the remaining 47% have been retouched only in one face, mainly the ventral one. Taking into account these numbers, the distribution of retouch shows that 48% of the bifaces made on flake supports were subject to some sort of retouch on their ventral surface. When their technical characteristics and position are considered in general, trimmed bifaces (where retouch is discontinuous or partial and affects only limited areas of one of the edges) predominate. Some cases are documented where the two edges of the same face have been subject to discontinuous retouch, while in others, a discontinuous retouch on one face is combined with a more continuous retouch on the other. When a normative, more standardized, continuous retouch does appear (41% of the retouched objects, n=7), its position may be inverse (n=3), alternate (n=2), direct (n=1) or bifacial (n=1). At ST69, hard-hammer retouch is mainly simple (35-55°) and marginal. In one artifact, simple retouch has been performed in order to obtain a notch morphology in the distal area.

A total number of 13 specimens have been included in the group of handaxes elaborated from nodule blanks, although in three cases the support has not been clearly distinguished and another could have been made on a thick tabular piece of nephelinite. A laterally fragmented piece, whose measurements and morphology have been approximately reconstructed, is also included in the general counts. Most of these handaxes were intensively exploited on both surfaces and thus no signs of cortex can be observed. In cases where cortical areas are preserved (n=2), these are limited to one-quarter of the upper face. The mean number of negative scars observed on both surfaces is 8.5 and the mean dimensions of the largest removals are 40.6 x 49.8 mm on the upper face and 40.9 x 51.63 mm on the lower face. These numbers show that as a general rule, detached flakes from these nodules were preferentially wider than longer. In 84% of the cases a centripetal reduction strategy has been identified, while the intersection angle formed by the two flaking surfaces is always simple. As in the case of handaxes made on flake blanks, simple centripetal removals facilitate both perimeter configuration and enhance biconvexities. The basal region of these artifacts tends to be sharply convex (75%), followed by thick (17%) and pointed (8%) shapes. Hard-hammer retouch has been observed in five pieces. Two handaxes show proximal retouch in the base, one of them continuous on both faces, while the other shows continuous retouch on one face and has been residually trimmed on the other. One handaxe shows continuous, alternate retouch in both faces. In the remaining two objects a continuous retouch can be observed on one of their faces. As in the artifacts produced from flake blanks, retouch is mostly simple and marginal, although one piece has abrupt retouch.

At ST69, the mean length of total cutting edges is 260 mm (mean maximum length = 146 mm) in the core handaxes and 319 mm (mean maximum length = 170 mm) in the flake handaxes. These numbers, which in both cases almost double the mean maximum length, stress that the configuration patterns carried out on these artifacts have succeeded in the goal of providing active cutting edges along the entire perimeter of the objects. From a sagittal view, cutting edges tend to be straight in 44% of the cases, and sinuous in the remaining 56%. From a horizontal view, handaxes are symmetrical in 66% of the cases and non-symmetrical in the remaining sum.

To calculate the refinement of the axes, we used Bordes' (1961) index of thickness relative to width (m/e). Figure 8.5 shows the percentage and number of specimens sorted by their refinement index. A total of 69% (n=33) of the total bifaces retrieved are thick (m/e < 2.35), while the remaining 31% (n=15) are thin (m/e > 2.35). No thin handaxes have been found in the group of artifacts made on nodule supports. The m/e index of these objects ranges from 1.56 to 2.11. When the assemblage of bifaces made on large flakes is taken into consideration, 43% of the artifacts fall within the range of thin pieces. Here three cleavers that show some of the highest refinement ratios are included. However, among the thick flake bifaces, a cleaver that has the lowest refinement ratio of the whole assemblage retrieved at this site (1.3) is included. In terms of morphology, these handaxes are in all cases pointed. In a number of cases deep or more marginal removals have been used to refine and enhance the distal point. Following Bordes' typology, 71.42% (n=30) of these artifacts fall in the range of amygdaloid handaxes, 26.19% (n=11) are elongated cordiforms and 2.38% (n=1) subcordiforms. Figure 8.6 shows the percentage of Bordes' types sorted by blank type.

At ST69 six cleavers, in which a transversal and distal natural cutting edge is observed, have been recovered. Striking platforms have only been preserved in two specimens. In these cases, butts are uni-facial and are located on the right proximal side. Bulbs are only observed in two cases and ventral surfaces are convex in all pieces but one, which is flat. Evidence of negative scars on ventral surfaces has been observed in some cleavers. Dorsal faces preserve no evidence of cortical areas and, at least in five artifacts, the distribution of negative scars and dorsal ridges show that these large flakes were obtained through centripetal reduction models. The base is always sharply convex or straight. Lateral continuous retouch has been observed in four cases, two alternate and one inverse. Retouch is mainly simple and abrupt. Natural edges, not always symmetrical, have been obtained through the intersecting planes of a previous centripetal removal and the removed large flake. Mean maximum length of the distal natural cutting edges is 84.2 mm (mean maximum width of these artifacts is 89.6 mm). From a typological point of view all of these specimens can be included in Type II of Tixier's (1956) classification, i.e. cleavers performed on large non-Levallois flakes.

Flakes and cores

Four large flakes have been retrieved along with the biface collection, three on basalt and one on nephelinite. The mean size of these artifacts is 81 x 125.5 x 27 mm and the mean weight is 322.5 g. Compared to the bifaces made on flake supports (whose mean size and weight are 170 x 92 x 41 mm and 747 g, respectively), these objects are smaller and lighter. Regular flakes are always wider than longer. Taking into account that as a general rule, flake handaxes preserve no traces of striking platforms or bulbs and therefore they have been measured using their morphological axis, it is possible that the same flaking pattern observed in simple flakes, which has always been measured using their technological axis, might also apply for the large flakes used as biface supports. At Peninj, production of flakes wider than longer for transformation into a variety of large tool forms has already been suggested (de la Torre 2005:498). The same striking pattern has also been reported for the production of large flake supports at several sites located in the Ethiopian Middle Awash valley (Clark and Schick 2000:201).

Flakes do not preserve cortex on dorsal surfaces or on butts. Therefore, all of them belong to Toth's (1982) Type 6. Ventral surfaces always preserve traces of bulbs and the striking platform may be uni-faceted (n=2), bi-faceted (n=1) or broken (n=1). The mean number of negative scars preserved on dorsal surfaces is five. The arrangement of dorsal ridges and negative scars shows traces of clear radial flaking in at least three cases. Although the collection of detached pieces is very small, a centripetal reduction model can be assumed as the preferred knapping process, while the initial flaking stages (represented by cortical or semi-cortical flakes) are not present in the survey sample. Taking this information into account, production of large and medium-sized flakes might not have been carried out at the site. One flake shows traces of continuous, abrupt, marginal and direct retouch on its distal edge. Typologically this piece can be considered a rather coarse transverse scraper.

Two basalt cores were identified in the archaeological sample. One of them is a laterally fractured cobble, measuring 164 x 117 x 53 mm and weighing 1,355 g, which has been centripetally exploited on one surface. Four negative scars were observed. Measurements of the largest negative scar are 66 x 108 mm. The other piece (153 x 97 x 52 mm and 782 g) has been centripetally exploited on both surfaces and no traces of face hierarchization have been observed. Thus, striking platforms have not been prepared *sensu stricto*, although no cortical butts could have been obtained due to the alternating exploitation of both surfaces.

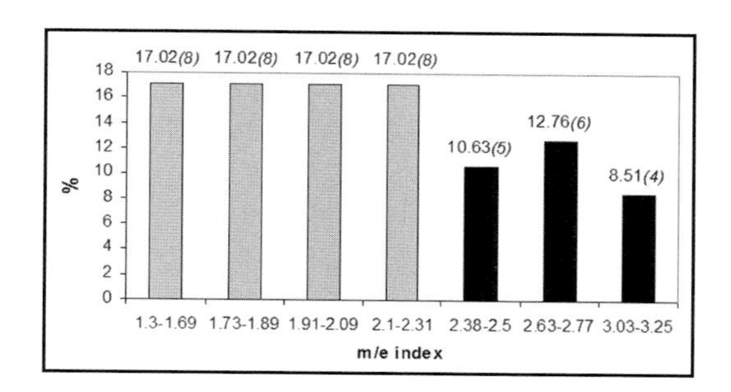

Figure 8.5 Percentage distribution of bifaces by their m/e (thickness relative to width), also known as the refinement index (Bordes 1961). Total number of specimens in italics.

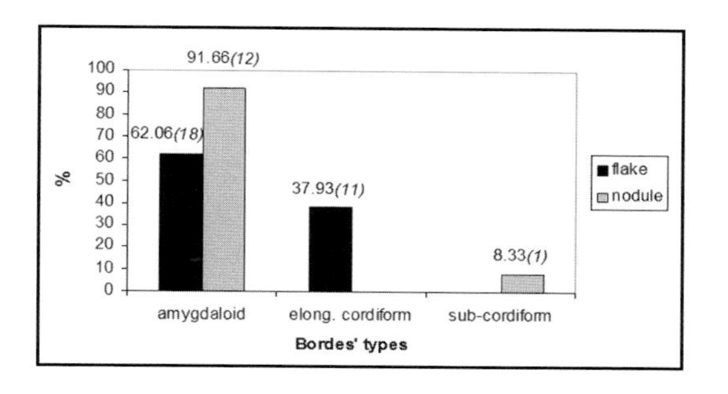

Figure 8.6 Typological classification of bifaces from ST69 (Bordes 1961) and percentage distribution by support type.

Conclusions and discussion

Survey carried out at ST69 has produced a small lithic collection made on local and semi-local volcanic rocks. Most of these artifacts are bifaces (mainly handaxes, although a small sample of cleavers is also represented), the hallmark of the Acheulean industrial complex. Available dating of the Upper Tuff member of Moinik Formation suggests an estimated age of 1.2 Ma for the lithic assemblage. The main technological and typological characteristics observed in the biface production at this site are the following:

1) Most of the heavy-duty artifacts have been made on thick, large and heavy flakes. Although striking platforms have not been preserved in most cases, it is probable (taking into account the characteristics observed in the small number of flakes that were not subject to bifacial configuration) that flake blanks destined for the production of

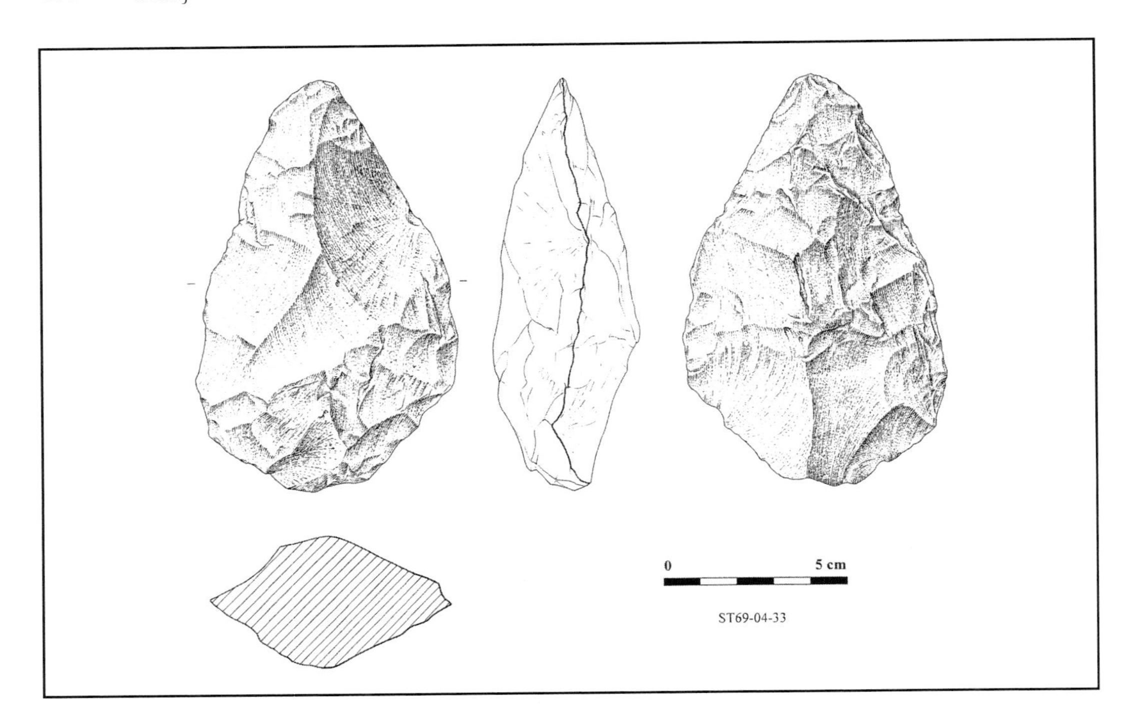

Figure 8.7 Amygdaloid handaxe on basalt nodule support. Drawing by L. Pascual, Aratikos Arqueólogos.

bifaces were obtained from side-struck flaking processes that allowed obtainment of pieces that were wider than they were long.

2) As can be inferred from the flake dorsal surfaces and from directionality of removals in bifaces produced from nodules and core types, centripetal reduction constitutes the major strategy carried out at this site and it has been observed in 79.6% of the sample analyzed. Radial flaking was thus the main goal of Acheulean knappers and it seems that they were efficiently obtaining large supports from centripetal exploitation. The majority of flakes preserve no cortex on their dorsal surfaces and the initial stages of the reduction process are absent from the sample. Although post-depositional forces might have played a role at this site, it seems apparent that knapping processes did not take place on site and that ST69 is a heterogeneous accumulation of large artifacts already configured or exploited elsewhere. The lack of large cores from which these large products should have been detached and the smaller by-products of the complete knapping sequences that would have been performed by hominids confirm this perspective.

3) A number of natural edges were secondarily transformed by retouch. Fifty-four percent of the sample shows traces of retouch. Trimmed objects predominate, although in some cases a more continuous retouch, mainly simple, has been observed.

4) Detached flake knapping and retouch has helped to remove striking platform thickness or to improve tip shape. In most cases simple flaking and retouch on flake and nodule supports have helped to increase cross-section

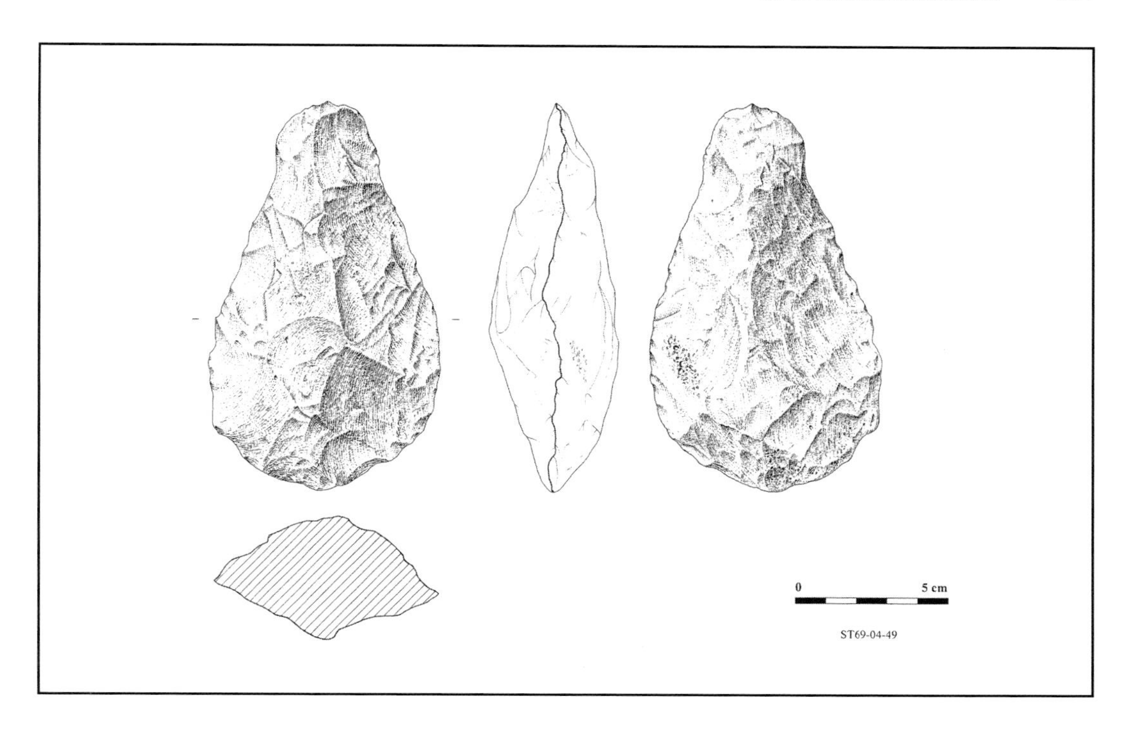

Figure 8.8 Amygdaloid handaxe on nephelinite flake support. Drawing by L. Pascual, Aratikos Arqueólogos.

convexities. Bifaces are rather symmetrical from the horizontal view and non-sinuous edges are abundant. Final shapes are quite standardized in typical pointed thick forms (amygdaloids followed by elongated cordiforms). The smaller pieces that must have been subjected to a more intensive reduction sequence (McPherron 2003), show the same morphological pattern.

The Acheulean industrial complex appears in the African archaeological record at about 1.7-1.5 Ma. Among the oldest sites reported to date between 1.7 and 1 Ma, it is worth mentioning Konso (Asfaw et al. 1992), Gona (Quade et al. 2004), PEEN1 and PEES 2 at Peninj (de la Torre 2005), EF-HR in Bed II at Olduvai Gorge (Leakey 1971), FxJj 63 at Koobi Fora (Isaac and Harris 1997), several localities at West Turkana (Roche et al. 2003; Roche and Kibunjia 1994), Sterkfontein (Kuman 1988), several sites at Gadeb (Clark and Kurashina 1979) and the Daka Member at Bouri Formation (Schick and Clark 2000). Other archaeological sites, traditionally ascribed to the Developed Oldowan but probably related to the Acheulean complex, can be found at Olduvai Bed II (Leakey 1971) or Melka Kunture (Chavaillon et al. 1979).

Leaving aside the discussion on the functional meaning of the morphological differences between Developed Oldowan and Acheulean *sensu stricto*, extensively treated by other researchers (Leakey 1976; Stiles 1979; Jones 1994), early Acheulean assemblages share the following traits (Clark 1994; Schick 1998): raw materials are preferentially local in origin and selective use of rocks according to their properties is not seen; hominids were able to obtain large

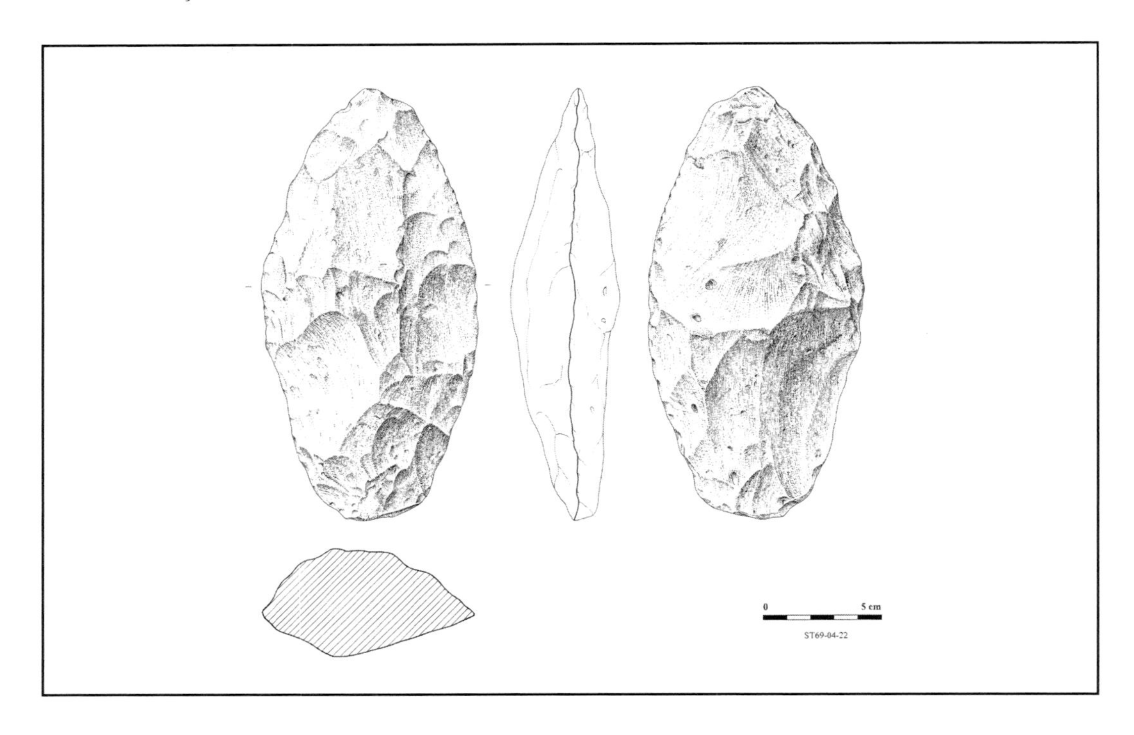

Figure 8.9 Amygdaloid handaxe on nephelinite flake support. Drawing by L. Pascual, Aratikos Arqueólogos.

and thick flakes (those in Konso, for example, reach 268 mm in length, similar to those from PEES2 at Peninj), although these supports have not been carefully produced and selected in order to improve final shape, as we can see in later periods (Texier and Roche 1995); biface general morphology shows no signs of a high degree of configuration and standardization: flake scars are few and usually not invasive or are limited to the edges, secondary hard-hammer retouch is crude and discontinuous, thick pieces predominate, edges tend to be sinuous, non-symmetrical artifacts are abundant and cortical areas in the base or the dorsal face are commonly observed.

At Peninj, Acheulean sites have been found in the North Escarpment and South Escarpment areas (Isaac 1965), dated to 1.3–1.2 Ma (see Chapters 9–10). The analysis of old and new collections retrieved at the North Escarpment (NE1) have been recently carried out (de la Torre 2005). At NE1, large retouched tools are made on large and heavy flakes, wider than they are long. It is evident that hominids here were able to obtain large flake supports, one of the basic technological traits related to the dawn of the Acheulean techno-complex. However, knapping sequences leading to the configuration of bifacial tools are rare. Once the large flake was detached, abrupt and partial retouch was generally carried out to reduce striking platform thickness. Retouch is in most cases marginal and is performed on one face, while volumetric traits related to bifacial reduction strategies are absent. Therefore, large artifacts can be classified in most cases as a variety of knife and heavy-duty scraper types. True handaxes (3.6%) and cleavers (14.4%) are scant.

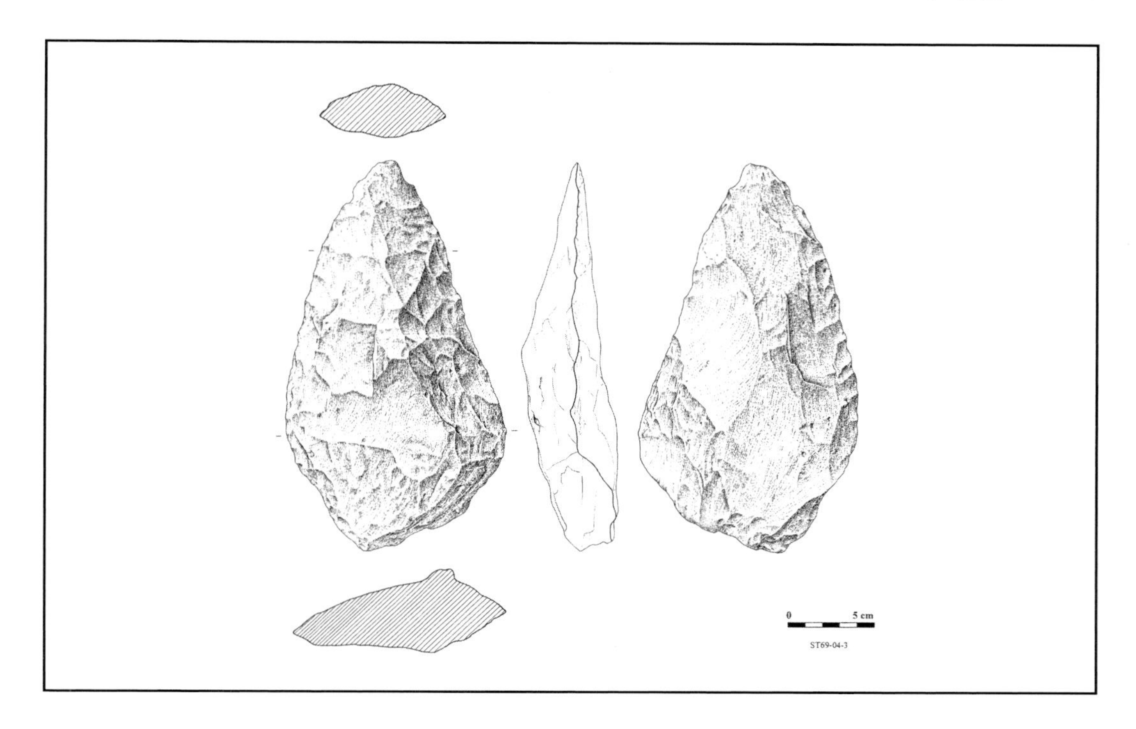

Figure 8.10 Elongated cordiform handaxe on nephelinite flake support. Drawing by L. Pascual, Aratikos Arqueólogos.

The technological characteristics observed in the lithic collection at ST69 are in agreement with some of the general patterns for the African Early Acheulean noted above. However, it is worth considering some traits which suggest that we are dealing with a sample where the concept of bifacial reduction and configuration has already been established. Active cutting edges in most of the perimeter of the blanks have been obtained mainly through invasive bifacial centripetal knapping and retouch from both core and large flake supports. Therefore, it seems evident that hominids were able to manipulate volumes rather than merely slightly transform peripheral areas of the edges. Although when found, stone hammer retouch is preferentially limited to marginal areas of the natural edges, an incipient and quite simple volumetric transformation of both faces that enhances cross-section biconvexities has been observed. Regardless of dimension, final forms are rather standardized. Pointed forms have been the main goal of the knapping processes and negative scars and/or retouch were frequently performed in order to shape distal tips. These procedures allowed hominids to obtain rather symmetrical bifaces with relatively straight edges along most of the perimeter of the blank, a feature that to us is a basic defining feature of large tool configuration during the Acheulean period.

While the appearance of the Acheulean concept can be mainly linked to the production of large flakes marginally shaped on the edges, the ST69 collection indicates that by 1.2 Ma (estimated) hominids had acquired the basic knowledge to repeatedly produce bifacial tools with a desired shape and therefore, had succeeded in controlling the basic principles of symmetry (Wynn 1979). However, it would not be until later stages (with selection of suitable raw materials, the use of predetermination techniques for the production of thinner supports and the use of soft hammer percussion) that bifacial configuration techniques would reach a higher control and refinement.

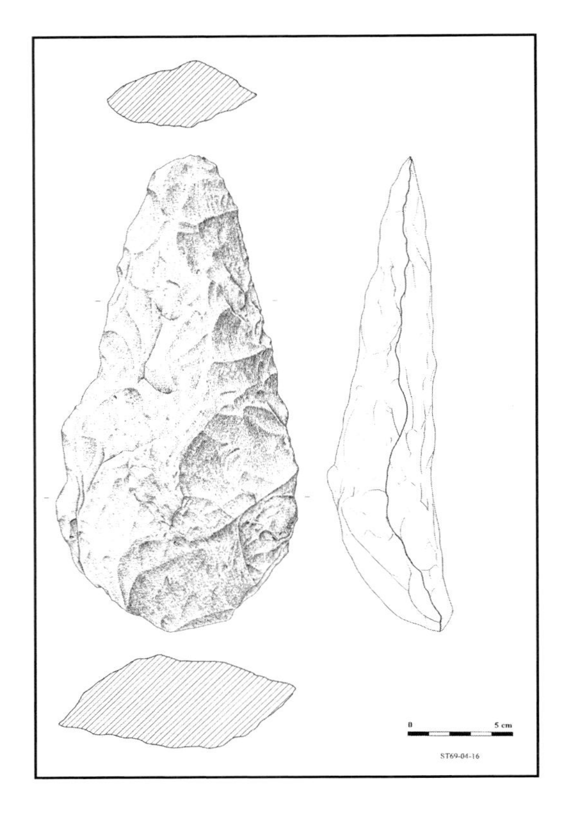

Figure 8.11 Elongated cordiform handaxe on basalt flake support. Drawing by L. Pascual, Aratikos Arqueólogos.

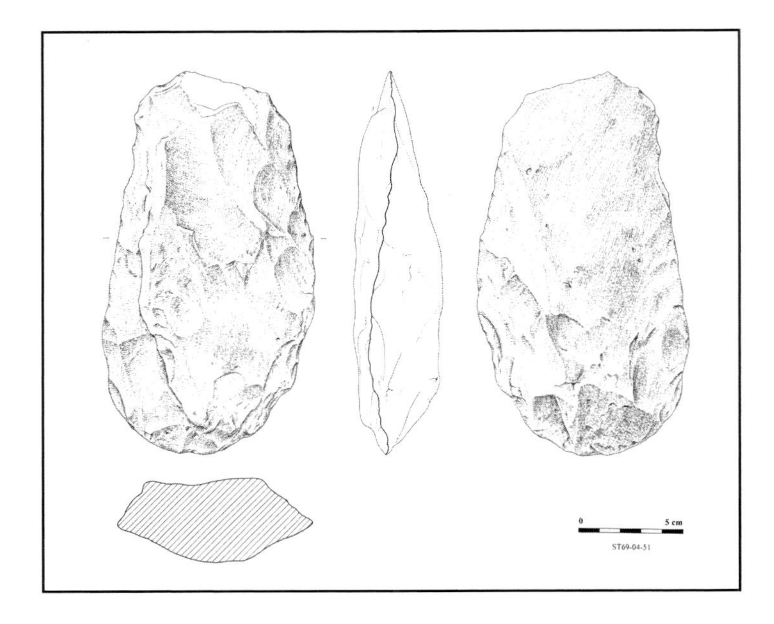

Figure 8.12 Nephelinite cleaver. Drawing by L. Pascual, Aratikos Arqueólogos.

9

The Acheulian Sites from the South Escarpment

Manuel Domínguez-Rodrigo, Jordi Serrallonga, Luis Luque,
Fernando Diez-Martín, Luis Alcalá, and Pastory Bushozi

Introduction

The South Escarpment is located close to Type Section (Maritanane; Figure 9.1). Isaac (unpublished notes) interpreted it as a deltaic front near a lacustrine shoreline. It is impossible to ascertain the exact paleoecological location of the lacustrine environments represented in the South Escarpment, given its lithological homogeneity. However, in areas where the Main Tuff appears devoid of overlying sediment, the top surface of the tuff shows traces of bioturbation and pedological maturation, implying the formation of soils. This means that these soils were created in the absence of frequent periodic lake transgressions, which is suggestive of an alluvial-riverine fan setting, more distant from the lake shore than was previously hypothesized by Isaac (see below).

Only four sites were discovered in the South Escarpment (Figure 9.2). PEES4 (PE for Peninj, ES for "Escarpe Sur") is the oldest one. It appeared under the Main Tuff in one of the few patches where the lower Humbu Formation appears exposed on the escarpment. It is a paleontological site with several bones, mainly of *Antidorcas* and *Gazella*. PEES3 is a small cluster of bones and stone tools found just 50 m away from PEES2. The latter is the largest archaeological site in the South Escarpment, with an impressive accumulation of heavy-duty tools in which handaxes and large knives dominate.

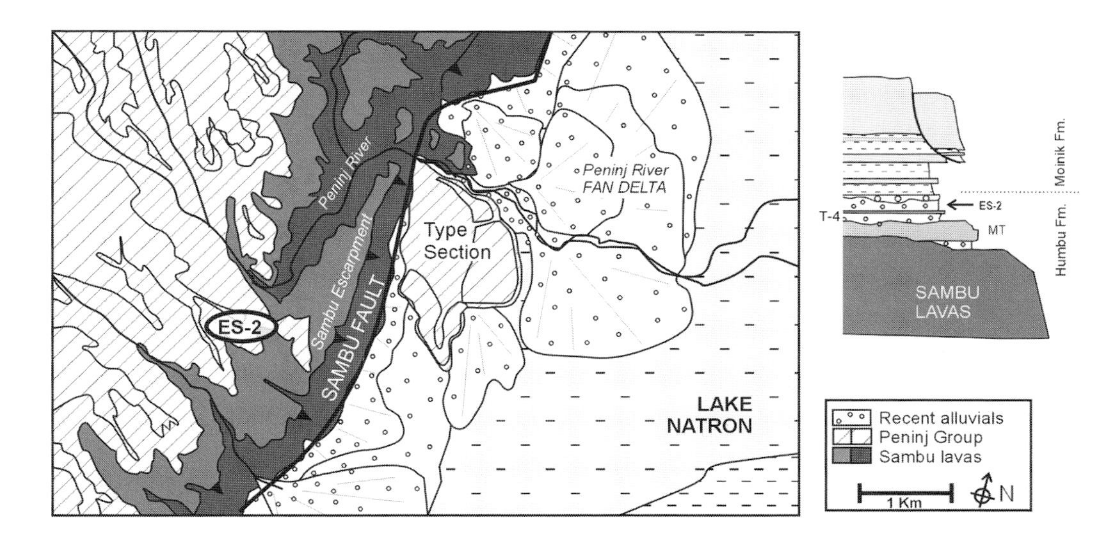

Figure 9.1 Geomorphological map of Peninj and location of the South Escarpment (ES) and the stratigraphic position of the main archaeological site (PEES2).

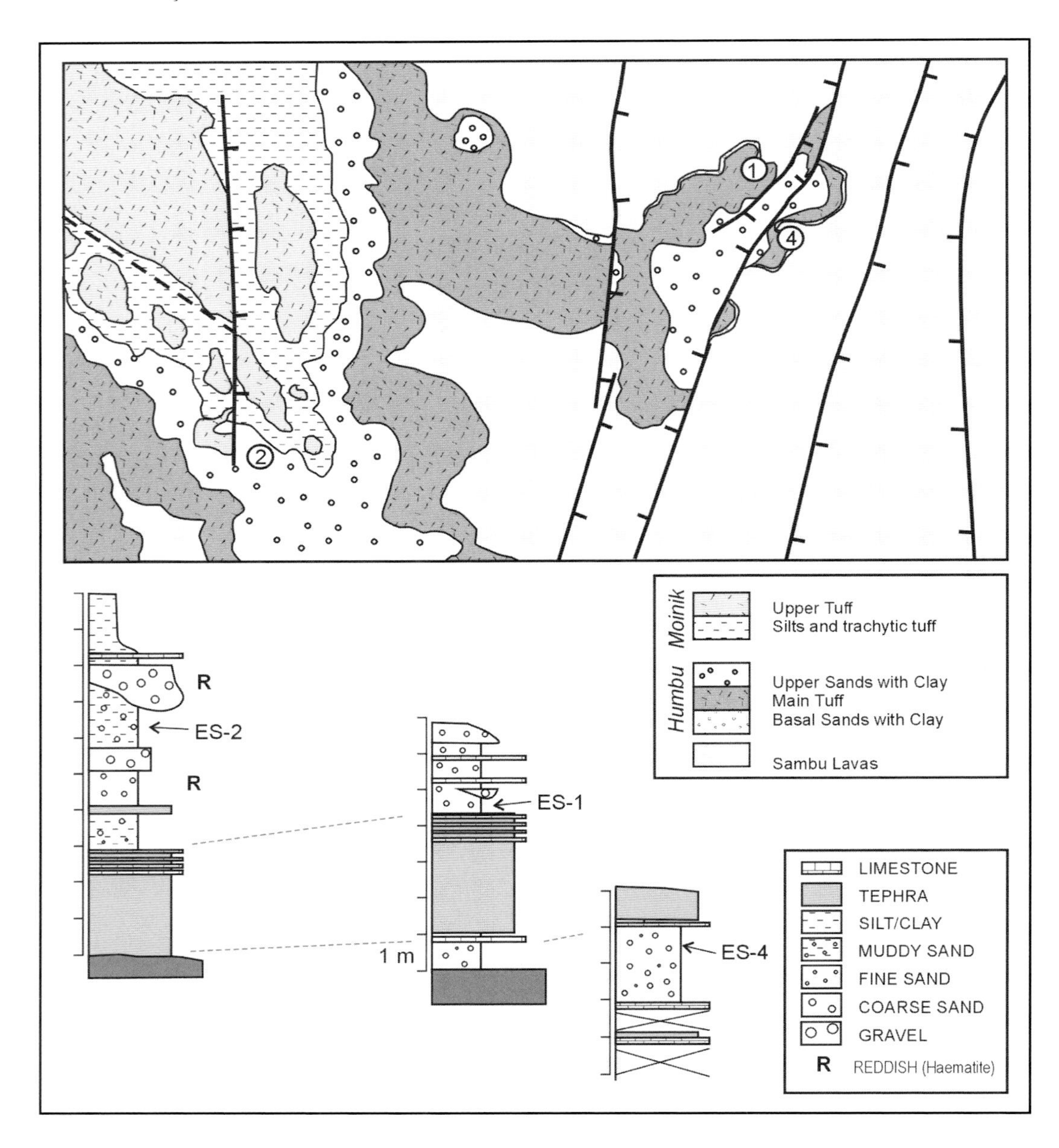

Figure 9.2 Location of the main sites discovered in the South Escarpment and their stratigraphic position.

Whereas PEES2 and PEES3 are set well apart from the escarpment edge, towards the western hills representing the Moinik Formation and modern alluvial sedimentation, PEES1 is situated about 300 m from PEES2 in the direction of the escarpment edge. Below follows a description of the geology, excavations and archaeological contents of PEES1 and PEES2, which were the only two sites that were excavated in extension (PEES2) and test-trenched (PEES1) in the South Escarpment.

The geology of the South Escarpment

The South Escarpment is located on the uplifted fault block of the Natron half-graben, 2.5 km southwest of the Type Section area (Figure 9.1). The Peninj Group stratigraphic record begins there with the upper part of the Basal Sand Member and the Main Tuff member. These deposits overlie the Sambu Lavas above the escarpment. Several faults run parallel to the main direction of the Rift Valley and dissect the sedimentary deposits. These faults with minor displacements gave rise to some steps and narrow horst/graben systems. Other gravitational faults affected the Peninj Group deposits, especially those belonging to the Moinik Formation. However, faulting did not strongly disturb the Peninj Group deposits and a wide ribbon of continuous outcrops can be followed along the rim of the escarpment. Outcropping sediments consist of sands, sandy clays and clays with interbedded tephra layers which span up to 30 m in thickness. The erosive retreat of the front of the Humbu and Moinik deposits over the more homogeneous lava rocks gave rise to this wide sedimentary exposure both in tectonic and fluvial slopes. Only some units of the Peninj Group member are well-preserved in the area and accordingly, few archaeological sites have been recorded in it (Figure 9.2). The PEES2 site is the most important of them, as will be discussed below.

The lowermost unit of the Humbu Formation, the Basal Sands with Clays (BSC), represents an infilling of the former Lake Natron basin up to the rim of the previous Sambu escarpment, before the lacustrine expansion. At the same time, erosion and valley incision in the escarpment gave rise to an irregular landscape that led to a greater overlap with the BSC sediments towards the west. The thickness of BSC over the Sambu escarpment rarely exceeds 1.5 m. To the west, the Main Tuff (MT) member directly overlies the Sambu lavas. The MT member shows a widely developed limestone layer rich in gastropods (Gabbia) at the base. Orange tephra layers including fine limestone levels also occur in the middle and are usually around 5 m thick. The top surface of the volcanic sediments is often weathered and is sometimes covered by strongly cemented coarse sandstone. In many of the areas around the South Escarpment, this paleosurface indicates a sedimentary hiatus that spans a considerable amount of time during the formation of the lower Upper Sand with Clay (USC) member of the Humbu Formation. Coarse and muddy sandstones that overlie this surface correlate with the middle and upper units of the USC member in the Type Section. The maximum thickness of USC in the area is 8–10 m. The most useful unit for stratigraphic correlation is the tephra layer T-4. This ash-fall deposit is the only one recorded in primary position in the USC in the area and it is easily linked to T-4 in the Type Section. The Moinik Formation silts and volcanic ashes were deposited abruptly and conformably over the Humbu Formation.

The most stratigraphically complete outcrop is located at the PEES2 site, where the entire ES sequence can be described (Figure 9.3): Over the Main Tuff member, thin layers of well-cemented coarse sands and gravels, usually containing tephra clasts, were deposited. In some places, silts and clays with pedogenic limestones can also be found. There are also nearly 2 m in thickness of greenish and poorly-sorted sandy clays, with muddy sandstone lenses, showing zeolites, calcite nodules and scattered small pyrite crystals. During this facies, a poorly-drained deltaic floodplain, crossed by several distributary channels, stretched over the landscape. The T-4 tephra layer, which is 0.15 m thick, locally overlies sandy clays. Three to four upward-grading beds of coarse to fine sandstones overlie the volcanic sediments. These are clayish at the top, showing crossed lamination, and span 2 m in thickness. These typical channel-fill deposits usually include gravel lenses. Locally, sands and gravels are reddened by hematite and show two thin iron oxide crusts at the top of the layers. Rootcasts are more frequent at the top of the sequence. Lateral migration of channels or proximal deltaic progradation with water table oscillations is suggested for this depositional environment.

The following facies (1–3 m thick) consists of greenish to brownish-gray massive muddy sandstones, rich in small pyrite growths, with interbedded coarse sandstone lenses, and grading thin silt beds where the PEES 2 site is found. The floodplain was probably highly vegetated near the distributary channels, as is interpreted from this facies. Coarse white to reddish channel-fill deposits, which are deeply erosive and show cross-lamination, become thinner laterally and only reach a maximum thickness of 2.3 m. These deposits cover the muddy sandstones. Flow direction was reported by Isaac (unpublished notes) as being to the northwest. At the top of the Humbu Formation, a massive, poorly-sorted, green to reddish sandstone, 1 m in thickness, is overlaid by a thin bed of well-cemented coarse sand. The lower section of the Moinik Formation was deposited conformably over these coarse sands, consisting of homogeneous laminated silts and dolomite/ calcite layers of lacustrine origin, plus erionitic tephra layers. Paleomagnetic data obtained by Thouveny and Taieb (1986, 1987) in the Upper Humbu Formation deposits suggest reversed polarity, younger than the Olduvai event.

Isaac (unpublished notes) carried out a preliminary environmental intepretation of the South Escarpment, based on the MSH (PEES2) site that generally coincides with our own observations, although some significant differences exist. Following Isaac's observations and comparing the South Escarpment with the sedimentary record in the Type Section, it can be argued that after the westward migration of the deltaic facies during the deposition of the Main Tuff, under a lacustrine environment, a period of eastward progradation began. Isaac suggests that during the deposition of the USC sediments, the South Escarpment exhibited various forms of lacustrine sub-environments and different lithofacies have been described by him in this regard (unpublished notes) and by Kaufulu (1983).

These lacustrine environments turned more fluvial in the upper part of the sequence. In this area, poorly-drained distributary channels, as evidenced by pyrite precipitates, covered the landscape. The PEES2 site was located in muddy sandstones with abundant evidence of rootcasts and massive muddy sandstone facies. PEES 2 is found a few meters to the north of the distributary channel which leads into the delta plain. Authigenic minerals like pyrite or zeolites suggest swampy, alkaline and saline conditions (Hay 1975). Abundant rootcasts were formed from plants growing on exposed interfluvial sand bars. Channel incision and an abundance of hematite (reflecting water table oscillations), localized calcite nodules, and a probable abundance of

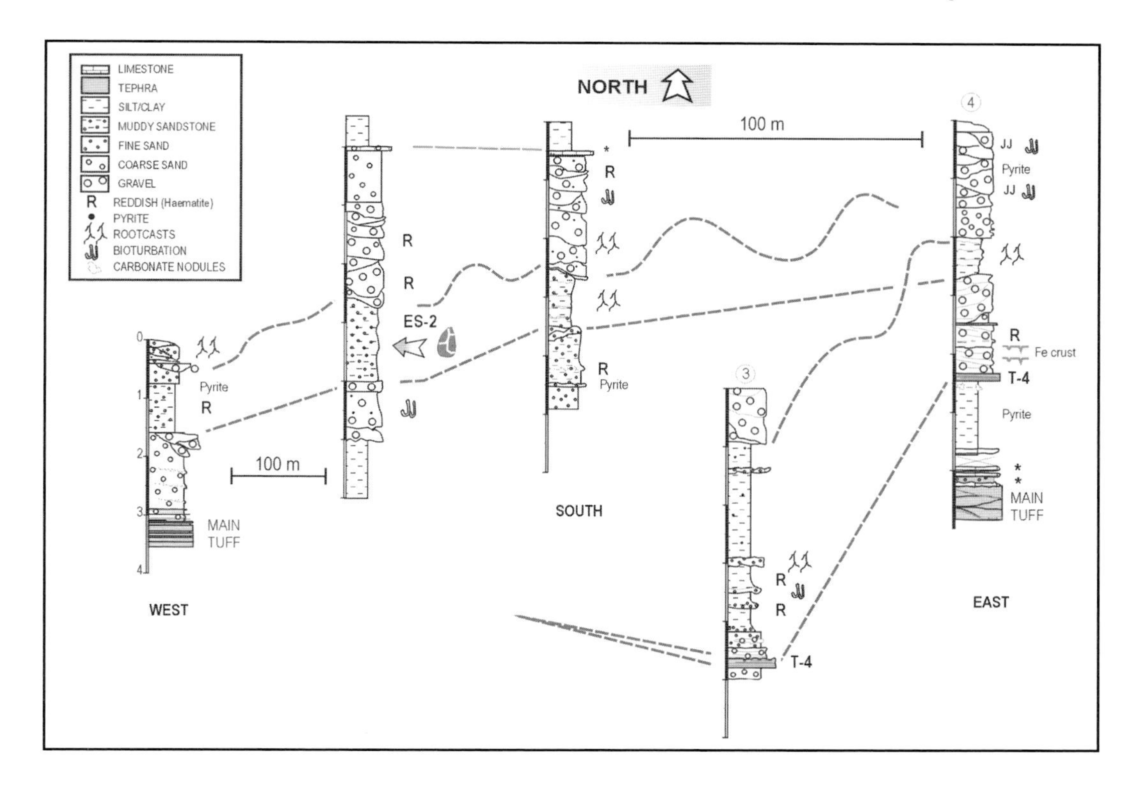

Figure 9.3 Stratigraphic column of the Upper Sands with Clays of the Humbu Formation in the South Escarpment.

vegetation (masking sedimentary structures) indicate that the environment was less lacustrine than Isaac previously suggested, and a deltaic flood plain crossed by distributary channels is the most probable environment during PEES2 site formation (Figures 9.3–9.4). PEES2 lies in the deltaic floodplain near a channel (Figure 9.4). Water alkalinity, weathering exposure, and pedogenic processes explain the chemical alteration of lithic artifact surfaces (in the form of an aureole in the section of the artifact, and evidenced by pyrite appearing around lithic tools).

The ES facies are comparatively more proximal and less lacustrine than contemporary deposits in the Type Section. The flat surface created after sub-basin infilling allowed the development of a large plain where channels migrated along the deltaic floodplain to reach the lake. There may have been temporary or seasonal swamp conditions, due to minor lake-level fluctuations in this wide flat relief. Sedimentary evolution in ES indicates at least two deltaic prograding sequences after the deposition of T-4, and coarse-grained facies generally increase over time.

Two meters of coarse-grained sandstone show a channel-fill deposit, 50 m wide, flowing towards the northwest (Isaac unpublished notes). This abrupt change, despite being highly localized, can be correlated with larger coarse sand channel deposits found at the top of the Humbu Formation in the Type Section. Later, the Moinik Formation overlies the Humbu Formation across

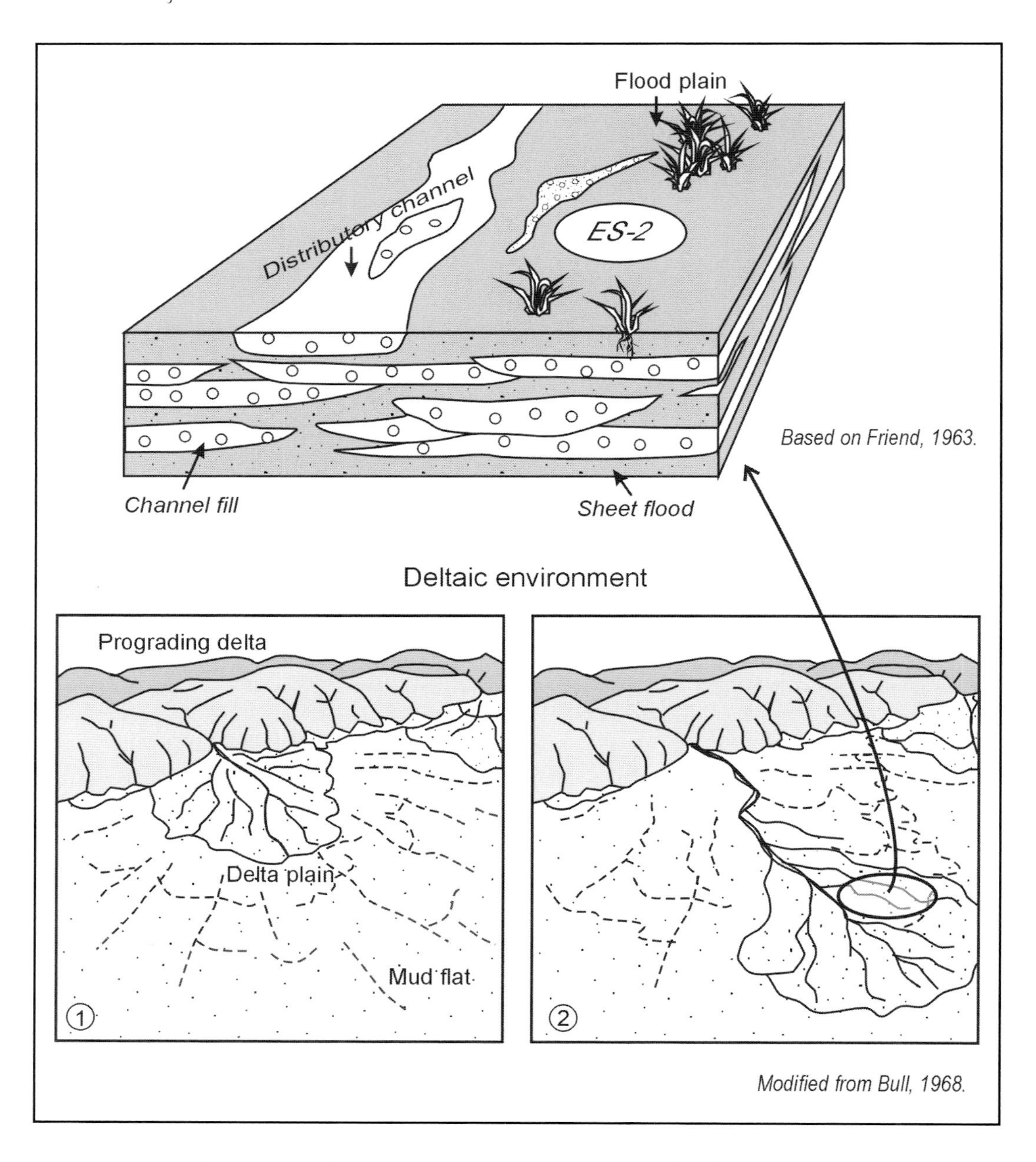

Figure 9.4 Reconstruction of the delta spreading into lacustrine mudflats where the PEES2 site was located, and detailed location of the site with respect to the nearest channel.

Figure 9.5 Exposure of the eroded archaeological level of PEES1 in the foreground, where the Main Tuff surface is visible.

the entire region. Previously, this was attributed to a change in water supply resulting from climate change. However, now it is interpreted as a consequence of minor Moinik tectonic ramp displacement (see Chapter 2).

The archaeology of PEES 1

This site was discovered in 1995 during the first field season. It is located directly on the first well-cemented coarse sands and gravels overlying the Main Tuff member (Figure 9.5). The stratum in which the site is found consists of 2.35 m of coarse to medium, grey to brownish sands with interbedded gravel lenses and well-cemented coarse sands. A calcite-crystal aggregate layer is also present. The upper part of the section is rich in rootcasts. Few *in situ* remains have been recorded across the large surface of the site due to recent erosion. Most stone tools and bones were found loose on the ground overlying the Main Tuff. Some of them were half-buried in loose, eroded sand. Most artifacts appeared extremely fresh, as is suggested by a lack of weathering (Figure 9.6); lava and basalt tend to wear after being exposed for a very short amount of time.

Most of the site surface was devoid of sedimentation, with the exception of the westernmost area, which was overlaid by the Upper Sands with Clays of the Humbu Formation. Most of the materials were found loose on the ground, as a result of deflation of the overlying sediments. Some pieces were found encrusted on the tuff surface and sealed by carbonate; these were extremely hard to detach from the soil. This suggests that some of these pieces were *in situ*. This was confirmed by placing two test trenches in the western area, one of which yielded one bone lying on the tuff surface at the bottom of the trench.

A grid was set up covering an area of 10 x 20 m and pieces were plotted spatially (Figure 9.7). Most of the spatial distributions may reflect erosion rather than the original position of the pieces. However, it is interesting to notice that whereas bones are scattered randomly over the grid, stone tools seemed to be much more concentrated. This could indicate a lack of association

Figure 9.6 Handaxe discovered at PEES1 made from a cobble.

between both types of materials or separate depositional histories.

The faunal remains belonged to four different individuals, represented by diverse anatomical parts (Tables 9.1-9.2). Most of the fragments were very small and showed intensive diagenetic fragmentation. Bone surfaces were generally good but not a single cut mark, percussion mark, percussion notch or any other hominid-imparted modification was observed. By contrast, several bones showed conspicuous and inconspicuous tooth marks. From a taphonomic point of view, no functional link can be established between the bones and the stone tools deposited at the site.

The stone tool assemblage is composed of 21 pieces (Table 9.3): 19 of basalt and two (flakes) of quartz. Four pieces are green phonolite. The nearest source of this raw material is a few kilometers away, in the Shirere Hills. A handaxe (Figure 9.6) and a centripetally-flaked core suggest that the kit can be defined as Acheulian. Two polyhedrons, one of them very elaborated (Figure 9.8), were also found. Most of the remaining pieces are either flakes or flake fragments. Only three flakes retained some cortex on

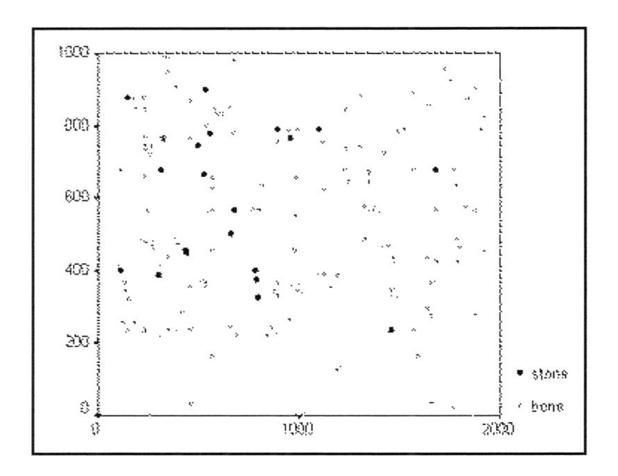

Figure 9.7 Spatial distribution of stone tools and bones at PEES1.

Figure 9.8 Polyhedral core found at PEES1.

Table 9.1 Taxa representation at PEES1

	MNI
Antidorcas recki	1
Connochaetes taurinus	1
Equus sp.	1
Ceratotherium cf. *simum*	1
Total	4

Table 9.2 Minimum number of specimens identified for each skeletal part at PEES1

	size 1-2	size 3	size 4-6
Skull	1	2	0
Vertebrae	0	0	0
Ribs	13	11	0
Pelvis	0	0	0
Scapulae	0	0	0
Humerus	1	3	0
Radius-ulna	3	6	0
Metacarpal	2	4	2
Femur	0	1	0
Tibia	1	4	0
Metatarsal	2	3	1
Carpal/tarsal	2	4	0
Phalanges	1	1	0
Indet.	19	34	0
Total	45	73	3

Table 9.3 Number of each type of artifact represented at PEES1

Artifact Type	Number
Flaked pieces	
core	1
chopper	1
polyhedron	2
handaxe	1
Detached pieces	
whole flakes	8
flake fragment	8
Total	**21**

their dorsal surface and one was almost completely cortical. Given the small sample and the intense erosion of the site, it is impossible to ascertain if the missing steps in the operational chain are due to hominid behavior or to postdepositional processes. This mini-site also might have been the result of a very brief occupational episode, whereas at larger sites the occupation might have been more prolonged.

This site is significant because it suggests that an Acheulian site could be contemporary with the Oldowan ST site complex in Maritanane. This was compared to a similar situation in Olduvai Bed II, where Oldowan sites are near the lake and Acheulian sites are away from the lake and in riverine contexts; this inspired the "Ecological Hypothesis of the Acheulian," which suggested that at Olduvai and Peninj, different site functionality could be attributed to differing ecological contexts (Domínguez-Rodrigo et al. 2005). However it is important to note that Acheulian sites may occasionally occur near lakes, bearing the same properties (e.g., a lack of faunal remains) as Acheulian sites occurring in fluvial contexts (see Chapter 8).

The archaeology of PEES2

This special site presented preservation conditions that contrasted with many classical Acheulian sites found elsewhere. Situated on the edge of a fluvial channel and with clear indications of a moderate to low-energy environment, it presented an opportunity to further understand some of the important accumulations of handaxes found in this type of site. The PEES2 Acheulian site was initially called MHS by Isaac (1965, 1967) and he later adopted the Sonjo name of Bayasi (Mturi 1991). It is situated on the South Escarpment to the southwest of the deltaic exposures of the Type Section (Maritanane). The Peninj Group overlies the basaltic Sambu basement where the site occurs, creating a deposit which is much thinner than at Maritanane. This implies that a small (20-50 m) escarpment and a proto-Peninj valley existed during Humbu Formation times (see above and Chapter 2). The PEES2 site is situated along an outcrop in the uppermost Humbu Formation beds. Isaac's initial excavation (according to his unpublished notes) located the archaeological horizon within a clayish sandstone close to the top of the Humbu Formation, in a narrow erosional gap that now serves as a livestock track. It was thought that the cluster of artifacts was discarded on a deltaic front in and among the distributary channels. The large handaxes and picks were interpreted as having been made elsewhere and carried in. The materials may have been rearranged but they were not transported far.

This Acheulian site was initially excavated by Isaac in 1964. Isaac (1965, 1967) notes that his team excavated a 15 m^2 (3 m x 5 m) step trench. This suggests that the topography of the outcrop containing the sequence of the Humbu and Moinik Formations must have been targeted. A total of 39 artifacts were retrieved *in situ* and 60 more were found on surface (Isaac 1965, 1967). Unpublished notes from the 1981 field season indicate that Isaac's team resumed excavations in this area, establishing a new 9 m^2 (3 m x 3 m) trench, laid at the west corner of the 1964 trench . This excavation located materials in a thick deposit: a total of 13 stone tools were found *in situ* and 97 were retrieved from the surface. Isaac's notes show that there might have been up to three levels, although materials were vertically dispersed over some 1.2 m of consolidated sand with clay. Isaac's interpretation of the site's formation was that the scatter of artifacts throughout the deposit suggested that they were introduced by hominids over a long interval of time.

Before beginning our work at PEES2, we tried to find Isaac's 1964 and 1981 test trenches from, but erosion over more than two decades had deleted any trace of their original location. Figure 9.9 shows the possible location of these trenches. A 1 m^2 gap inside the front line of the outcrop suggested an artificial origin (Figure 9.9A), although it could be a natural gully. A variation of the 1981 trench shape is shown in Figure 9.9B. However in Isaac's field notes, the location of the 1981 trench seems to be closer to the aforementioned livestock path. Alternatively, his trenches might have been further from the modern outcrop wall than ours were (Figure 9.9C). This seems most likely, since when we found the outcrop it was eroding and showed some patches of sediments between the livestock trail and the level we excavated (Figure 9.10). Therefore, it is very likely that Figure 9.9C represents the closest approximation to the original location of Isaac's trenches with respect to ours. Here we focus on our own excavations, since the 1981 materials were not located in Tanzania and could not be analyzed and included in our study.

In 1996, 2000, and 2002 our team excavated this site, opening a 12 m^2 grid along the erosive wall of the outcrop where the site was located, and a 4 m^2 grid into the wall of the outcrop, following a gap created by a small gully that cut across the wall. The sands and clays were extremely consolidated and the use of picks was inefficient, given the hardness of the rock. Excavation was therefore slow and took three field seasons to complete. At the Moinik/Humbu Formation transition, consolidated sediment was dug with the aid of electric hammers and careful attention was paid so that artifacts could be spotted. Sediment was sieved. Artifacts were plotted in 1996 with a theodolite and in 2000 and 2002 with a laser theodolite.

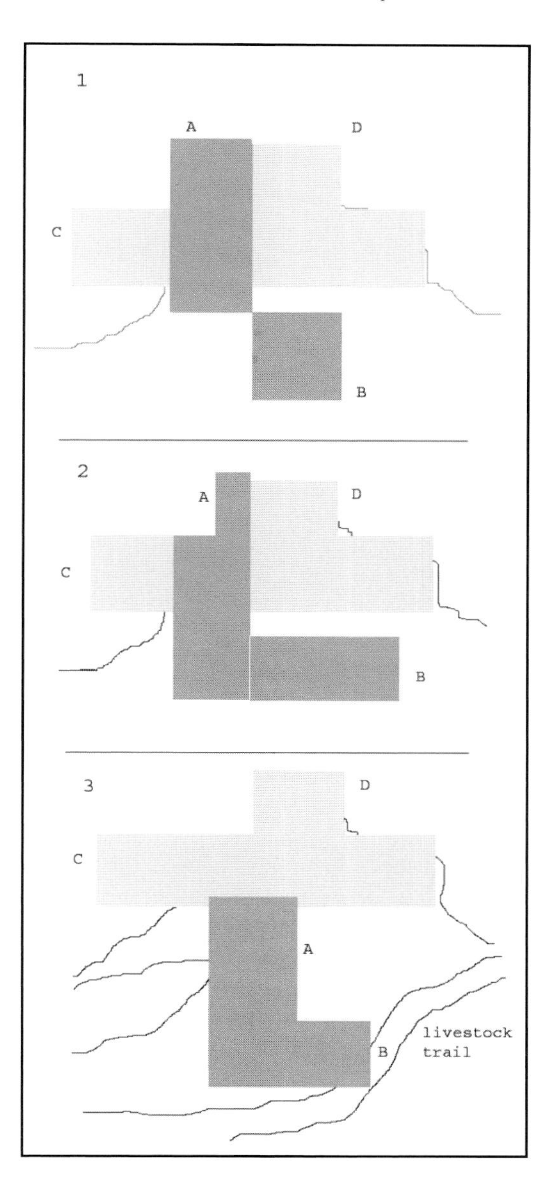

Figure 9.9 Three hypothetical diagrams for the location of Isaac's former trenches at PEES 2 (dark color) and our trenches (light color). See text for explanation.

Figure 9.10 Outcrop where PEES2 was located in the year it was found by our team (1995).

Two meters below the modern surface of the outcrop, we began locating artifacts in high numbers. All the archaeological remains were concentrated over about 60 cm in depth in the northernmost part of the excavation, where two archaeological levels appeared visible. In the southern part of the excavation, archaeological materials spanned about 30 cm in depth, and if two separate levels existed here, they were impossible to distinguish; instead we observed only one (Figure 9.11). The archaeological level was easily pinpointed because some artifacts appeared on the front of the outcrop due to erosion, still *in situ* (Figure 9.12).

In the 4 m^2 grid cutting into the outcrop, the entire stratigraphic sequence of the Moinik and Humbu Formation deposits was exposed at the site for the first time, enabling us to determine the vertical distribution of artifacts. In contrast with Isaac's observations, we did not find a 1.2 m-deep vertical scatter of materials. Only one polyhedron was found at the top of the Green Sands with Clay sequence, where the site is *in situ*. If we include that in the sequence excavated, then we could claim that the vertical distribution of stone tools is greater than 1.2 m. However, if we exclude that isolated find, the remainder of the stone tools was located at the mid-bottom of the sequence, in one or two archaeological levels as described above, spanning about 60 cm in depth. This site yielded very few bone fragments. A leopard canine and three unidentifiable bone fragments were discovered in the excavated areas. The lack of fauna cannot be attributed to differential preservation: a few meters from the site, the outcrop yielded several points where bones were found.

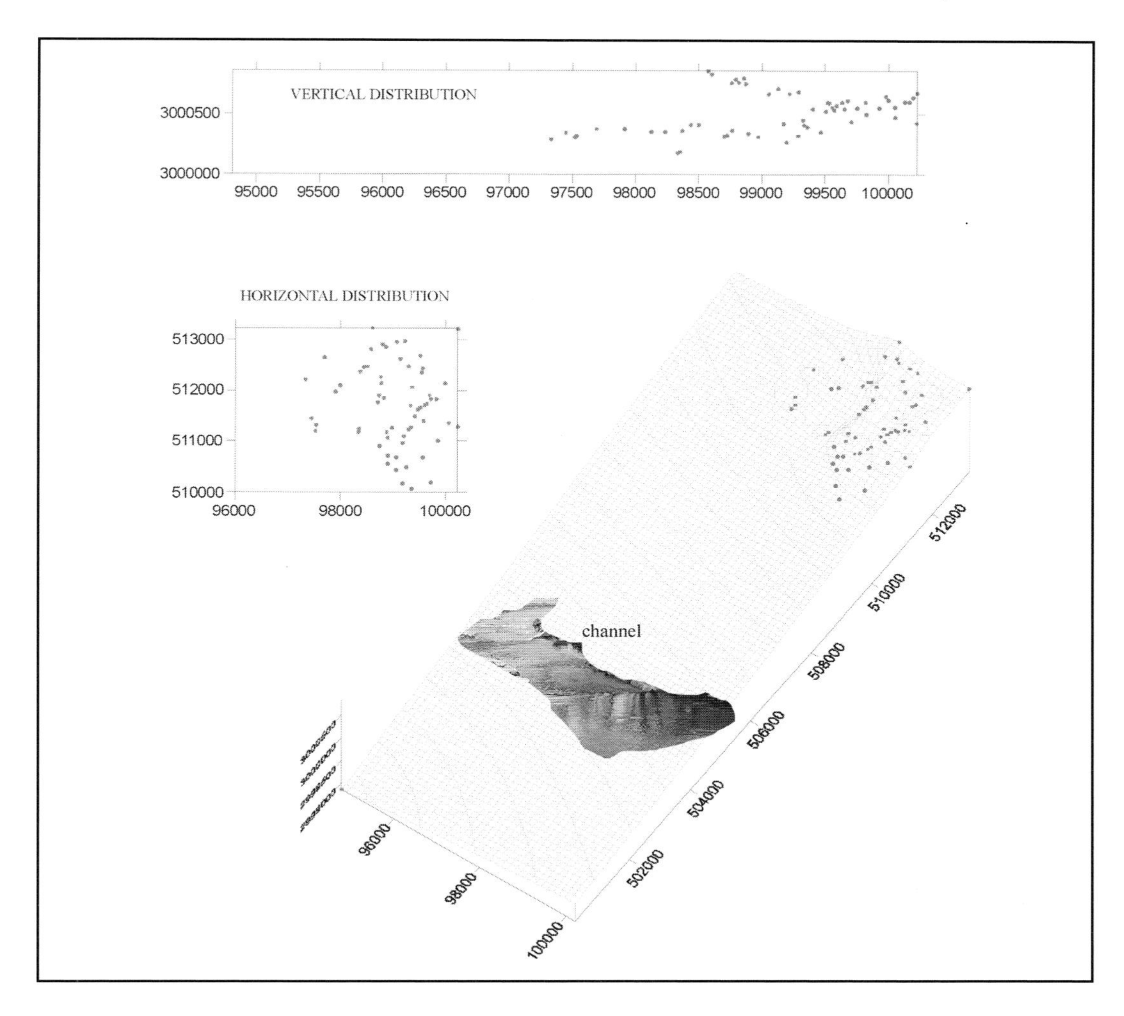

Figure 9.11 Horizontal and vertical distribution of lithic artifacts at PEES2. Note the proximity of the channel to the site.

Artifacts, comprising large and small pieces of basalt and quartz, showed no preferential orientation. Most of the artifacts, especially those from the lowest level, appeared flat on the surface, although some were dipping. Quartz flakes and quartz debris were found together in one of the squares, suggesting a lack of transport or other disturbance. The clayish context and the presence of debris and pieces of all sizes, together with a lack of polishing or abrasion, suggest a high-integrity assemblage. However, previous exposure of the pieces to high levels of humidity, probably during a high water table, affected the mostly basalt pieces by deteriorating their surfaces, giving the black and hard basalt a powdery, whitish texture in the outside 2 mm of the piece, when seen in section.

Figure 9.12 Handaxe appearing through erosion in the profile of PEES2, indicating the location of the archaeological level.

Table 9.4 Number of specimens of each lithic category found in PEES2

Flaked pieces	
Polyhedrons/cores	15
Handaxes/large knives	53
Detached pieces	
Complete flakes	91
Broken flakes/debris	121
Pounded pieces	
Hammerstones	2
Transported pieces	
Manuports	10
Total	**292**

A total of 171 stone pieces, plus 121 small fragments, were found (Table 9.4). Small flakes are the most represented category, although in contrast with other sites, they only constitute slightly more than half of the lithic assemblage. Serrallonga used for his doctoral research (in progress) a slightly different approach to flake types than Toth's (1982, 1987) system. He divided flakes according to the amount of cortex on the dorsal surface, irrespective of the presence or absence of cortex on platforms:

Type A: Cortex covers the entire dorsal surface.
Type B: Cortex covers between 50% and less than 100% of the dorsal surface.
Type C: Cortex covers approximately 50% of the dorsal surface.
Type D: Cortex covers between 25% and less than 50% of the dorsal surface.
Type E: Cortex covers less than 25% of the dorsal surface
Type F: Dorsal surface shows no cortex.

Figure 9.13 shows the distribution of these types in the PEES2 flake collection, showing that non-cortical flakes dominate the assemblage. This, together with the lack of completely cortical flakes, suggests modification of cores and polyhedral forms prior to their transport into the site. However, curation by hominids extends beyond that. When we count negative flake scars on the core/polyhedron set, a minimum of 60 scars were obtained, which would require a similar number of flakes to explain their exploitation at the site. However, if we include all the handaxes and large heavy-duty knives, these alone show a total of 263 scars. It is clear that there is an underrepresentation of flakes in relation to the number of negative scars on the flaked pieces of the assemblages.

This could indicate taphonomic post-depositional disturbance, but the sandy-clayish context together with the lack of observed polishing or abrasion, the orientation of artifacts, the preservation of phytoliths (see below) and the presence of all size categories, including debris, indicates otherwise. It is very likely that most large flaked pieces entered the site in the condition which is observed upon their retrieval. This is especially true of the handaxe/knife assemblage. If the length and breadth dimensions of the flakes and the negative scars of all flaked pieces are plotted (Figure 9.14), we see that the negative scars in the handaxe/knife assemblage tend to have larger dimensions, which is not reflected in the flakes, very few of which match the dimensions of those scars. This was previously taken as evidence of flakes being the result of hominids flaking cores and polyhedrons on-site, but not handaxes, which might have entered the site already crafted (Domínguez-Rodrigo et al. 2005). Some pieces from the large flake group (n=17) showed a very acute angle between the platform and the dorsal surface, and in most of the specimens of this size group, the dorsal surfaces showed no overlap with previous negatives scars. We interpreted this as evidence of flakes being obtained from the handaxes or heavy-duty knives. Half of the largest flakes represented in Figure 9.14 belong to this category. This is suggestive of handaxes edges being retouched and reused.

The last interpretation is supported by the fact that most of the handaxes have a wavy outline along sections of their edges (Figure 9.15), and frequently some step fractures are associated with this outline. This indicates use. The heavy-duty activities for which handaxes were used left conspicuous traces on the edges; this is further substantiated by the high number of pieces with (sometimes extensively) retouched edges.

The handaxe/knife assemblage is composed of flaked pieces produced from large flakes obtained from large boulders. These pieces do not show all of the retouch usually observed in bifaces; rather, they have the typical morphology of large heavy-duty knives. However, the standardization of the shape, the fact that in several specimens the retouch is bifacial and the fact that Isaac (in his unpublished notes) interpreted them as unifacial and bifacial handaxes, compels us to defend the use of both terms in this site interchangeably. They frequently have been retouched uni-facially, creating the typical pattern that characterizes PEES2: a large flake is used as a blank in which the thickest part, usually corresponding to the platform of the original flake, is opposite the thinnest part of the flake

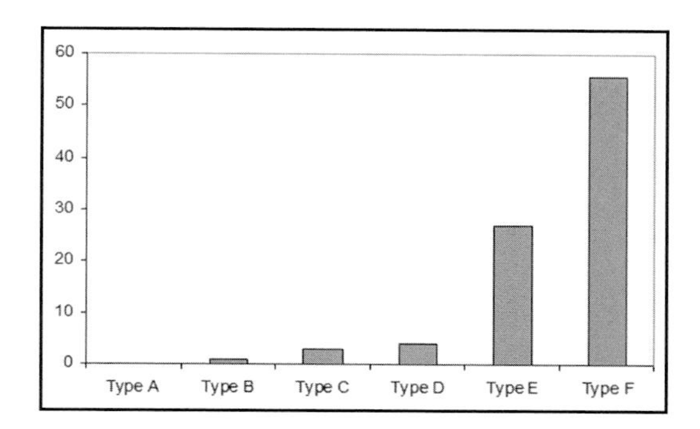

Figure 9.13 Distribution of flake types at PEES2 according to the amount of cortex on the dorsal surface. See text for explanation.

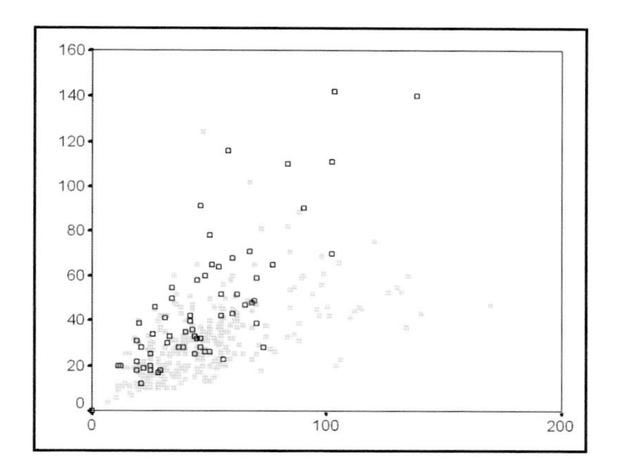

Figure 9.14 Distribution of flake (open squares) sizes and negative scar sizes of flaked pieces (closed squares). Length (x axis) and breadth (y axis) measured in mm.

where the edges appear shaped and retouched (Figure 9.15). In PEES2, large cutting tools were sought; a large edge in a heavy tool is continuously reproduced. This pattern contrasts with that observed in PEEN1, where pick-shaped and pointed artifacts were systematically knapped. However, some pointed pieces also appear in the PEES2 assemblages (Figure 9.16). Sometimes, the piece is flaked more extensively to create a double edge (Figure 9.17).

Figure 9.15 Typical handaxe/knife from PEES2 with edge occurring opposite the platform.

Figure 9.16 Pointed handaxe, bifacially flaked, from PEES2.

The large blanks from which handaxes were made were not produced from large cores at the site. In our surveys, we have not found any boulder of the size required to flake these tools within a radius of a few kilometers. The closest source of raw material observable today is located 5 km away at the North Escarpment. Our experimental research shows that several of the large blanks used for elaborating handaxes break easily during the knapping process. It is therefore unsurprising that hominids were probably creating blanks at the raw material source and transporting them in a fairly complete state to the site.

All handaxes are larger than 100 mm, the largest being nearly 240 mm (Figure 9.18). The narrowest handaxe is 60 mm and the broadest is 140 mm. These are very large, heavy artifacts (all together over 50 kg), which clearly indicate their importance, given their accumulation at the site. Hominids must have transported them over large distances, used them and discarded them at the site. Given the minimal flaking on the smaller flaked pieces, it can be argued that handaxes were the main tools used at PEES2. During excavation, they appeared frequently in small clusters (Figures 9.19–9.20), whereas detached pieces were spread across the entire area. This specific clustering of handaxes could also suggest differential treatment of these pieces versus the other elements in the assemblage.

The paleosol at PEES2 was sampled around the stone tools, within the artifact-bearing horizon, to analyze and compare its phytolith contents with the phytolith samples extracted from the artifacts. This was designed as a control to determine the amount of contaminants adhering to the working edge of the stone tools. By contaminants, we mean the phytoliths and other old residues naturally deposited in the soil before or during the sedimentation process, which might have accidentally adhered to stone tool surfaces, thereby mimicking the residues originated from artifact use.

Phytoliths were found preserved on three of the stone tools. Their morphotypes contrasted with those of phytoliths recovered from the artifact-bearing paleosol, which were mainly grasses (Domínguez-Rodrigo et al. 2001b). The specimens found on the artifacts were polyhedral (Figure 9.21) and different in shape from those extracted from the paleosol sample (Figure 9.22), implying different sources of phytoliths for each (see below). Therefore, we strongly argue that phytoliths from the surrounding soil did not contaminate the edge of the artifacts. The marked difference between the types and proportions of phytoliths from the paleosols versus from the artifacts clearly shows that the phytoliths preserved on the edge of the artifacts were due to utilization rather than a result of postdepositional processes.

The functional difference between phytolith representation in the soil sample and the artifacts is further supported by fact that the distinctive phytoliths on the artifacts were located on the tool surfaces and not on the rest of the sediment matrix which was adhered to them. Furthermore, phytoliths have not been found on the sharp surfaces of the edges but were located on the internal working edge of the artifacts. This supports adsorption as a result of utilization, as observed in experimentally-used stone tools (Anderson 1980; Juan et al. 1996; Hardy and Garufi 1998). The surface of the external part of the edges and the

Figure 9.17 Bifacial handaxe with double edge from PEES2.

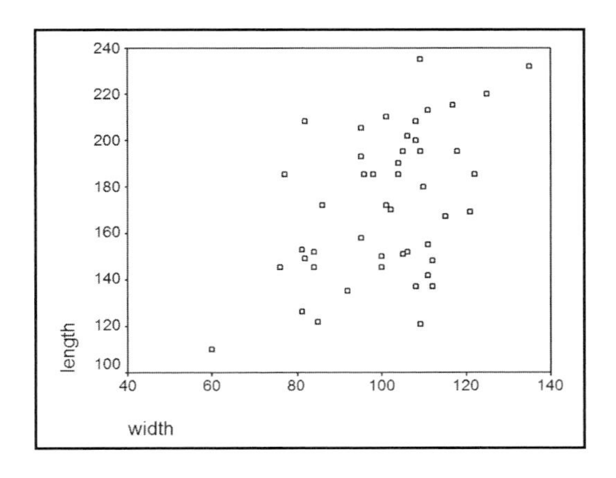

Figure 9.18 Length and breadth of handaxes from PEES2.

Figure 9.19 Clustering of handaxes from PEES2.

Figure 9.20 Detail of the best-preserved cluster of handaxes during excavation.

intense friction these areas undergo during use leads to phytolith preservation only on the internal section of edges. Phytoliths adhering to artifacts from soil contamination would not show such a preferential location.

Calcium oxalate phytoliths with polyhedral forms were identified on the artifacts, some of them *in situ* associated to plant tissues, and some *ex loco* as silt particles. This type of phytolith is very common in woody tissues, and its size, quantity and distribution have been used as indicators of type of wood by some researchers (Chattaway 1936; Barakat 1995). The plant remains found on the Peninj tools are formed by calcium oxalate phytoliths in association with parenchyma tissues, silified tracheids and fibers. These types of plant remains are documented in some Leguminosae (especially in several species of *Acacia*) and in Salvadoraceaeas well as in leaf tissue of dicotyledons. *Acacia* sp. is the only legume species identified at Peninj through pollen analyses during Upper Humbu times (Domínguez-Rodrigo et al. 2001a). *Salvadora persica* is also present. The presence of parenchyma tissues is consistent with the interpretation of a good preservation environment and therefore, a low-energy depositional context, as was suggested above by the lithological analyses. Cortical cells have been found on one of the flakes, strongly suggesting that the flake may have been used for a task involving the removal of cortical fibers from branches, whereas the handaxes are more likely to have been used for chopping wood.

From these data, our interpretation of the analyzed tools is based on the following reasoning. The phytoliths found on the tool surfaces are indistinguishable from those reported by other researchers as belonging to *Acacia*. Given the heavy-duty activities performed by most of the analyzed handaxes, as reflected in the damage

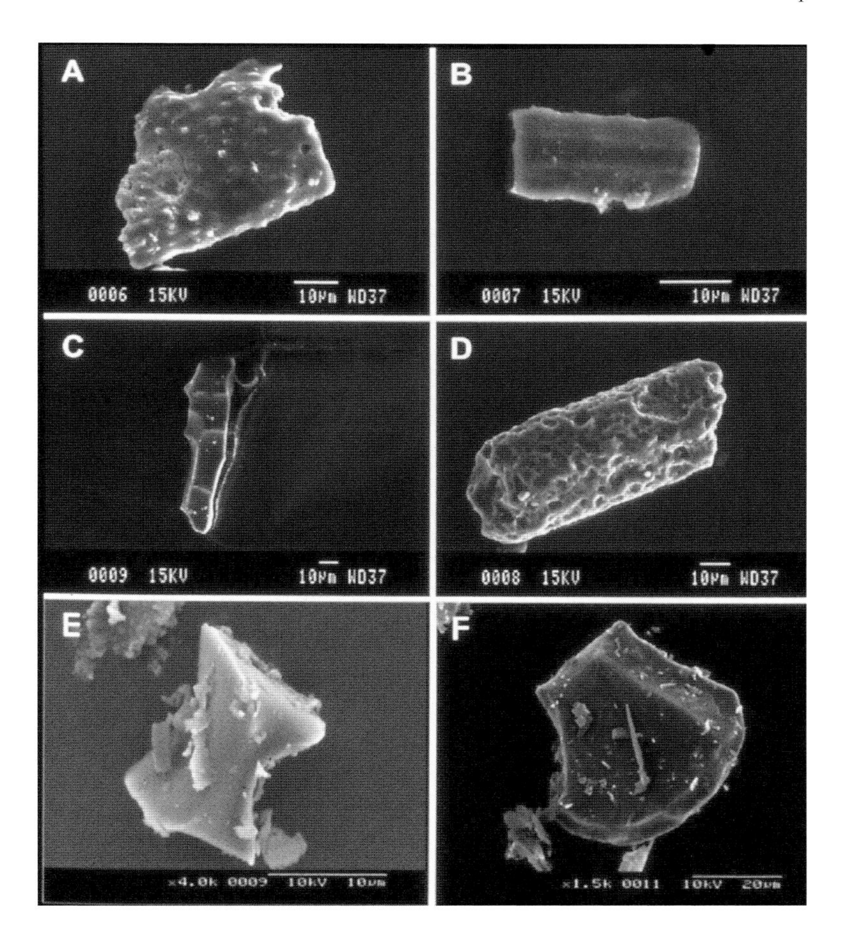

Figure 9.21 (above) SEM images of the phytolith morphotypes distinguished in the soil study. a) opaque platelet phytolith; b) short cell phytolith from Poaceae; c) elongate phytolith with a faceted surface; d) phytolith eroded bydissolution; e) saddle phytolith from Poaceae; f) fan-shaped bulliform phytolith (keystones) from Poaceae.

Figure 9.22 (left) Polyhedral phytoliths found on the inner edges of handaxes at Peninj (Domínguez-Rodrigo et al. 2001).

Figure 9.23 The PEES2 excavation.

Figure 9.24 A portion of the excavated lower level at PEES2 showing two clusters of stone tools, viewed from above.

Figure 9.25 (left) Excavation of the two archaeological levels at PEES2.

Figure 9.27 (above) Handaxe in the process of being excavated.

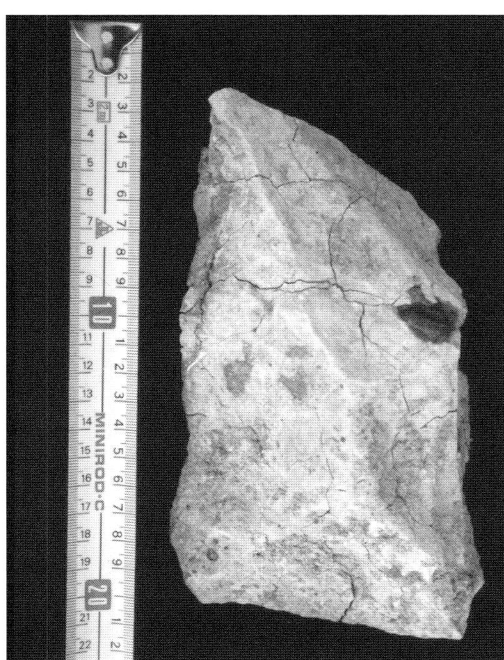

Figure 9.26 (left) Cleaver found at PEES2.

patterns on the edges, it seems that *Acacia* rather than *Salvadora* was targeted by hominid woodworking activities. *Salvadora* branches as well as dicotyledon leaves do not require heavy-duty tools for processing and manipulation, as has been documented in modern human populations in Africa, nor would they produce the heavy damage documented on some handaxe edges. This would also explain the complex behavior linked to transport of handaxes over long distances, as is also documented in flakes, strongly suggesting that the flakes may have been used for a task involving the removal of cortical fibers from branches, whereas the handaxes are more likely to have been used for chopping wood.

This interpretation has an important bearing on our understanding of hominid behavior. It has been suggested that hominids at that time could not have been hunters, because they lacked the means to capture animals. According to this study, rudimentary spears could have been one type of wooden tool that humans were making 1.3-1.2 million years ago. This could have enhanced their adaptation as hunters to open environments, and gives us further insight into the intelligence of hominids at that time.

10

The Acheulian Sites from the North Escarpment

Manuel Domínguez-Rodrigo, Fernando Diez-Martín, Luis Luque, Luis Alcalá, and Pastory Bushozi

Introduction

The North Escarpment of Peninj is situated at the foot of the Sambu volcano. It embodies a set of longitudinal alluvial sediments from Early Pleistocene time belonging to the Humbu and Moinik formations. Isaac (1965, 1967) discovered a large concentration of handaxes in this area, which he excavated. He called the site RHS, which together with the MSH site in the South Escarpment were the main archaeological discoveries made by the first paleoanthropological expedition to Peninj. After the brief excavation in 1964, research at the site was not resumed until 1981, when Isaac re-excavated the site and changed its name to Mgudulu (Mturi 1987). The results of this recent excavation were never published, given Isaac's sudden death. The age of these sites (initially thought to be about 1.5 Ma) made them, together with EF-HR, the oldest Acheulian sites known for a while (Isaac and Curtis 1974).

The North Escarpment was again surveyed in 1995 and 2001 by us.[1] In this survey, several paleontological and archaeological localities were found. The density of artifacts on the landscape is greater than anywhere else in Peninj (see Domínguez-Rodrigo et al. 2005). This could be due to a higher occupation of the area by hominids given its higher ecological productivity (as suggested by Downey and Domínguez-Rodrigo 2002) or to better preservation factors. In contrast, faunal remains have been poorly preserved. In 2002, extensive excavations took place at PEEN1, which enabled us to study the artifacts and their context and even to put Isaac's (1964) materials into context. Of the three archaeological windows to the past at Peninj, the North Escarpment is the furthest from the lake, formed in a riverine habitat in which the Peninj river was the main hydrological source.

Geology of the North Escarpment

Geomorphology and geography

The archaeological sites of the North Escarpment region (EN for "Escarpe Norte") are located on the upthrown fault block of the Natron basin half-graben. This is located on the southern hillsides of the Oldoinyo Sambu volcano, whose lava flows underlie the Peninj Group sedimentary deposits north of the ES and ST areas. The Peninj Group outcrops are shown near the Peninj river valley and the erosive front over the Sambu tectonic escarpment (E35°53.5', S2°14.7'; see Figure 10.1). These exposures are the northern limits of our research area. It is also the northern limit of the structural surface that constitutes the Upper Tuff of the Moinik Formation. Sediments in the area are very sandy, not well-compacted and are intensely eroded in deep gullies and trenches bordering the erosive front as well as in small hills in the surroundings. The geological substrate underlying the Peninj Group deposits are the Sambu lavas or Hajaro lavas, as Isaac (1965) previously suggested.

The EN deposits are slightly tectonically affected. Nevertheless, our geological review of the area and subsequent sedimentary correlation analysis show that EN sediments were

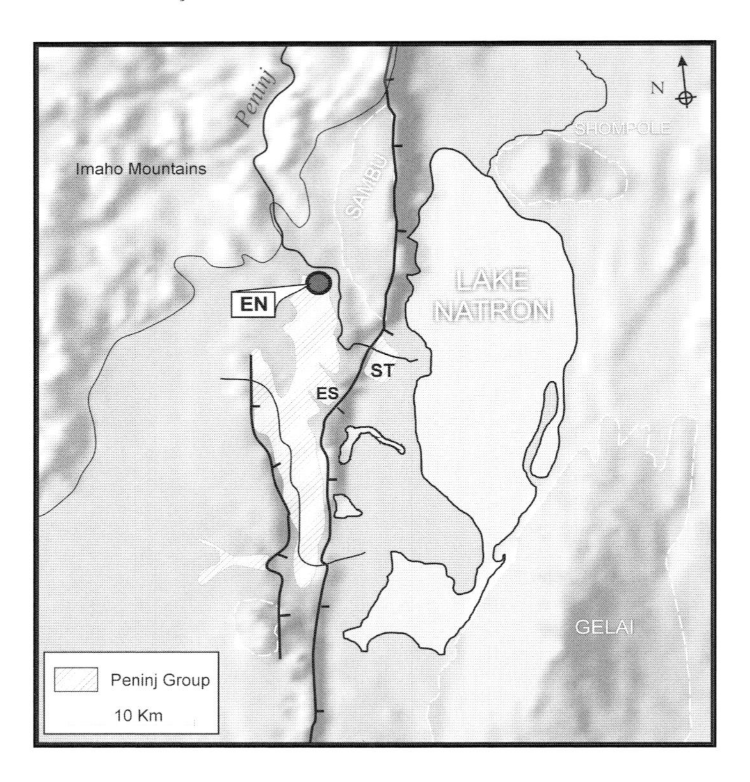

Figure 10.1 Map of the northwest Lake Natron sub-basin, indicating the areas where sites have been excavated. North Escarpment (EN) sites are located on the Sambu Escarpment near the Peninj river valley.

deposited in a small, depressed area, probably due to tectonic subsidence. Sediment distribution and sedimentary features suggest that EN sedimentary deposition occurred in a marginal and partially isolated sedimentary sub-basin, at least during the deposition of the lower Humbu formation. This sub-basin subsidence would be conditioned by the activity of faults both parallel and oblique to the main rifting structural directions. The area has undergone active sedimentation over a long period both during and before the formation of the Peninj Group. As Isaac noted, the Pre-Peninj river flowed through this area during an eruptive pause in the flow of the Sambu lavas, as well as during the Hajaro times (Isaac 1965). The proximity to Oldoinyo Sambu (2,045 m) and the precambrian Imaho mountains (1,700 m) caused a high relief modelled by water flow. This water supply was probably much greater than in southern contemporary alluvial environments. In contrast, the sedimentary rate seems to have been lower, probably due to a less effective sedimentary source. The space to accommodate these sediments must also have been smaller than in areas at the foot of the Sambu tectonic escarpment.

Previous works in the area are restricted to Isaac's surveys and excavations during the 1960s. Isaac (1965) argued that the Basal Sand member was deposited outside the main escarpment. The Basal member of the Peninj Group was deposited overlying a fresh surface of Hajaro lava reaching more than 15 m of thickness in the westernmost outcrops. These deposits were part of an alluvial fan built by a Proto-Peninj river that circulated from the Precambrian hills onto the lava plain. A site rich in the Acheulian industry was called RHS and was correlated to the

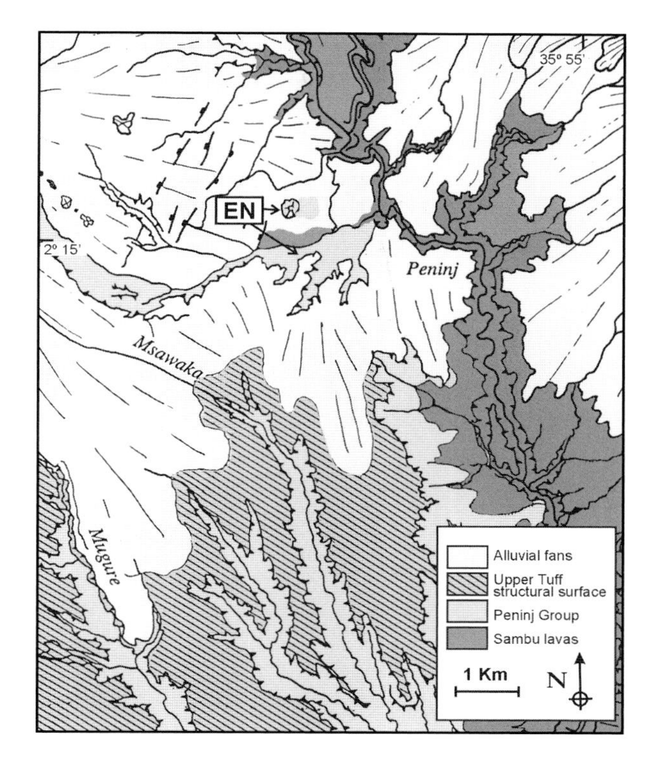

Figure 10.2 Geological map of the North Escarpment area.

Humbu Formation over the Main Basaltic Tuff member. South of this area, Isaac notes that the Upper Series Clays (Moinik Formation) mark a change to saline lacustrine conditions, and 5 km west from the Type Section this part of the sequence is interbedded with gritty, poorly-sorted alluvial and deltaic formations. The Upper Tuff of Moinik also changes its features from very pure vitreous tufas in the east to altered, sandy sediment with quartz in the west, where it rapidly disappears. All of these observations have been confirmed by our own research and field observations.

Geology of the EN sites

The lithological facies outcropping in the North Escarpment area are quite different from those of the South Escarpment and the Type Section. The upper and younger sedimentary units of the Peninj group cover the main part of the exposures (Figure 10.2). This unit corresponds to the Moinik Formation and appears very homogeneous; it is very thick (up to 18 m) and is comprised of graded, interbedding sequences of white, coarse to fine, poorly-sorted quartz sandstones, with erosive lower surfaces in a set of channel-fill deltaic deposits. Rootcasts, bioturbation and pedogenetic calcite nodules are frequent. The top of the Peninj Group deposits laterally corresponds to the Upper Tuff structural surface, but few remains of this tephra layer are preserved there. A few kilometers southward, sands decrease and an erionitic ash layer develops, reaching up to 6 m in thickness.

All of these coarse sediments lie unconformably over more heterogeneous older deposits, which are thinner, finer, reddish and sometimes sandy. A large erosive discordance

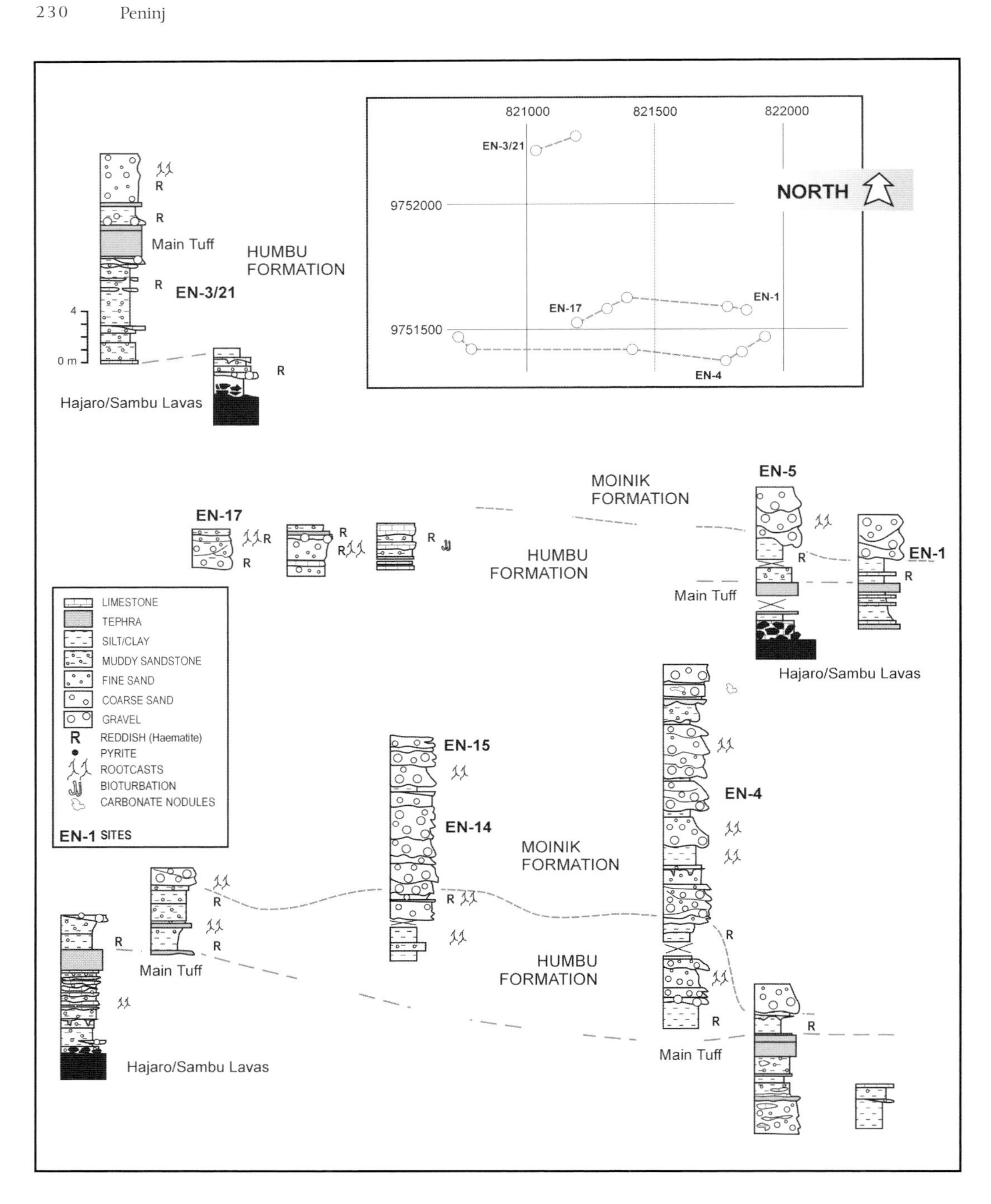

Figure 10.3 Stratigraphic sections near PEEN1 showing the clear differences in Humbu Formation preservation between northern and southern areas of the North Escarpment. Note the southward increase in thickness of the coarse sands of the Moinik Formation unit above the erosional surface.

separating both units created a significant paleo-relief that is clearly visible in the EN area.

There are two different areas of sedimentary exposures in the EN region (Figure 10.2): a plain north of the Peninj river where small relict hills still remain, and an erosive front on the southern slope of the river valley. The former shows Humbu Formation deposits, including BSC, Main Tuff and USC members. Southern exposures show mainly Moinik arenaceous deposits and some scarce Humbu Formation outcrops, including the PEEN1 site (Isaac's RHS; Figures 10.3-10.4).

The Northern exposures (Figure 10.4) show the Humbu Formation lying unconformably over Sambu or Hajaro lavas. Silts and sandy silts lie on fresh lava boulders or flows, previously interpreted as Hajaro lavas (Isaac 1965). Lava blocks and boulders are in a brown clayey matrix. Overlying the lavas, fine-grained muddy sands and silts from the Basal Sand member show several graded interbedding sequences. At the base of each sequence a lag of coarse sands is commonly found. Tuffaceous reddish to brownish sands and sandy silts and clays are found in the middle, and limestone rich in root-casts at the top. The thickest exposure in the isolated hill shows 5.5 m of Basal Sands that includes four sequences grading from muddy sands to sandy silts and clays, with limestone on the top. Some folded sediments in the upper part could be attributed to seismic liquefaction.

The Main Tuff member is present in the middle of the hill. Limestone rich in the gastropod *Gabbia* is 0.1 m thick, and also shows some fish remains and ostracods. The Tephra layer is 1.6 m thick and is rich in pumice and coarse pyroclasts. In addition, USC starts with sedimentation of well-cemented coarse sands and gravels, overlaid by sandy clays which are red in color, bioturbated and weathered. A 0.2 m-thick tephra layer covers sands and sandy clays with similar features to T-1 tuff of Type Section. About 2.8 m of red, coarse muddy sands rich in rootcasts and intensely weathered top the outcrop. Lithic remains are abundant but no *in situ* remains have been found (PEEN21). Fossil bones are very rare and are heavily weathered (PEEN3).

The PEEN1 area (Figure 10.4) shows a few outcrops where underlying lava flows are exposed. In all of these, blocks of lava overlie lava flows. Big blocks are embedded in limestone and clay matrix. Unconformable contact between the Humbu and Moinik Formations is usually coated by recent colluvia but some scarce outcrops are found. For this reason, the PEEN (RHS) area provides one of the better exposures for understanding stratigraphic relationships between different units of the Peninj group in the northern area. The Humbu Formation has been partially eroded in the area because of the erosional discordance with the Moinik Formation. In some places, Basal Sands are observable, and near the PEEN1 site, they consist of 1.45 m of red, graded sandy clays, rich in pedogenic calcium carbonate, rootcasts, and burrows, which intercalates with an analcimic tephra layer, a few centimeters thick, as well as with finely-laminated tuffaceous silts at the base. In other exposures of the area, the Basal Sands consist of more than 4 m of silts and fine sands, as well as poorly-sorted muddy sands rich in pedogenic carbonates, and calcretes and rootcasts.

The Main Tuff member shows a reduced thickness, thinner than 0.45 m. Tephra is rich in pumice, especially at the top of the layer. In this case, the USC consist of red sandy clays that include a thin layer of red analcimic tephra, rich in pyroclasts. Middle layers show a great abundance of sandy, porous, highly-burrowed calcretes and limestone nodules, interbedded with red clays. In the upper part, 0.8-1 m of highly homogeneous red sandy clays tops the sequence, and is overlaid by an erosive unconformity.

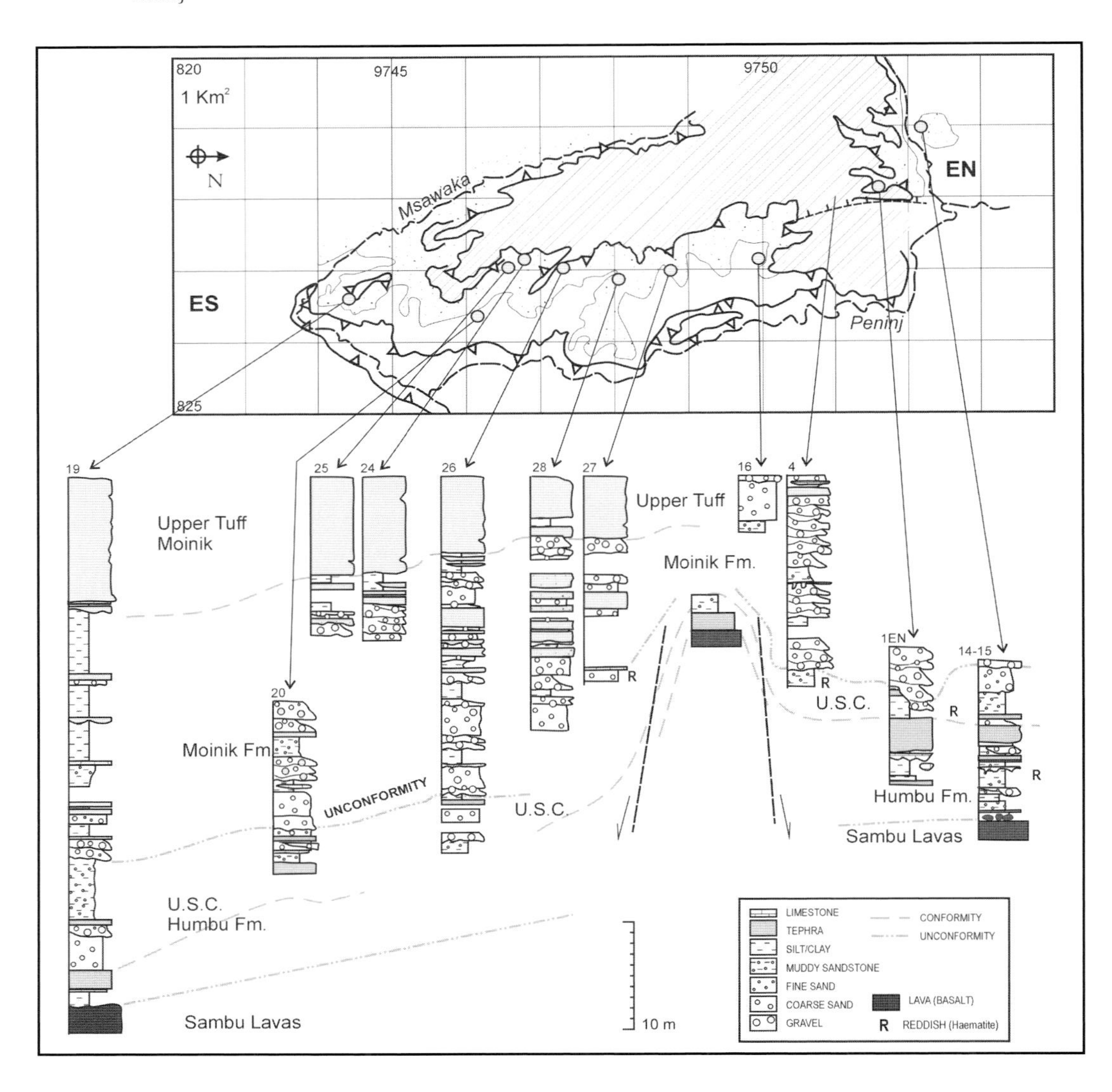

Figure 10.4 Stratigraphic sections and south-north correlation between sedimentological units of the Humbu and Moinik Formations above the Sambu Escarpment. Note that the coarse-grained deltaic facies changes to the lacustrine clayey facies in the south.

Over the erosive unconformity that limits the Humbu fine to coarse sediments, the Moinik Formation exposures start up to the top of the sequence. Moinik deposits consists of 15-20 m-thick sandstones that infill and cover the paleorelief caused by erosion of the underlying Humbu Formation. Sandstones are formed by graded interbedding sequences, from well-cemented gravels at the base to poorly-sorted, sometimes laminated coarse sands, and fine sediments like silts and clayey sands rich in rootcasts, mud-cracks and limestone nodules at the top. These sediments are interpreted as channel-fill deposits in an alluvial deltaic environment. At least six main grading sequences overlap in the largest exposures, each of the sequences partially eroding the underlying one. The basal part seems to include the coarsest sediments while the upper part seems to show the most "edaphized" ones. Basal layers consist primarily of coarse gravels and sands, which include: red clay clasts coming from the underlying Humbu Formation, boulders of analcimic tephra (also from Humbu), and large and weathered lithic remains. These features clearly show the strongly erosive processes acting on the Humbu Formation. Sands and gravels become finer as one moves up the sequence. Some of the upper sand layers are slightly tuffaceous and erionitic. A number of archaeological sites have been recorded in this deltaic sequence in different areas.

Western exposures (Figure 10.4)
A number of good outcrops, but no *in situ* lithic remains, have been found in the erosive front south of the Peninj river valley and West of PEEN1. In this area, the Humbu exposures are rare but Moinik sediments are widely represented. Where the Humbu Formation appears, it is common to find some remains on surface. Basal sands in this area are as thick as 4 m, showing some sandstone sequences rich in limestone and calcretes as well as some layers which were deformed by liquefaction processes. The Main Tuff member is 0.9 m thick and rich in pumice. The thickness of the USC member depends on erosion, but it is generally 2-4 m thick and it consists of sandy clays and red, gravel-rich sands. A tephra layer is also included in this member.

Southern Exposures (Figure 10.5)
Exposures of the Peninj Group along the erosive front located upon the Sambu Escarpment are widely represented (Figures 10.3, 10.5). The thickness of each sedimentary unit depends mainly on paleorelief and tectonic factors. A network of horsts and grabens seems to affect this upper part of the Moinik Ramp (as was previously described in Chapter 2). One of these faults is located east of EN and it was previously reported by Isaac (unpublished notes). It runs approximately north to south, parallel to the main rift structures, but other faults running orthogonally can be deduced from stratigraphic analysis. In all these exposures, the Humbu Formation is poorly represented because of erosion and recent sediment coverage. Locally, the Main Tuff member directly overlies the Sambu lavas, showing the presence of a marked paleorelief that probably limits the southern extension of the Humbu Formation at least during its initial phases (Basal Sands). Muddy sands of Humbu Formation are more clearly visible southward, where a fine Basal Sands facies reaches 1.5 m in thickness. The Moinik Formation, always lying unconformably over Humbu, thickens towards the south, from 17-19 m in thickness near PEEN1 to 36 m in thickness in the Msawaka valley, at the other side of the valley where the PEES sites are located (as is visible in the PEES2 site). The most notable feature of the Moinik facies is the clear transition from pure arenaceous deltaic facies in the north to more clayey lacustrine ones in the south. Deltaic influence can be traced up to 4 km south of EN, but it is temporarily dominant along the Moinik Formation deposition, up to 7 km southward. Tephra layers are more frequent as the depositional energy is lower, with finer sediments of lacustrine origin. At the top of the sequence, the Upper Tuff of Moinik increases its thickness from 6 to 12 m over a few km, changing from greenish arenaceous composition to yellowish pure erionitic composition. Sites are

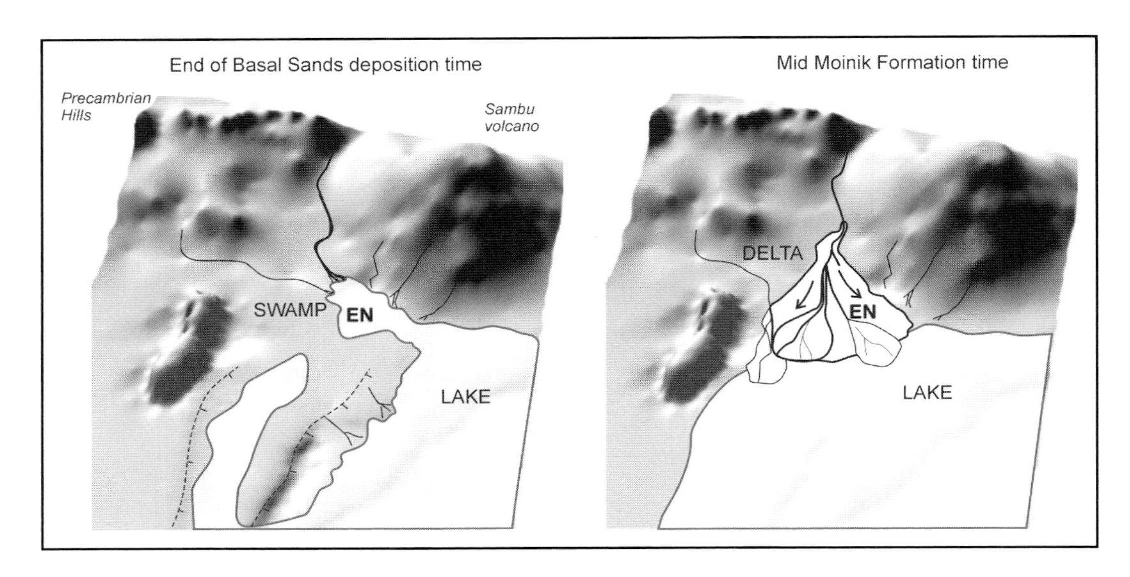

Figure 10.5 Paleoenvironmental interpretation of the North Escarpment area divided into phases of the Peninj Group deposition. During the Basal Sand member formation a swampy area developed, probably due to sinking lateral to the main Rift direction. The second phase shows the deltaic environment in the west/northwest area, grading into a lake in the most subsiding areas southward.

scarce in this area but some bones have been found embedded in tephra ash layers, showing the partially terrestrial environment.

Paleoenvironmental interpretations

The North Escarpment area shows the presence of the same sedimentary units that characterize the Peninj Group deposits. Sedimentary evolution of this area seems to be parallel to others like those of Type Section or South Escarpment. As in the rest of the basin, sedimentary processes are strongly conditioned by tectonics. A network of horsts and grabens not only created a paleorelief, but also partially isolated some areas from those in the basin. In this case, tectonics associated with the Moinik ramp, plus an oblique faulting, gave rise to a depressed area near the Oldoinyo Sambu volcano and Precambrian hills that acted as a local trap for sediments. At the time of deposition of the Basal Sands, the area was isolated form the southern lacustrine environments and probably joined them eastward from the main basin (Figure 10.5). Swamps, mud flats and alluvial plains were developed during these times in the PEEN area, being deeper or more commonly flooded in the north. The Main Tuff member is well-represented in the whole area and was deposited in a lacustrine environment with nearby freshwater sources, as suggested by the presence of fish remains. This member's reduced thickness may be explained by the greater distance from it to the southern volcanic cone, but its richness in pumice and coarse pyroclast contradicts this hypothesis. Contemporary eruptive activity of the Oldoinyo Sambu volcano may be an explanation. The USC show strongly weathered surfaces. Oxidization and pedogenesis are common processes, and changes in the water table can be interpreted. The lower rate of accumulation in comparison with other areas is an reflection of these type of processes. Sedimentary facies include alluvial and mudflat facies and the time of deposition correlates with T-1 and T-2 deposition in ST. These facies probably extend over the whole area and are connected with the rest of the basin.

At the base of the sequence the Moinik Formation incorporates a number of materials from the Humbu Formation: clay boulders, tephra and lithics. Facies are purely deltaic, expanding southward and showing different moments of progradation and retreat along its deposition time (Figure 10.5). Multiple channel-fill deposits in erosive upward-grading sequences are typical of these braided systems. Water table oscillations gave rise to carbonate migration and precipitation in the lower surfaces of channel-fill deposits, where permeable coarse sands are in touch with compact, fine and sometimes clayey deposits. High-energy deposition does not allow the formation of fine ash-fall layers. For this reason, the northern deposits lack the tephra layers that are present in the southern exposures. The unconformity with the underlying formation becomes more conformable south of the Msawaka River.

Erosive processes in the north and the deposition of unconformable deltaic sediments, as well as of conformable lacustrine sediments to the south, suggest differential movements of the Moinik ramp, leaving higher parts above the base level susceptible to erosion, and submerged southern parts in a lake environment. A fast deltaic progradation from the Pre-Peninj river is contemporary to sedimentary aggradation in the ES area (see Chapter 2).

Archaeology of the EN1 Site

Geological characteristics of PEEN1

The Peninj Group has a different aspect at the North Escarpment from that of the Type Section. The Humbu Formation overlies the Sambu basalt basement directly. This basement shows traces of erosion. A stratum of lava conglomerate overlies the basalt basement, indicating the beginning of the sedimentary process that created the Basal Sands with Clays of Lower Humbu. Over one meter of sands and clays with laminated silt levels and limestone nodules overlies the previous stratum. This has a more clayish aspect towards the north, which suggests that there must have been some relief separating this subbasin from the lacustrine basin in the south, and that the river stream must have also been accompanied by swampy/lacustrine habitats as the main sedimentary sources. A level of sandy carbonated nodules was deposited on top, suggesting a partial transgression of the lake in this area. Overlying the Basal Sands with Clays, the Main Tuff can be found.

The PEEN1 section shows both Humbu and Moinik Formation facies (Figures 10.6–10.7). The Humbu Formation was deeply eroded before the deposition of the Moinik Formation. This is clearly visible in the site section, where the height of the Humbu strata varies by 0.8 m over a width of just 7 m. A few m to the south, it reaches more than 1.5 m in thickness. Humbu consists of red to brownish massive clays, which are partially stained brown below the tephra layer correlated with T-1. This layer is laterally eroded and filled by Moinik gravels and sands. Directly over the erosive discordance, well-sorted red fine gravels are deposited, including lithics and thin levels of quartzitic gravels. This layer also includes red clay blocks (intraclasts), boulders of analcimic tephra (from Humbu), small blocks of well-cemented sandstone, and abraded, rounded fossil bones. Sandy limestone strongly cemented the basal layers. Overlying sediments include coarse sand and gravel bars that show crossed lamination and ripples, and which locally include some pedogenetic limestone calcretes, sometimes rich in rootcasts. Nearby exposures show a homogeneous

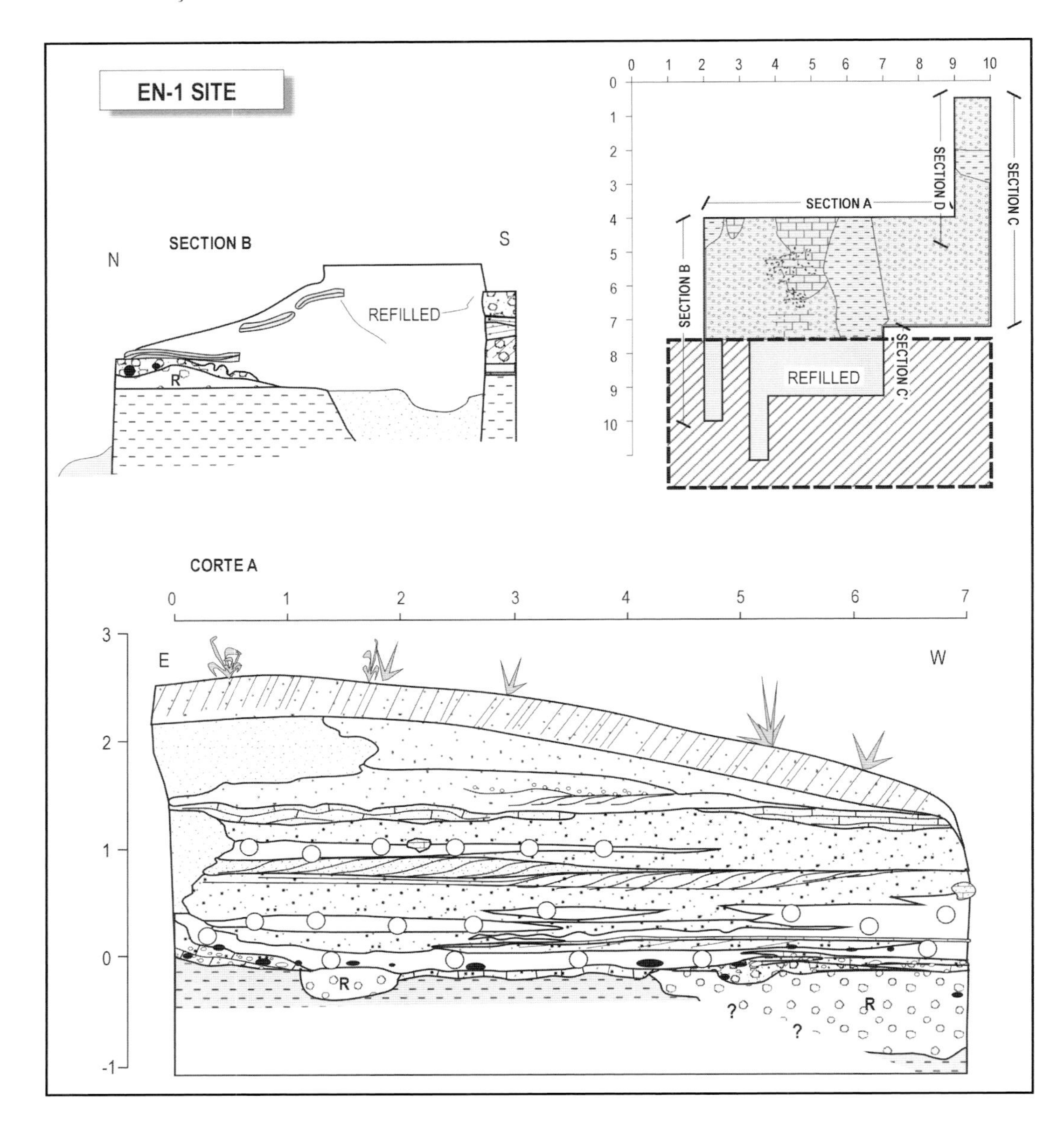

Figures 10.6 (above) and 10.7 (right). PEEN1 site sections showing their sedimentological features. In this area, the erosional surface between the Humbu and Moinik Formations is clear. The first Moinik deposits are full of reworked Humbu materials such as basaltic tephra blocks, red mud intraclasts, or fossil and lithic remains. Backfilled trenches excavated by Isaac are also observed in the sections.

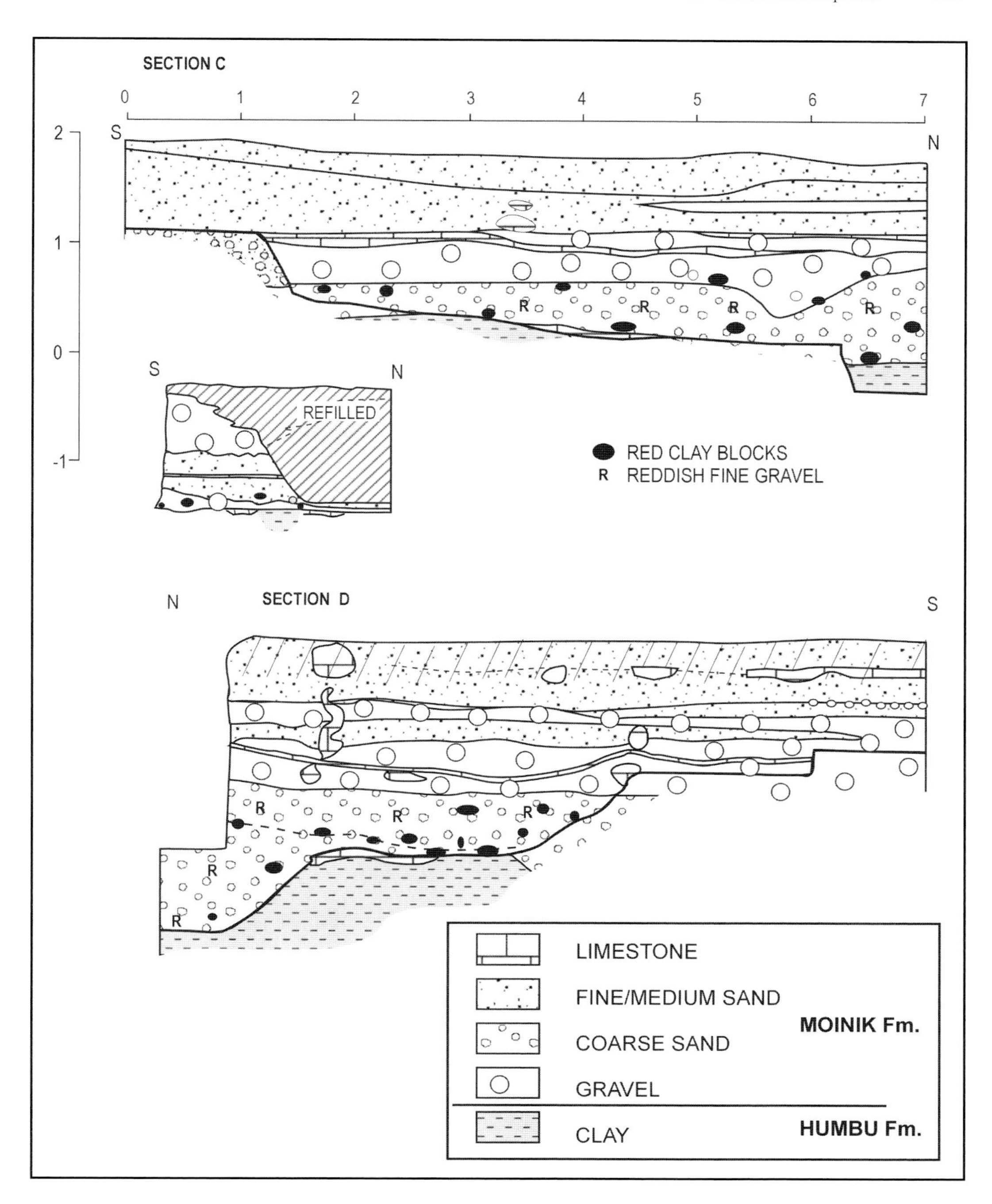
SECTION C
0
1
2
3
4
5
6
7
2
1
0
-1
S
N
R
S
N
REFILLED
RED CLAY BLOCKS
R REDDISH FINE GRAVEL
N
SECTION D
S
R
LIMESTONE
FINE/MEDIUM SAND
MOINIK Fm.
COARSE SAND
GRAVEL
CLAY
HUMBU Fm.

Figure 10.8 Main excavation area of PEEN1.

succession of channel-fill deposits up to the structural surface on the top of the Moinik Formation.

Excavation of the site

Isaac's (1965) initial excavation of the site (8 m x 6 m trench) - produced 215 lithic items. His excavation in 1981 produced more artifacts but his field notes do not contain the total number of materials retrieved from the site that year. Our excavation began by making a detailed topography of the area and spatially plotting all artifacts found on surface. Then we expanded Isaac's excavation by opening an additional 38 m^2 (Figure 10.9). The site is embedded between fluvial deposits. This supports Isaac's interpretation of the site as a sandy bank close to a channel, although a substantial part of the site is found within a channel itself. Figure 10.8 shows that most materials lie on a slope to the west, with a vertical distribution within a 10-cm interval in the north of the trench. In the south of the excavation, archaeological materials appear distributed in a vertical deposit spanning 1 m (Figures 10.9-10.10). The reason seems to be that in the southern part of the excavation there is a gravel level belonging to the channel, in which artifacts are vertically distributed, in contrast with the northernmost area of the excavation (Figure 10.11). Two depositional dynamics seem to have affected the site: the first is in the channel (i.e. the southern part of the excavation), which is a high-energy environment, given the gravel lithology and the size of fossils that the channel contains; the second is along the channel edge (i.e., the northern part of the excavation), with moderate to high-energy depositional processes that account for sedimentation.[2] This is reflected in some archaeological materials.

Our team retrieved 197 lithic artifacts from the surface and 155 *in situ* in the excavation. An initial study of this lithic assemblage was made by de la Torre (2005). Not all these artifacts were available for the detailed study presented in this chapter and the analysis below is based only on those 152 pieces which were in our hands at the

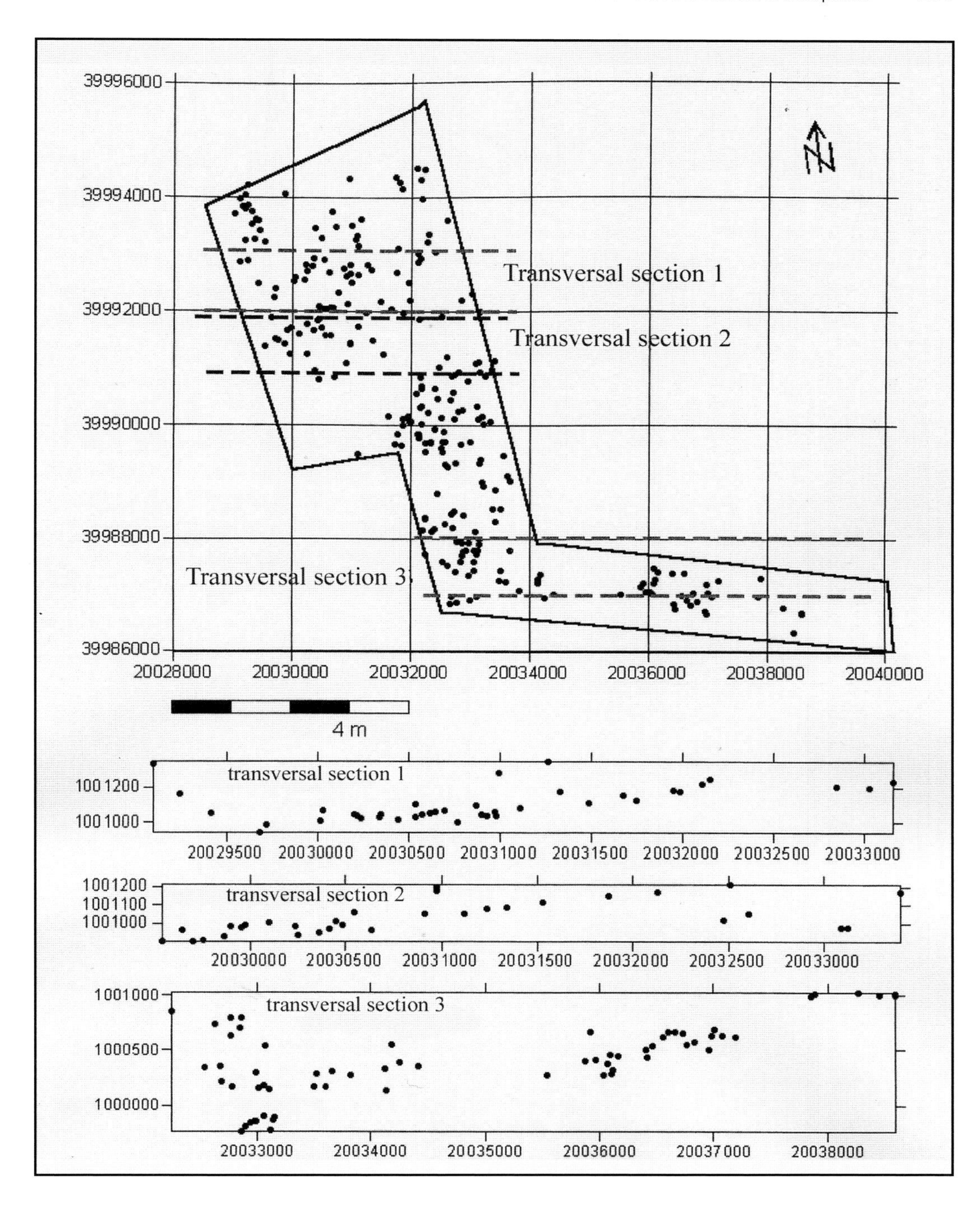

Figure 10.9 Horizontal and vertical distribution of artifacts at PEEN1 by section (E-W) (redrawn from de la Torre 2005).

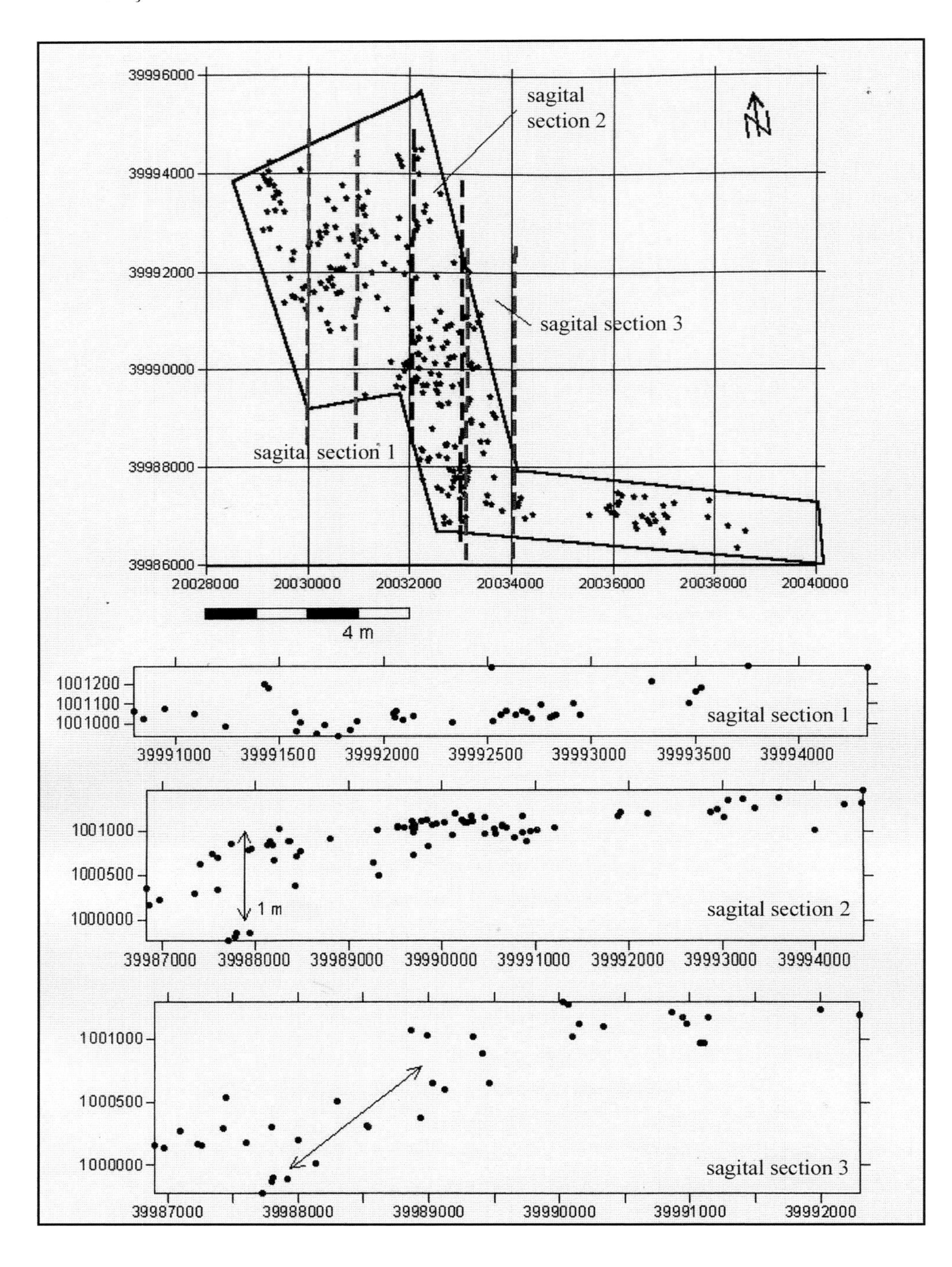

Figure 10.10 Horizontal and vertical distribution of artifacts at PEEN1 by section (N-S) (redrawn from de la Torre 2005).

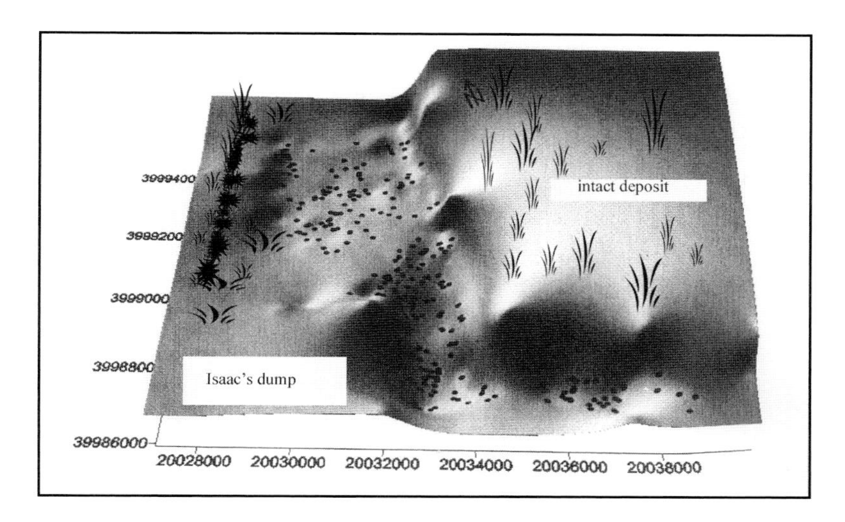

Figure 10.11 Isometric reconstruction of artifact distribution at PEEN1. Note the somewhat horizontal distribution in the background (attributable to the lower edge of the former river overbank), and the dipping slope representing the channel in the foreground (redrawn and modified from de la Torre 2005).

moment of analysing the site.[3] A total of 126 bone fragments were also retrieved *in situ*. The taphonomic analysis of bones is particularly informative. Most of the bones are fragments smaller than 3 cm (Figure 10.12), all of which show intensive abrasion and polishing; these are signs of size-sorting selection and long-distance fluvial transport. No functional relationship whatsoever can be claimed for the spatial association of bones and stones. Bones were deposited in the site as another sedimentary particle, given their similar size to the gravels and clasts geologically associated with them. These transported bones are found across the entire excavated surface.

A high frequency of abrasion is also observed in some stone tools. A total of 58.7% of lithic artifacts show some degree of abrasion. Of these, 43% show intensive abrasion. This abrasion cannot be from subaerial weathering, which often occurs with basaltic materials, since it also affects quartz pieces. More than 75% of quartz pieces show moderate to high abrasion. Like the bones, the abraded lithics appear across the entire excavated surface. In sum, serious post-depositional disturbances affected the site, transporting away small pieces and importing clasts, bones and perhaps lithics from other sources. Discriminating between local and allochthonous lithic pieces is not easy; local origin can only be convincingly argued for the heavy-duty tools, based on their large size and marginal abrasion.

Analysis of lithic artifacts from PEEN1

The 152 lithic artifacts, together weighing 41.6 kg, were studied by the authors. This sample includes pieces recovered both from survey and excavation. Figure 10.13 shows the distribution of the sample by artifact category. Detached pieces are the most abundant objects, representing 67% of the collection. Here three different categories have been considered: flakes, retouched flakes and large/heavy duty flakes (including non-transformed blanks and supports that have been transformed by retouch into large

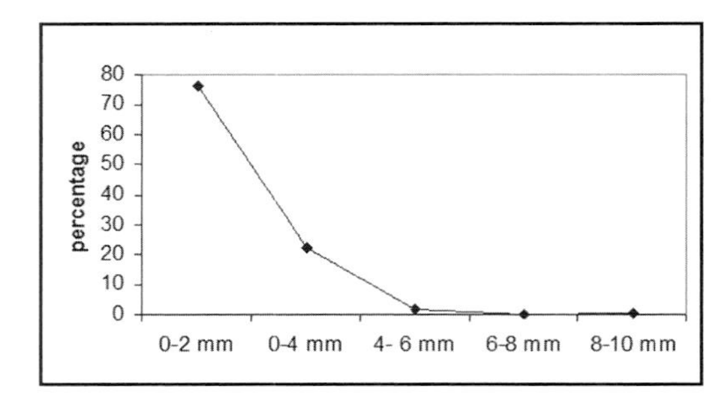

Figure 10.12 Bone specimen size distribution at PEEN1.

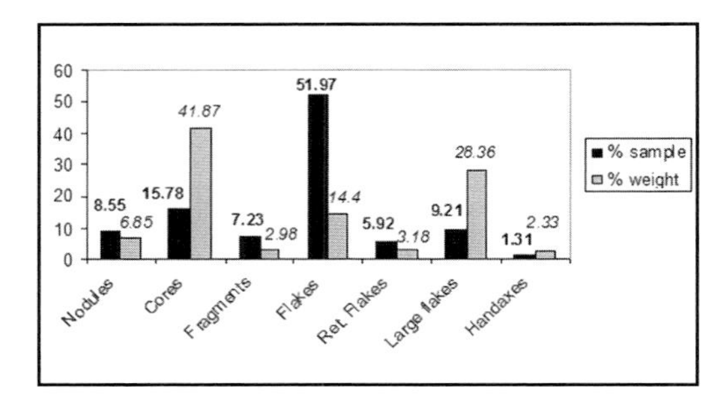

Figure 10.13 Percentage representation of technical categories, sorted by number of artifacts by category (bold type) and weight (in italics).

tools). Cores and core fragments represent 23% of the collection. In the category of nodules, both natural non-transformed small nodules and hammerstones have been included. When technical categories are sorted by their weight, cores (41.87%) and large flakes (28.36%) show the highest percentage values, while flakes comprise only 14.4% of the total weight.

Three different types of raw materials have been used in the knapping processes carried out at the North Escarpment sites: basalt, nephelinite and quartz. As can be seen in Table 10.1, basalt is the predominant raw material. It is worth noting that within this igneous rock type, a wide range of qualities and textures are included. Low-quality basalts predominate; these include coarse-grained rocks and rocks with porous textures, abundant fissures and weakness planes. Although some fine-grained basalts also occur, as a general rule hominids at PEEN1 were using low-quality basalts (ill-suited for conchoidal knapping) in considerable proportions. This is corroborated by the abundant number of step fractures recorded in detached pieces (24% of these objects show step fractures).

Table 10.1 also shows artifact categories sorted by raw material type. Obviously, basalt has the highest percentage values in most categories, although it is worth mentioning that quartz produced the majority of fragments. At PEEN1 most of the quartz materials retrieved are coarse-grained. Therefore, it is understandable that this type of raw material predominates in fragments and hammerstones (50% of nodules involved in percussion activities are quartz objects) and that it is not represented in the small and large tool configuration sequences. However, these limitations did not prevent quartz from being repeatedly used in production sequences, as can be deduced by the relatively high percentage of quartz flakes retrieved. Although nephelinite is the best raw material (usually having a fine-grained and homogeneous texture), it is only marginally represented in the collection. The low percentage of nephelinite and the abundance of poor-quality rocks might mean that at PEEN1 hominids were having problems in finding and/or selecting good rocks for their knapping strategies and that most of these activities were carried out with ad hoc raw materials found in the vicinity of the site.

With regard to the degree of abrasion observed on the external surfaces of the objects, four weathering groups have been established. W1 indicates fresh to slightly abraded surfaces;

Table 10.1 Distribution of technical categories by raw material

Category	Basalt (%)	Nephelinite (%)	Quartz (%)	Total
Nodules	11 (84.61)	-	2 (15.38)	13
Cores	18 (75)	2 (8.33)	4 (16.66)	24
Fragments	4 (25.36)	1 (9.09)	6 (54.54)	11
Flakes	53 (67.08)	8 (10)	18 (22.7)	79
Ret. flakes	8 (88.88)	1 (11.11)	-	9
Large flakes	12 (85.71)	2 (14.28)	-	14
Handaxes	1 (50)	1 (50)	-	3
Total (%)	107 (70.39)	15 (9.86)	30 (19.73)	152

Percentage representation by tool type category is provided in parentheses.

W2 indicates objects with signs of slight to medium abrasion; W3 indicates pieces clearly abraded and, finally, W4 indicates objects with heavily patina-covered surfaces. As can be seen in Figure 14, most artifacts have been weathered to a certain degree. Only 18% (n=12) of the total number of artifacts included in the W1 category are completely fresh, while the number of pieces that have been subjected to intense weathering (W3-4) constitute 24% of the sample. This observation is in agreement both with the contextual environment of the site (a fluvial channel) and with the idea that processes related to fluvial forces or alternatively, to prolonged periods of sub-aerial exposure, might have played a role in the modification of lithic pieces and the formation of the site. Therefore the sample retrieved might have been the result of various depositional events; some pieces might be related to *in situ* deposition (although small waste resulting from knapping processes is not present in the collection), while others might have been due to the accumulation of materials displaced down the channel from relatively nearby spots (de la Torre 2005:476).

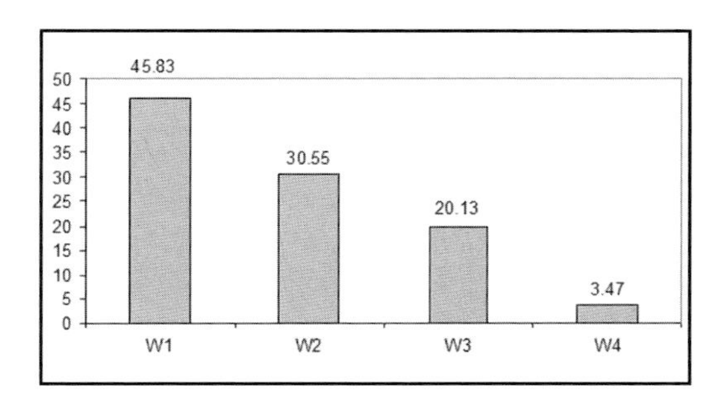

Figure 10.14 Distribution of lithic artifacts by weathering group.

Nodules and hammerstones

At PEEN1 a total number of 13 nodules have been retrieved. Most of these objects (61.53%) are small, natural, unmodified basalt nodules. However, there are five pieces (three basalts and two quartzes), which show traces of percussion on their surfaces. Mean dimensions of these hammerstones are 104 x 77 x 50 mm, and the mean weight is 571 g.

Flake production and reduction sequence

In this section the technical categories related to core exploitation and flake production are included: i.e., cores, core fragments, flakes, flake fragments and retouched flakes. Large flakes used as supports for their transformation into large-format tools will be considered alongside handaxes in the section devoted to large tool configuration chains.

Table 10.2 Dimensions of cores from PEEN1

	minimum	maximum	mean	std. deviation
length	46	180	99.37	34.57
width	36	147	81.66	23.83
thickness	28	92	62.41	16.47
weight	61	2,090	726.91	550.26

Reduction strategies

Twenty-four cores were retrieved at PEEN1. Table 10.2 shows measurements of the objects included in this technical category. Most (90%) of the cores have been moderately exploited, showing traces of reduction on all their surfaces, although only one exhausted core has been observed (a core fragment that shows a sequence of small negative scars). A total of 86.36% of the cores show no signs of cortex on their upper surfaces (59% retain no cortex on the lower surfaces).

Conversely, 13.6% retain some cortex on their upper surfaces and 41% in their lower surfaces (the most abundant cortical presence is two-quarters of the surface covered by cortex). The mean number of negative scars per core is 8.6.

A systematic classification of the core reduction strategies at PEEN1 is problematic. A number of basalts (the main raw material used in this category) are low-quality rocks and identification of the exact direction of flaking is sometimes ambiguous. At the same time, some pieces have been heavily altered by diagenetic processes and weathering, making the analysis of the flaking patterns followed difficult. Some other pieces have been fragmented and the interpretation of their original volumetric shapes has only been possible by interpretive reconstruction. Taking into account all these problems, the classification of cores has been carried out mainly by the interaction of two types of attributes: faciality (unifacial, bifacial, trifacial, and multifacial exploitation of the natural volumes), and scar directionality or polarity (unipolar, bipolar, multipolar and centripetal).

The different combinations of faciality and polarity have been summarized in four basic exploitation models (Figure 10.15), whose percentage distribution is shown in Figure 10.16.

E1. Unipolar simple. This consists of the repeated exploitation of natural planes of the cobble using the same horizontal plane as the striking platform. Cores exploited in one, two and three different natural planes or faces have been included here.Longitudinal negative scars are observed and sub-parallel flakes are obtained. Five cores (one in quartz and four in basalt rocks), which show the lowest degree of exploitation and the highest proportion of cortical areas retained, have been included in this type. One of these pieces might be a large flake fragment or a tabular piece of basalt. The mean number of negative scars in E1 type cores is 5.25.

E2. Unipolar bifacial. This type is comprised of cores that are exploited in two faces, using a portion of the same natural edge as the striking platform. The sequence of flaking may be alternate between each face, although it is common that exploitation takes place first on one face and then on the other. Therefore, previous negative scars on one face are used as the striking plane to exploit the other. Three basalt cores have been included in this flaking type. One of them is a rather heavy nodule (155 x 147 x 92 mm), clearly exploited in

Table 10.3 Measurements of flakes sorted by their dimensional module

	minimum	maximum	mean	std. deviation
17-40 mm				
length	17	40	28.55	6.9
width	16	61	32.18	9.95
thickness	5	29	11.7	4.95
weight	2	44	12.81	10.91
41-80 mm				
length	41	77	53.64	9.9
width	32	107	54.67	18.05
thickness	9	37	18.32	6.14
weight	18	244	61.5	46
>80 mm				
length	85	132	98.28	15.71
width	60	123	84.14	21.61
thickness	15	39	32.14	8.29
weight	93	498	296.14	134.98
all flakes				
length	17	132	50.53	23.94
width	16	123	50.21	23.69
thickness	5	39	32.66	8.95
weight	2	498	75.98	105.4

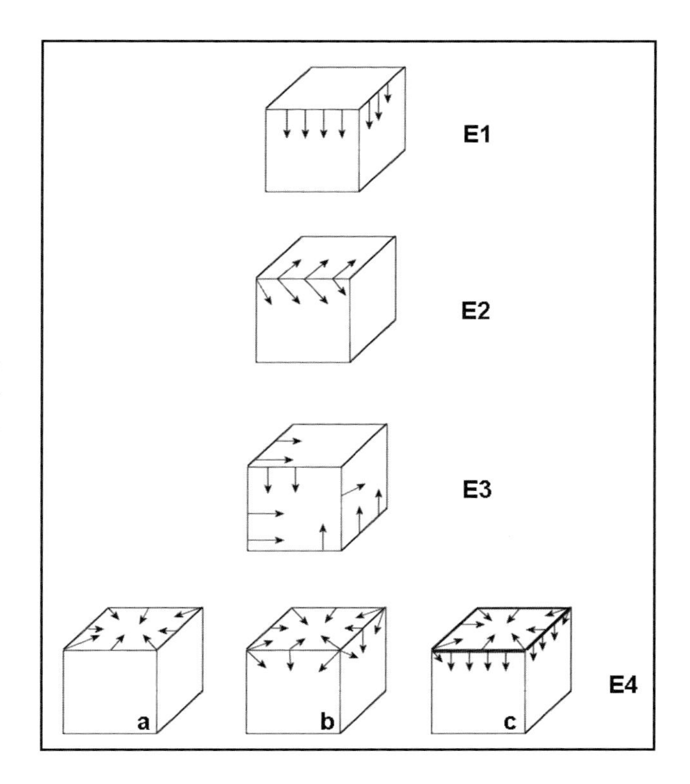

Figure 10.15 Schematic representation of reduction strategies documented at PEEN1: E1, unipolar simple; E2, unipolar bifacial; E3, multipolar; a) E4, unifacial discoid; b) bifacial discoid; and c) predetermined.

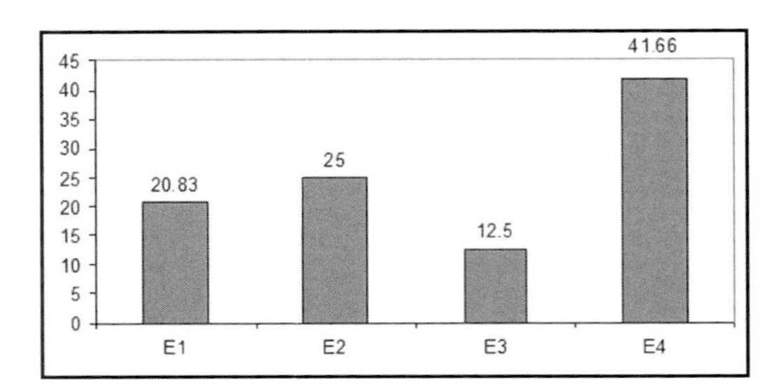

Figure 10.16 Percentage distribution of core types.

a unipolar bifacial fashion, that has produced eight medium-sized flakes (measurements of the largest negative scar are 81 x 77 mm). The mean number of negative scars in this core type is 7.

E3. Multifacial/multipolar. Commonly known as polyhedral cores, this type consists of cores that show negative scars in more than three faces and more than three striking planes. Six nodules (three basalts, two quartzes and one nephelinite) have been included in this reduction model. One of the pieces exploited in multiple planes is a large and heavy nodule from which a large flake (128 x 110 mm) has been obtained in the horizontal plane, and which shows other negative scars in the transversal plane with multiple directionalities. Although this is the only possible example of a core for the production of large flake blanks in the studied sample, it is not evident that the detachment of the main flake follows a planned or organized reduction pattern. Another object, which could be considered a very coarse trihedral pick, has been included in this core type. The mean number of negative scars is 8.5.

E4. Discoid. This reduction model is comprised of cores in which a radial flaking pattern has been recognized. Radial exploitation constitutes the most abundant reduction type in the core sample analyzed. The mean number of negative scars observed in centripetal cores is 10.75. Three different types of radial reduction have been included here (Terradas 2003): First, unifacial discoid cores, where centripetal exploitation is performed on one side. The other side shows no signs of radial flaking or preparation of the striking planes. Three cores have been considered as this type: two basalts and one nephelinite. Second, bifacial discoid cores, in which centripetal exploitation can be observed on two faces. Six cores have been included in this group: one quartz and five basalts. Three cores clearly share the patterns that commonly define the bifacial non-hierarchized discoid flaking method (Boëda 1993) and the other two could show hierarchization of both surfaces, one being the preparation plane subordinated to the radial flaking performed on the main surface. Third, predetermination cores, in which all the characteristics that define the production of predetermined flakes can be observed (Boëda 1995). One basalt core, from which a preferential medium-sized flake was detached (81 x 56 mm), has been included in this group.

A sample of 11 core fragments has also been retrieved; including three artifacts whose support would have been a flake or a flake fragment. Quartz fragments predominate. The mean measurements are 66 x 46 x 27 mm and the mean weight is 113 g. The mean number of negative scars is five. Although it is difficult to reconstruct the reduction patterns which produced these fragments, at least in two cases a radial organization of scars has been observed.

Detached products

Detached objects (flakes, flake fragments and retouched flakes) constitute 58.55% of the total sample studied and 71.77% of the technical categories included in the exploitation operational chains. Figure 10.17 shows the dimensions of non-fragmented flakes; note that flake dimensions vary widely (from 17 to 132 mm in length, for example). In order to carry out an accurate metric and descriptive study of such a varied sample, and in order to better understand the different knapping strategies to which detached products can be related, plain flakes have been included in three different groups by length (Table 10.3): M1 (17-40 mm), M2 (41-80 mm) and M3 (>81 mm). Here, length has been taken as the reference measurement criterion of each module because mean length of all flakes is greater than mean width. At EN1, M2 flakes are the most abundant pieces (51.13%), followed by M1 flakes (36.36%) and M3 flakes (12.5%). M1 includes four pieces of debris (conventionally, debris is considered to be very small flakes less than 20 mm long), and 27 small flakes; M2 includes 45 medium-sized flakes and M3 includes 11 large flakes and fragmented large flakes. Previous analysis by de la Torre (2005:481) has interpreted this dimensional diversity as the result of three different knapping processes involved in the production of Acheulean tools. In the small flake group, a number of retouching flakes (the waste resulting from façonnage processes carried out in the configuration of large tools) would be included. Medium-sized flakes would include a number of flakes resulting from the preparation processes carried out on large cores exploited for the production of large flake supports. Large-sized flakes would include potential supports to be transformed into large configured tools. A number of parameters have been considered to determine to what extent technical traits can help us to distinguish which flakes in the sample belong to various processes related to production and configuration of large flake tools, versus simply to core exploitation. These parameters are: number of flake dorsal scars, thickness of flakes and butts, and negative scar dimensions.

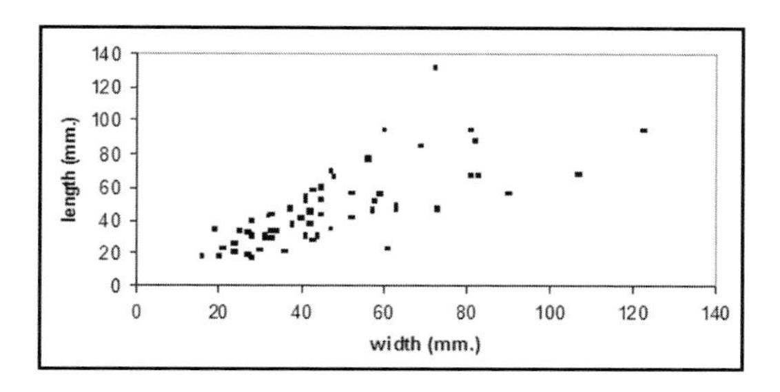

Figure 10.17 Size scatterplot showing length and width of flakes (including retouched flakes).

The number of dorsal scars in flakes ranges from one to 10 (mean=3.37). Four groups have been established based on the number of scars observed on dorsal faces: 1 (one dorsal scar); 2 (two to three scars); 3 (four to five scars); 4 (six or more scars). Figure 10.18 shows the distribution of flakes by dimensional module (described above) and by dorsal scar group. In all three modules, flakes which have two to three dorsal scars are the most abundant, representing always more than 66% of each group. It appears that regardless of the dimensional range considered, flakes at PEEN1 are the results of rather organized production processes, with the performance of various knapping sequences in the cores from which they were obtained. However, it also seems clear that the percentage of dorsal scars, and therefore the complexity of knapping patterns, is greater in medium-sized and large flakes. With regard to flakes with one single dorsal scar (and with no dorsal ridges), Dag and Goren-Inbar (2001) have suggested that dorsally plain flakes (when flat and bearing uni-faceted butts) are in many cases the by-products of configuration activities, this is to say, flakes obtained in retouching processes. In the sample studied here, non-cortical flakes with no dorsal ridges are more abundant within the small flake group (reaching 26.6%), although flakes with the same

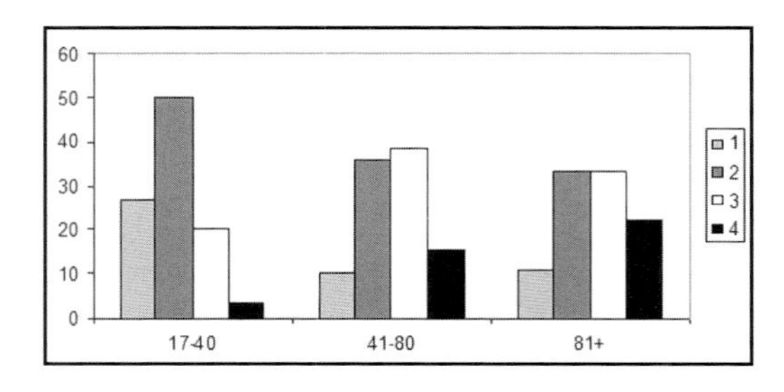

Figure 10.18 Percentage distribution of flakes sorted by dorsal scar number: 1= one dorsal scar; 2= two to three; 3=four to five, and 4= six and more.

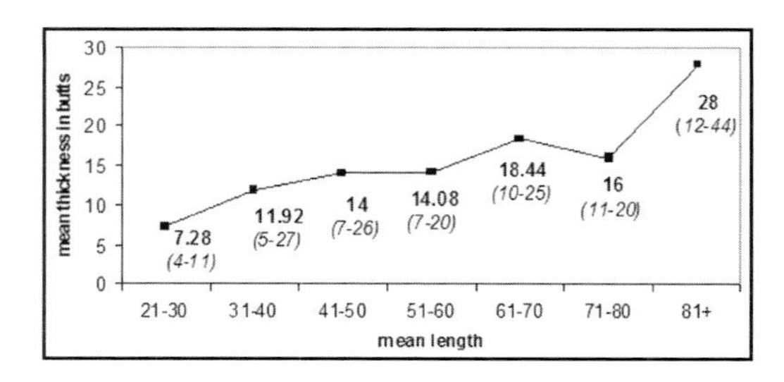

Figure 10.19 Mean thickness of striking platforms in flakes, sorted by seven dimensional groups (minimum and maximum values of each group are in italics).

dorsal patterns have been observed even in large flakes.

Along with the mean values for thickness presented in Table 10.3, flake thickness has been studied in accordance with two different complementary parameters: carination index (CI) and thickness of the striking platform. According to Laplace's (1974) carination index, 47.72% of flakes are thick (CI between 1 and 2.23) and 52.27% are thin (CI >2.23). When sorted by their dimensional module, thick or carinated flakes represent 47% of M1, 47% of M2 and 55% of M3. Figure 10.19 shows mean thickness of striking platforms in flakes sorted by seven dimensional groups. Here, it is apparent that butt thickness tends to increase in accordance with dimension, reaching very high values in the group of large flakes. However, as a general rule, striking platforms tend to be rather thick when flakes larger than 40 mm are taken into consideration.

In sum, all the parameters considered above show that we are dealing with a varied sample of detached products whose traits cannot be definitively attributed to a particular flaking process. It seems plausible that within the smallest flake group there are several by-products that are the result of retouching activities carried out in the configuration of large tools. In addition, most of the retouching negative scars whose length was measured in large retouched flakes fall between 31-35 mm. It is difficult to ascertain that larger negative scars in some of these tools are the result of retouching activities; for example, one large flake could be classified as a pick, since its large negative scar (62 x 53 mm) seems to have been the result of a flaking process rather than a retouching process. Although flat flakes predominate in the small and medium flake groups, medium-sized and in particular, large flakes, tend to be rather thick in their butt area. This means that as a general rule hominids were applying strength in their percussion impacts on heavy basalt blanks (which is particularly obvious in the case of large flakes), and that points of percussion were quite invasive in the flaking process. Although at PEEN1 there was a clear knapping strategy leading to the production of large flakes suitable for their transformation into large retouched tools, available data cannot empirically confirm if medium-sized flakes are in fact by-products related to this process. As we have already seen, large cores for the production of large flakes are almost absent in the analyzed sample and therefore it is impossible to reconstruct in detail the techniques performed to obtain these supports. Negative scars measured in the core sample retrieved (Table 10.4) show that at PEEN1 hominids were

Table 10.4 Measurements of negative scars observed in the core sample

	minimum	maximum	mean	std. deviation
length	17	128	48.05	22.18
width	17	110	47.33	16.69

mainly producing medium-sized flakes (46.15% of the negative scars measured in cores, and 51.13% of the flake collection) and small flakes (43.58% and 36.36%), with uni-faceted butts (79.31% of these scars would have had a plain striking platform). With this data at hand, it is easier to suggest that most of the small and medium-sized flakes are the result of a reduction process of medium-sized cores rather than the by-products of a structured target of large flake production. This does not mean that large flakes could have not been obtained in the spot, but available data and the absence of cores for the production of large flakes support the idea that the flake collection is mainly the result of a core exploitation pattern aimed at producing medium-sized flakes.

With regard to other descriptive parameters, although clear discrimination of cortical areas is problematic here (since occasionally, surface abrasion and raw material quality make it difficult to distinguish between cortical and non-cortical surfaces), at PEEN1 most flakes fall within the last stages of the reduction sequence, i.e., Toth's (1982) Type VI (Figure 10.20). Cortical and semi-cortical products are very rare; only 2.28% of flakes belong to the initial flaking stages (Types I, IV). The predominance of non-cortical products may indicate that the initial stages of knapping did not take place on-site, or that some sort of taphonomic bias might have accounted for this observation, considering the depositional environment. Striking platforms are mainly non-cortical and uni-faceted butts predominate, while cortical platforms are scarce (Figure 10.21). These

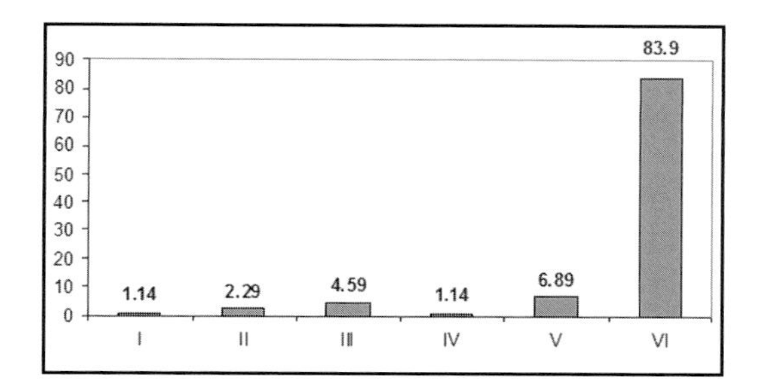

Figure 10.20 Percentage distribution of flake types according to Toth's (1982) model.

data are in agreement with the flaking patterns observed in the core negative scar sample, although here cortical butts seem to be overrepresented: uni-faceted butts in core scars represent 79.31%, followed by cortical (17.24%) and bi-faceted butts (3.44%). In sum, flakes at PEEN1 come from later stages of the reduction sequence, where striking platforms were prepared in some way. Traits observed in butts suggest that striking platforms used mainly previous extractions as the platform from which to detach the following flake, and that more complex platform preparation systems (bi-faceted and multi-faceted) are not very numerous (17.56%).

Small format tool configuration

At PEEN1, there are nine flakes and flake fragments that show evidence of retouch on their edges. Another six flakes (7.5% of the plain flake category) have probably been trimmed, but here retouch is isolated and we cannot with total confidence assign them to the group of small format

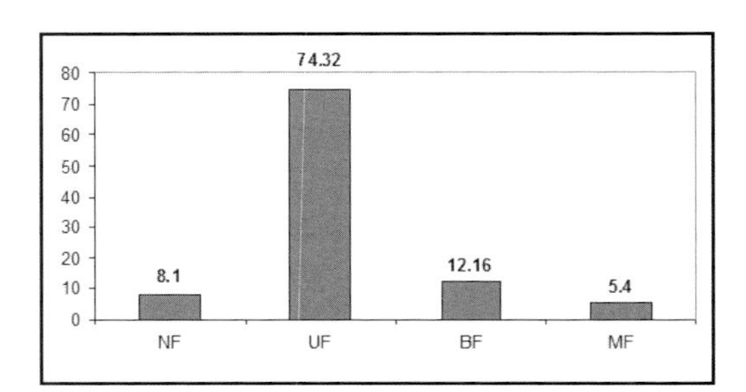

Figure 10.21 Percentage distribution of striking platform types in flakes. NF, non-faceted or cortical; UF, uni-faceted or plain; BF, bi-faceted or dihedral; MF, multifaceted.

retouched objects. Therefore, they have been included in the plain flake section. Table 10.5 shows mean measurement values of the retouched flakes. On average, retouched flakes are larger and heavier than plain flakes, although the retouched flake sample is too small to say that this dimensional pattern would have influenced the selection of products for their transformation into retouched tools. However, it is worth mentioning that at least two of the objects included in this category might be retouched fragments from large flakes. Retouched flakes share similar technical patterns to those observed in plain flakes. Most of these pieces show no traces of cortex on their dorsal surfaces or butts (78% of pieces are Toth's Type VI), although one Type II and one Type III flake have been found. In those instances where the striking platform has been preserved, five butts are uni-faceted, two are cortical and one is multi-faceted. The mean number of previous scars on dorsal surfaces is 3.55 (interval 1-5) and the mean edge length is 82.25 mm. The mean striking platform thickness is 23.3 mm. The location of retouch is mainly right- (n=4) and left-sided (n=2), although distal and proximal retouches have also been recorded. Retouch is mainly continuous, abrupt (n=5) or semi-abrupt (n=4) and deep (n=6). An equal number of pieces show exterior, interior or bifacial retouch. From a typological perspective, retouched objects can be defined mainly as scrapers (67%: three side scrapers, two transverse and one proximal), although a denticulate, a notch and a trimmed flake have also been observed.

Large format tool reduction sequence

Large retouched flakes. A sample of 14 large flakes, the majority of which have been transformed by retouch, has been studied from PEEN1. The raw material is mainly basalt (85.7%), followed by nephelinite. In all cases, the maximum dimension of the flake (length or width) is always larger than 100 mm. As the mean dimensional values show (Table 10.6), large flakes tend to be wider than longer. Ventral surfaces show no signs of bulb in 71.42% of these large flake tools. Striking platforms are quite thick, showing a mean length of 40.5 mm. Removed butts by later retouch are the most abundant (46.14%), followed by uni-faceted (30.76%), non-faceted (15.38%) and broken butts (7.69%). Dorsal surfaces are mainly non-cortical (84.61%); in 15.38% cortex is present on two-quarters of the dorsal surface and only one flake (7.69%) has a completely cortical dorsal surface. The mean number of negative scars on the dorsal surface is 4.18. In eight cases, the arrangement of scars on dorsal surfaces shows that these flakes were obtained from radial reduction strategies. Mean cutting edge length is 220 mm.

Identification of retouch is problematic on these large flake tools. Retouch is always coarse and cannot always be unambiguously distinguished from scars due to exploitation. Mean length of measured negative scars is 37.11 mm (interval 31-62 mm) and mean width is 48.33 mm. (interval 32-59 mm). Three large flakes, whose edges have been damaged by pseudo-retouch or show signs of abrasion produced by use, have been considered to be non-retouched.

Table 10.5 Measurements of negative scars observed in the core sample

	minimum	maximum	mean	std. deviation
length	40	121	66.66	27.46
width	37	89	58.66	19.43
thickness	14	39	26	8.66
weight	32	425	147.44	137.02

Table 10.6 Dimensions of large flakes from PEEN1

	minimum	maximum	mean	std. deviation
length	77	185	122.57	37.95
width	80	260	129.78	48.85
thickness	39	73	53.37	12.64
weight	429	1,650	844	311.37

Here a very large and heavy cortical flake (131 x 260 x 63 mm and 1650 g) has been included. Another large and thick flake has been considered to be a rather coarse trihedral pick, showing large side scars that do not seem to be the result of retouch. Therefore, a trihedral distal tip might be the result of exploitation carried out on a very thick flake rather than a specific design. Retouch has been performed in the remaining 10 large flakes, one of which could be a cleaver whose transversal edge has been damaged. Retouch is transverse (n=5) and sided (n=4), discontinuous (n=6), abrupt (n=6) and semi-abrupt (n=3), deep (n=6) and exterior (n=6). From s typological point of view, these large flakes are cannot be considered true bifacial tools, and share no technological or morphological similarities with the two handaxes retrieved at PEEN1 or with the biface collection studied at ST69. Large flakes were residually transformed by a coarse and abrupt retouch that in many cases was carried out to reduce butt thickness. Abrupt retouch on the striking platform area is usually opposed to a natural cutting edge that, in some cases, has been marginally trimmed. Therefore, some of the traits defining large bifacial tools are lacking in the studied sample; these include volumetric transformation due to bifacial reduction and retouching in the perimeter of the blank, symmetry on the horizontal plane, and transversal biconvexities. As other researchers have already suggested (de la Torre 2005:498), that these objects could be typologically defined as knives (*sensu* Kleindienst 1962) rather than normative bifacial tools.

Handaxes. Two handaxes have been retrieved at PEEN1. The first one has been made on a basalt nodule blank, is fairly weathered, measures 144 x 70 x 39 mm and weighs 450 g. This object has been extensively exploited and does not retain any cortex. Thirteen negative scars have been observed, the largest being 38 x 56 mm. Hard-hammer retouch has been carried out on both surfaces and along the entire perimeter of the piece (the length of cutting edges is 250 mm), although it particularly enhances the distal tip. Retouch is marginal and mostly simple. From a sagittal view, the cutting edge is sinuous while from a horizontal view this tool is non-symmetrical. Thickness relative to the width index (m/e=1.79) shows that this handaxe is thick and from a typological point of view, it can be classified as amygdaloid (Bordes 1961).

The other handaxe has been created from a nephelinite nodule. It measures 130 x 8 x 46 mm and weighs 524 g. No cortical areas have been observed on either surface and 10 negative scars have been counted (the largest removal being 67 x 28 mm). The lower end is thick. As in the other object, hard-hammer retouch, mainly simple and marginal, has been performed on both faces. The distal tip is very pronounced. The cutting edge (200 mm of total length) is sinuous and the frontal morphology is rather symmetrical. The refinement index (1.78) shows that this handaxe is thick and it falls within Bordes' amygdaloid type.

Technological patterns at PEEN1: Conclusions and discussion

At PEEN1 previous and modern excavation and survey programs have enabled us to retrieve a fairly abundant lithic collection that can help reconstruct the technological behaviors of the hominids at this site 1.3–1.2 Ma. The lithic sample studied for this chapter (152 objects) shows the following characteristics:

1) Lithic objects were mainly produced from volcanic rocks (80%), although metamorphic raw materials have also been recorded. The majority of the knapping was performed on a variety of local basalts, most of which are coarse-grained and are poor materials for conchoidal fracture. However, some fine-grained basalts and good-quality nephelinite were also used. Quartz is generally low-quality (coarse-grained) and would have been a difficult raw material to exploit. Therefore, at EN1 hominids were knapping low-quality rocks in high proportions, which might account for some of the morphological and technological traits observed in the collection.

2) Most of the lithic artifacts can be attributed to exploitation processes (cores, flakes and retouched flakes). The production of different types of flakes for a variety of purposes would have been the main goal of the knapping activities performed at the site. Cores are relatively abundant. Reduction strategies preferentially tend towards centripetal exploitation in different degrees of complexity: unifacial, bifacial discoid and predetermined. Multipolar reduction models are also well-represented. However, most of the cores were exploited to produce small and medium-sized flakes. The core sample studied here includes only one heavy and large core, from which a large flake was obtained. With the data we have at hand, it is impossible to unambiguously reconstruct which reduction strategies were performed to obtain the abundant large flake supports recovered at the site.

3) Plain flakes and flake fragments are the most abundant technical category. Most of these products retain no cortex on their dorsal surfaces and show some degree of preparation in their striking platform areas. As flakes become larger, the butt area becomes thicker. The patterns observed in dorsal scars and dorsal ridges suggest that these products come from fairly organized exploitation models, in which radial flaking (as confirmed by core type distribution) might have played a significant role. However, first-generation flakes are lacking from the sample. It is difficult to explain this bias by post-depositional disturbance (which should not have such an impact on qualitative parameters), and probably some sort of behavioral aspects should be considered here. A small group of large flakes and large fragmented flakes have been included in this category. Their dimensions suggest that these pieces probably were blanks to be transformed into large tools. Medium-sized retouched flakes are scarce. Most of them have been transformed into different types of scrapers.

4) Two types of large tools have been observed. Two pointed and thick handaxes were retrieved that show all the patterns related to configuration and volumetric transformation of bifacial large tools.[4] On the other hand, at PEEN1 knapping processes also led to the production of very large flakes that were partially transformed by abrupt and coarse retouch. Hominids did succeed in a systematic production of these large supports, although taking into account that cores for the production of large flakes are underrepresented, it is possible that most of these knapping processes took place off-site. In most instances, this retouch is performed in order to reduce butt thickness, although partial transformation of natural edges is also observed. Retouched areas are usually opposed to natural cutting edges. Volumetric transformation of these objects, far from normative bifacial tools, is absent.

In sum, taking into account the chronological framework and the technological characteristics reported here, PEEN1 stands as one of the earliest sites where traits related to the early stages of the Acheulean techno-complex are observed. Hominids were able to obtain very large flakes that were partially transformed into a variety of heavy-duty knife types, although the small collection of true handaxes performed on nodule supports could suggest that they also were succeeding in manipulating volumes into bifacial tools, the trademark of the Acheulean techno-complex. Considering that hominids were producing true handaxes at PEEN1, it is difficult to say (if only looking at the sample analyzed) if the coarsely transformed large flakes are the result of a specific and desired "less-effort" configuration strategy.

Taking into account the array of complex technological behaviors displayed, we are inclined to suggest that these large flake tools were intentionally planned. While the reduction sequence related to the production of large flakes is an important part of the knapping strategies performed at PEEN1, hominids were repeatedly producing small and medium-sized flakes that only on rare occasions were transformed into retouched tools. Technological traits observed in the flake sample and in the cores from which they would have been detached show that knapping processes were rather elaborated. Hominids successfully controlled a variety of discoid reduction methods, to the point where they even were able to produce some predetermined flakes. The characteristics defining exploitation and configuration processes stated above stress that the basic principles involved in the dawn of the Acheulean techno-complex had been successfully performed at PEEN1. However, the lack of functionality between stone tools and bones at the site, together with the unexplained existence of standarized pointed large flakes (Isaac 1967; de la Torre 2005) makes the economic behavior of hominids at the site difficult to reconstruct.

Endnotes

1 Recently, De la Torre et al. (2008) offer a biased account of our research in Natron in which they claim that our project lent "minor attention to the record from Bayasi MHS and Mugulud-RHS". They only refer to the time (two field seasons for Mora and three for de la Torre) during which they participated in our project. Research on the escarpments were conducted before and after their ephemeral participation in the Natron project. As this bookgoes to press, work in the escarpments is still being led by one of us (FDM). Truthfully, a similar area was excavated in the escarpments compared to excavations in Maritanane as can be seen in this book.

2 This is not clearly defined in de la Torre et al. (2008) work, leading them to produce a statement of confusion as to how the site was formed and post-depositionally altered. Their report does not include the geological study shown here, which is essential to understanding the site formation history.

3 Some materials had not been handed over by the time this study was conducted. This explains why his sample is larger than that presented here.

4 De la Torre et al. (2008) claim that no bifaces were made by Peninj hominids, since the distribution of volume is "asymmetrical, retouching is non-invasive, and central volumes of blanks remain unmodified". Although a large part of the heavy-duty tools crafted by the Peninj hominid were as defined by de la Torre et al. (2008), in every Acheulian site, completely bifacial tools exist with invasive knapping, central volumes of blanks modified and with a symmetry axis. Dozens of clear bifaces were found in ST-69 (see Chapter 8), only slightly younger than PEEN1, and bifaces exist in PEES 2 (see Chapter 9:Figures 9.16–9.17), and PEES1 (Chapter 9:Figure 9.6). De la Torre et al. (2008) did not have access to the collections from PEES1 and PEES2 excavated by us, thereby using a partial information for their statements.

Figure 10.22 Upper face of a hierarchized bifacial discoid core on basalt (Type 4c).

Figure 10.23 Preparation face of the core .

Figure 10.24 M1 flakes (17-40 mm) without ridges on the dorsal face.

Figure 10.25 M2 flake (41-80 mm).

Figure 10.26 M3 flake (81> mm).

Figure 10.27 Side scraper on a step-fractured flake.

Figure 10.28 Large retouched flake (knife) on basalt. Abrupt retouch has been performed on the distal edge.

Figure 10.29 (right) Large retouched flake (knife) on basalt. Simple retouch has been performed on the proximal area in order to reduce the striking platform thickness.

Figure 10.30 Large retouched flake (knife) on nephelinite. There is abrupt retouch on the left side of the flake, and a natural non-modified edge on the right side.

Figure 10.31 Amygdaloid handaxe on basalt.

Figure 10.32 (left) Amygdaloid handaxe on nephelinite.

11

Conclusions

Manuel Domínguez-Rodrigo, Luis Alcalá and Luis Luque

The archaeology project at Peninj contributes to the understanding of early human behavior with two behavioral phenomena so far suspected but undocumented for Lower Pleistocene times. One of them is related to Acacia woodworking and the existence of a technology based on organic materials, which could have enhanced predatory strategies by early hominids. The data drawn from PEES2 are most valuable and shows that one of the (very likely multiple) tasks for which large knives and handaxes of the early Acheulian were intended was to produce wooden implements. The other phenomenon is the existence of butchery places, which were repeatedly used by Lower Pleistocene hominids.

Ethnographic observations of large animal kill sites of modern Hadza hunter-gatherers in Africa have shown that butchery sites can show dense concentrations of bone refuse near water sources. Taphonomic indicators of these butchering sites can be extremely diagnostic. Given that carcass remains are not widely shared in these *loci*, bones from the same individual tend to form a spatially discrete cluster. When several individuals are obtained simultaneously or at different intervals, bone clusters from different animals are independent or minimally overlapping. Importantly, anatomical sections frequently are abandoned articulated. Bones discarded at home bases rarely appear articulated, whereas this is common in kill sites. Bones from carcasses deposited over multiple-year intervals will show differential subaerial weathering.

Peninj preserves some 1.5–1.2 million-year-old butchering sites in the ST Site Complex. One of them, ST4, is characterized by the partial semi-articulated distribution of different anatomical units of an extinct giraffe, with spatially differentiated distribution of taxa, which resembles the distribution of bones in Hadza kill sites and might be indicative of the place having acted as a butchering spot. The spatial association of bones and stone tools would support this (given the presence of cut marks and percussion marks), as well as the low density of artifacts and the presence of clusters of articulated bones. Several anatomical units of at least three different animals have been accumulated and manipulated by hominids in a spatially restricted area at the bottom of the deposit at the ST4 site. These animals were not deposited simultaneously but during repeated occupations of the site. The evidence from ST4 supports the hypothesis that lower Pleistocene hominids were processing carcasses and not just dismembered bones. Those hominids, at 1.5–1.2 Ma, created sites which are indistinguishable from modern foragers' "near-kill locations," and which might have served similar purposes. If so, planning and curation must already have reached a sophisticated level at that time. The existence of butchery (near-kill) sites, woodworking sites, and central places (at FLK Zinj in Olduvai), suggests that hominid behavior at that time was already quite varied and that the archaeological record must be explained using two complementary variables: ecological and behavioral factors.

Peninj also indicates that hominids enjoyed primary access to fleshed animal carcasses, without prior intervention by carnivores. This suggests a highly open and competitive environment where primary access to flesh had to be gained through predatory strategies, not by passive scavenging. Archaeological and taphonomic evidence from Peninj indicate that the scavenging hypothesis is not defensible anymore, at least not for hominids at 1.5-1.2 Ma. Hunting must be seriously considered.

Now that most of the older classical "hominid-created" sites from Olduvai Bed I have been revised to show minimal hominid involvement (Domínguez-Rodrigo et al. 2007), every single site where there is evidence of hominid carcass manipulation must be carefully scrutinized and treasured. At the time of publication, the only evidence prior to 1.5 Ma of a hominid-accumulated bone assemblage with clear evidence of butchery is the FLK Zinj site from Olduvai. Additional evidence of butchery appears at other sites in the form of isolated bones bearing cut or percussion marks. Both Peninj and Swartkrans (Pickering et al. 2004) appear as the oldest most solid evidence of hominid carnivory after Olduvai's FLK Zinj. The fact that only three sites embody most of the evidence (with Kanjera and BK pending publication) for hominid butchery for a period covering over one million years since the beginning of the archaeological record shows how little we know of early human behavior.

Peninj could yield further surprises in the future. This book describes the archaeology of Lower Pleistocene but remains belonging to the MSA (Figure 11.1) and LSA are far more abundant and appear scattered in thousands all over the landscape. Their lack of stratigraphic context, since they were deposited in the absence of sedimentary processes (evidenced by several lithics bearing a thick patina from exposure) is a handicap in their study. However they remain to be properly studied, and this should be considered in future research at Peninj.

Figure 11.1 Some surface collections of MSA lithic implements at Peninj.

References

Anderson, P. C.

1980 A testimony of prehistoric tasks: Diagnostic residues on stone tool working edges. *World Archaeology* 12:181-194.

Asfaw, B., Y. Beyene, G. Suwa, R. Walter, T. White, G. WodeGabriel, and T. Yemane

1992 The earliest Acheulean from Konso-Gardula. *Nature* 360:732-735.

Ayeni, J. S. O.

1975 Utilization of waterholes in Tsavo National Park (East). *E. Afr. Wildl. J.* 13:305-323.

Baker, B. H.

1958 Geology of the Magadi area. *Geol. Surv. Kenya Rept.* 42:79.

1963 Geology of the area South of Magadi. *Geol. Surv. Kenya Rept.* 61:27.

1986 Tectonics and volcanism of the Southern Kenya Rift Valley and its influence on rift sedimentation. In *Sedimentation in the African Rifts*, edited by L. E. Frostick et al. Geol. Soc. Spec. Pub. 25:45-58.

Baker, B. H., and J. Wohlenber

1971 Structure and Evolution of the Kenya Rift Valley. *Nature* 229:538-542.

Baker, B. H., and J. G. Mitchell

1976 Volcanic stratigraphy and geochronology of the Kedong-Ologersailie area and the evolution of the South Kenya Rift Valley. *Jour. Geol. Soc. London* 132:467-484.

Baker, B. H., L. A. J. Williams, J. A. Miller, and F. J. Fitch

1971 Sequence and geochronology of the Kenya rift volcanics. *Tectonophysics* 11:191-215.

Baker, B. H., P. A. Mohr, and L. A. J. Williams

1972 Geology of the Eastern Rift System of Africa. *Geol. Soc. Amer. Spec. Paper* 136:1-67.

Barakat, H.

1995 Middle Holocene vegetation and human impact in central Sudan: charcoal from the Neolithic site at Kadero. *Vegetation History and Archaeobotany* 4:101-108.

Barba Egido, R., and M. Domínguez-Rodrigo

2005 The taphonomic relevance of the analysis of long limb bone shaft features and their application to implement element identification: study of bone thickness and the morphology of the medullary cavity. *Journal of Taphonomy* 3:111-124.

Behrensmeyer, A. K.

1978 Taphonomic and ecologic information from bone weathering. *Paleobiology* 4:150-162.

Bell, K., and J. Keller (eds.).

Carbonatite Vulcanism. Oldoinyo Lengai and the Petrogenesis of Natrocarbonatites. Springer Verlag.

Berggren, W. A., F. J. Hilgen, C. G. Langereis, D. V. Kent, J. D. Obradovich. I. Raffi, M. E. Raymo, and N. J. Shackleton

1995 Late Neogene Chronology: New Perspectives in High-Resolution Stratigraphy. *GSA Bulletin* 107(11):1272-1287.

Binford, L. R.

1978 *Nunamiut Ethnoarchaeology*. Academic Press, New York.

1981 *Bones: Ancient Men, Modern Myths*. Academic Press, New York.

1984 *Faunal Remains from Klasies River Mouth*. Academic Press, New York.

1985 Human ancestors: changing views of their behavior. *Journal of Anthropological Archaeology* 4:292-327.

1986 Reply to Bunn and Kroll. *Current Anthropology* 27:444-446.

1988a Fact and fiction about the Zinjanthropus Floor: Data, arguments and interpretations. *Current Anthropology* 29:123-135.

1988b The hunting hypothesis, archaeological methods and the Past. *Yearbook of Physical Anthropology* 30:1-9.

Binford, L. R., and C. K. Ho

1985 Taphonomy at a distance: Zhoukoudian, "the cave-home of Beijing Man"? *Current Anthropology* 26:413-439.

Binford, L. R., L. G. L. Mills, and N. M. Stone

1988 Hyena scavenging behavior and its implications for the interpretation of faunal assemblages from FLK 22 (the Zinj floor) at Olduvai Gorge. *Journal of Anthropological Archaeology* 7:99-135.

Birt, C. S., P. K. H. Maguire, M. A Khan, H. Thybo, G. R. Keller, and J. Patel

1997 The influence of pre-existing structures on the evolution of the Southern Kenya rift valleys - evidence from seismic and gravity studies. *Tectonophysics* 278(1-4):211-242.

Blasco Sancho, F.

1995 *Hombres y Fieras. Estudio Zooarqueológico y Tafonómico del Yacimiento del Paleolítico Medio de la Cueva de Gabasa 1 (Huesca).* Monografías de la Universidad de Zaragoza. Zaragoza.

Blumenschine, R. J.

1986 Early Hominid Scavenging Opportunities. *Implications of Carcass Availability in the Serengeti and Ngorongoro Ecosystems.* B.A.R. International Series, 283, Oxford.

1988 An experimental model of the timing of hominid and carnivore influence on archaeological bone assemblages. *Journal of Archaeological Science* 15:483-502.

1989 A landscape taphonomic model of the scale of prehistoric scavenging opportunities. *Journal of Human Evolution* 18:345-371.

1991 Hominid carnivory and foraging strategies, and the socio-economic function of early archaeological sites. *Philosophical Transactions of the Royal Society (London)* 334:211-221.

1995 Percussion marks, tooth marks and the experimental determinations of the timing of hominid and carnivore access to long bones at FLK Zinjanthropus, Olduvai Gorge, Tanzania. *Journal of Human Evolution* 29:21-51.

Blumenschine, R. J., and H. T. Bunn

1987 On theoretical framework and tests of early hominid meat and marrow adquisition. A reply to Shipman. *American Anthropologist* 89:444-448.

Blumenschine, R. J., and T. C. Madrigal

1993 Long bone marrow yields of some African ungulates. *Journal of Archaeological Science* 20:555-587.

Blumenschine, R. J., and C. W. Marean

1993 A carnivore's view of archaeological bone assemblages. In *From Bones to Behavior: Ethnoarchaeological and Experimental Contributions to the Interpretations of Faunal Remains*, edited by J. Hudson, pp. 271-300. Southern Illinois University, Illinois.

Blumenschine, R. J., and F. T. Masao

1991 Living sites at Olduvai Gorge, Tanzania? Preliminary landscape archaeology results in the basal Bed II lake margin zone. *Journal of Human Evolution* 21:451-62.

1995 Landscape ecology and hominid land use in the lowermost Bed II Olduvai basin. Proposal submitted to the National Science Foundation.

Blumenschine, R. J., and C. R. Peters

1998 Archaeological predictions for hominid land use in the paleo-Olduvai Basin, Tanzania, during lowermost Bed II times. *Journal of Human Evolution* 34:565-607.

Blumenschine, R. J., and M. M. Selvaggio

1988 Percussion marks on bone surfaces as a new diagnostic of hominid behavior. *Nature* 333:763-765.

Blumenschine, R. J., J. A. Cavallo, and S. D. Capaldo

1994 Competition for carcasses and early hominid behavioral ecology: a case study and a conceptual framework. *Journal of Human Evolution* 27:197-213.

Blumenschine, R. J., C. W. Marean, and S. D. Capaldo

1996 Blind tests of inter-analyst correspondence and accuracy in the identification of cut marks, percussion marks, and carnivore tooth marks on bone surfaces. *Journal of Archaeological Science* 23:493-507.

Bocherens, H., D. Billiou, A. Mariotti, M. Pathou-Mathis, M. Otte, D. Bonjean, and M. Toussaint

1999 Palaeoenvironmental and palaeodietary implications of isotopic biogeochemistry of last interglacial Neanderthal and mammal bones in Scladina Cave (Belgium). *Journal of Archaeological Science* 26:599-607.

Bocherens, H., P. Koch, A. Mariotti, D. Geraads, and J. Jaeger

1996 Isotopic biogeochemistry ($\delta^{13}C$ and $\delta^{18}O$) of mammalian enamel from African Pleistocene hominid sites. *Palaios* 11:306-318.

Boëda, E.

1993 Le débitage discoïde et le débitage levallois récurrent centripète. *Bulletin de la Société Préhistorique Française* 90:392-404.

1994 *Le concept Levallois: Variabilité des méthodes*. CNRS, Paris.

1995 Levallois : A volumetric construction, methods, a technique. In *The Definition and Interpretation of Levallois Technology*, edited by H. Dibble and O. Bar-Yosef, pp. 41-68. Prehistory Press, Madison.

Bonnefille, R.

1984 Palynological research at Olduvai Gorge. *Research Reports* 17:227-243. National Geographic Society.

Bordes, F.

1961 *Typologie du Paléolithique ancien et moyen*. Imprimeries Delmas, Bordeaux.

Bradbury, J. W.

1980 Foraging, social dispersion and mating systems. In *Sociobiology Beyond Nature/Nurture*, edited by G. W. Barlow and J. Silverberg, pp. 189-207. Westview Press, Colorado.

Brain, C. K.

1981 *The Hunters or the Hunted*. Chicago University Press, Chicago.

Brooker, R. A., and D. A. Hamilton

1990 Three liquid inmiscibility and the origin of carbonatites. *Nature* 346:459-461.

Bunn, H. T.

1981 Archaeological evidence for meat-eating by Plio-Pleistocene hominids from Koobi Fora, Kenya. *Nature* 291:574-577.

1982 Meat-eating and Human Evolution: Studies on the Diet and Subsistence Patterns of Plio-Pleistocene Hominids in East Africa. Ph. D. Thesis. University of California, Berkeley.

1983 Evidence on the diet and subsistence patterns of Plio-Pleistocene hominids at Koobi Fora, Kenya, and at Olduvai Gorge, Tanzania. In *Animals and Archaeology: Hunters and their Prey*, edited by J. Clutton-Brock, pp. 21-30, B.A.R. International Series, 163, Oxford.

1991 A taphonomic perspective on the archaeology of human origins. *Annual Review of Anthropology* 20:433-467.

1994 Early Pleistocene hominid foraging strategies along the ancestral Omo River at Koobi Fora, Kenya. *Journal of Human Evolution* 27:247-266.

1995 Reply to Tappen. *Current Anthropology* 36:250-251.

1996 Reply to Rose and Marshall. *Current Anthropology* 37:321-323.

Bunn, H. T., and J. A. Ezzo

1993 Hunting and scavenging by Plio-Pleistocene hominids: nutritional constraints, archaeological patterns, and behavioural implications. *Journal of Archaeological Science* 20:365-398.

Bunn, H. T., and E. M. Kroll

1986 Systematic butchery by Plio-Pleistocene hominids at Olduvai Gorge, Tanzania. *Current Anthropology*, 27:431-452.

1988 A reply to Binford. *Current Anthropology* 29:123-149.

Bunn, H. T., E. M. Kroll, and L. E. Bartram

1988 Variability in bone assemblage formation from Hadza hunting, scavenging and carcass processing. *Journal of Anthropological Archaeology* 7:412-457.

1991 Bone distribution on a modern East African landscape and its archaeological implications. In *Cultural Beginnings: Approaches to Understanding Early Hominid Life Ways in the African Savanna*, edited by J. D. Clark. U.I.S.P. Monographien Band 19:33-54.

Bunn, H. T., J. W. K. Harris, G. Isaac, Z. Kaufulu, E. M. Kroll, K. Schick, N. Toth, and A. K. Behrensmeyer

1980 FxJj 50: An early Pleistocene site in northern Kenya. *World Archaeology* 12:109-136.

Butzer, K. W., G. L. Isaac, J. L. Richardson, and C. Washbourn-Kamay

1972 Radiocarbon Dating of East Africa Lake Levels. *Science* 175(4027):1069-1076.

Byrne, G. F., A. W. B. Jacob, J. Mechie, and E. Dindi

1997 Seismic structure of the Upper Range mantle beneath the Southern Kenya rift from wide-angle data. *Tectonophysics* 278(1-4):243-260.

Capaldo, S. D.

1995 Inferring hominid and carnivore behavior from dual-patterned archaeological assemblages. Ph. D. Thesis. Rutgers University, New Brunswick.

1997 Experimental determinations of carcass processing by Plio-Pleistocene hominids and carnivores at FLK 22 (Zinjanthropus), Olduvai Gorge, Tanzania. *Journal of Human Evolution* 33:555-597.

1998 Methods, marks and models for inferring hominid and carnivore behavior. *Journal of Human Evolution* 35:323-326.

Capaldo, S. D., and R. J. Blumenschine

1994 A quantitative diagnosis of notches made by hammerstone percussion and carnivore gnawing in bovid long bones. *American Antiquity* 59:724-748.

Capaldo, S. D., and C. R. Peters

1995 Skeletal inventories from wildebeest drownings at Lakes Masek and Ndutu in the Serengeti ecosystem of Tanzania. *Journal of Archaeological Science* 22:385-408.

Casanova, J.

1986 East African Rift Stromatolites. In *Sedimentation in the African Rifts*, edited by L. E. Frostick. *Geol. Soc. Spec. Pub.*, 25:201-210.

1987 Stromatolites et hauts niveaux lacustres Pléistocénes du bassin Natron-Magadi. *Tanzanie-Kenya Sci. Geol. Bull.* 40(1-2):135-154.

Casanova, J., and C. Hillaire-Marcel

1987 Chronologie et paléohydrologie des hautes niveaux quaternaires du bassin Natron-Magadi (Tanzanie-Kenya) d'après la composition isotopique (^{18}O, ^{13}C, ^{14}C, U/Th) des Stromatolites littoraux. *Sci. Geol. Bull.*, 40(1-2):121-134.

1992 Late Holocene hydrological history of Lake Tanganyika, East Africa, from isotopic data on fossil stromatolites. *Palaeogeography, Palaeoclimatology, Palaeoecology* 91(1-2):35-48.

Cavallo, J. A.

1998 A Re-Examination of Isaac's Central-Place Foraging Hypothesis. Ph. D. Thesis. Rutgers University, New Brunswick.

Cavallo, J. A., and R. J. Blumenschine

1989 Tree-stored leopard kills: expanding the hominid scavenging niche. *Journal of Human Evolution* 18:393-399.

Cerling, T. E., and R. Hay

1986 An isotopic study of paleosol carbonates from Olduvai Gorge. *Quatern. Res.* 25:63-78.

Cerling, T. E., J. Quade, Y. Wang, and J. R. Bowman

1989 Carbon isotopes in soils and paleosols as ecology and paleoecology indicators. *Nature* 361:138-139.

Cerling,T .E., J. M. Harris, S. H. Ambrose, M. G. Leakey, and N. Solounias

1997 Dietary and environmental reconstruction with stable isotope analyses of herbivore tooth enamel from the Miocene locality of Fort Ternan, Kenya. *Journal of Human Evolution* 33:635-650.

Chattaway, M. M.

1936 Anatomy of the family Sterculiaceae. *Phil. Trans. Roy. Soc. London*, Series B, 554:313-366.

Chavaillon, J., N. Chavaillon, F. Hours, and M. Piperno

1979 From the Oldowan to the Middle Stone Age at Melka Kunturé (Ethiopia): Understanding cultural changes. *Quaternaria*, 21:87-114.

Church, A. A., and A. P. Jones

1994 Hollow natrocarbonatite lapilli from the 1992 eruption of Oldoinyo Lengai, Tanzania. *Jour. Geol. Soc. London* 151:59-63.

Clark, J. D.

1994 The Acheulean industrial complex in Africa and elsewhere. In *Integrative Paths to the Past*, edited by R. Corruccini and R. Ciochon, pp. 451-469. Prentice Hall, New Jersey.

1996 Reply to Rose and Marshall. *Current Anthropology* 37:323.

Clark, J. D., and H. Kurashina

1979 Hominid occupation of the East-Central highlands of Ethiopia in the Plio-Pleistocene. *Nature* 282:33-39.

Clark, J. D., and K. Schick

2000 Overview and conclusion on the Middle Awash Acheulean. In *The Acheulean and the Plio-Pleistocene deposits of the Middle Awash valley, Ethiopia*, edited by J. de Heinzelin, J. D. Clark, K. Schick, and H. Gilbert, pp. 193-202.

Crane, K.

1981 Thermal variations in the Gregory Rift of southern Kenya. *Tectonophysics* 74:239-262.

Crossley, R.

1979 The Cenozoic stratigraphy and structure of the Western part of the Rift Valley in Southern Kenya. *Jour. Geol. Soc. London* 136:393-405.

Curtis, G. H.

1967 Notes on some Miocene to Pleistocene potassion/argon results. In *Background to Evolution in Africa*, edited by W. W. Bishop and J. C. Clark, pp. 365-367.

Dag, D., and N. Goren-Inbar

2001 An actualistic study of dorsally plain flakes: a technological note. *Lithic technology* 26:105-117.

Damnati, B.

1993 Sedimentology and geochemistry of lacustrine sequences of the Upper Pleistocene and Holocene in intertropical área (Lake Magadi and Green Lake Crater): paleoclimatic implications. *Jour. African Earth Sci.* 16(4):519-521.

Damnati, B., M. Taieb, and D. Williamson

1992 Laminated deposits from Lake Magadi (Kenya Climatic contrast effect during the maximum wet period between 12,000 y 100,000 years. *Bull. Soc. Geol. France* 163(4):407-414.

Darracott, B. W., J. D. Fairhead, and R. W. Girdler

1972 Gravity and magmatic surveys in Northern Tanzania and Southern Kenya. In R.W. Girdler (Ed.), East African Rifts. *Tectonophysics* 15(1/2):131-141.

Darwin, C.

1871 *The Descent of Man*. Random House, New York.

Dauvois, M.

1976 *Précis de dessin dynamique et structural des industries lithiques préhistoriques*. Fanlac, Périgueux.

Dawson, J. B.

1962 The Geology of Oldoinyo Lengai. *Bull. Volcanol.* 24:349-387.

1964 Carbonatitic volcanic ashes in Northern Tanganyika. *Bull. Volcanol.* 27:81-92.

1989 Sodium Carbonatite extrussions from Oldainyo Lengai, Tanzania: Implications for Carbonatite complex génesis. In *Carbonatites: Génesis and Evolution*, edited by K. Bell, pp. 255-277.

Dawson, J. B., and D. G. Powell

1969 The Natron-Engaruka Explosion Crater Area, Northern Tanzania. *Bull. Volcanol.* 33:791-817.

Dawson, J. B., H. Pinkerton, D. M. Pyle, and C. Nyamweru

1994 June 1993 eruption of Oldoinyo Lengai, Tanzania: Exceptionally viscous and large carbonatite lava flows and evidence for coexisting silicate and carbonate magmas. *Geology* 22:799-802.

Denys, C.

1987 Micromammals from the West Natrón Pleistocene deposits (Tanzania Biostratigraphy and paleoecology. *Sci. Geol. Bull.* 40(1-2):185-202.

Dibble, H. L.

1988 Typological Aspects of Reduction and Intensity of Utilization of Lithic Resources in the French Mousterian. In *Upper Pleistocene Prehistory of Western Eurasia*, edited by H. L. Dibble and A. Montet-White, pp. 181-197. The University Museum, Philadelphia.

Domínguez-Rodrigo, M.

1994a *El Origen del Comportamiento Humano*. Tipo, Madrid.

1994b Dinámica trófica, estrategias de consumo y alteraciones óseas en la sabana africana: Resumen de un proyecto de investigación etoarqueológico (1991-1993) *Trabajos de Prehistoria* 51:15-37.

1996 A landscape study of bone conservation in the Galana and Kulalu (Kenya) ecosystem. *Origini* 20:17-38.

1997a Meat-eating by early hominids at the FLK 22 Zinjanthropus site, Olduvai Gorge, Tanzania: an experimental approach using cut mark data. *Journal of Human Evolution* 33:669-690.

1997b A reassessment of the study of cut mark patterns to infer hominid manipulation of fleshed carcasses at the FLK Zinj 22 site, Olduvai Gorge, Tanzania. *Trabajos de Prehistoria* 54:29-42.

1999a Flesh availability and bone modification in carcasses consumed by lions. *Palaeogeography, Palaeoclimatology and Palaeoecology* 149:373-388.

1999b Meat eating and carcass procurement by hominids at the FLK Zinj 22 site, Olduvai Gorge, Tanzania: A new experimental approach to the old hunting-versus-scavenging debate. In *Lifestyles and Survival Strategies in Pliocene and Pleistocene Hominids*, edited by H. Ullrich, pp. 89-111. Edition Archaea, Schwelm, Germany.

2001 A study of carnivore competition in riparian and open habitats of modern savannas and its implications for hominid behavioral modelling. *Journal of Human Evolution* 40:77-98.

2002 Hunting and scavenging by early humans: the state of the debate. *Journal of World Prehistory* 16:1-54.

Domínguez-Rodrigo, M., and R. Barba

2006 New estimates of tooth marks and percussion marks from FLK Zinj, Olduvai Gorge (Tanzania): the carnivore-hominid-carnivore hypothesis falsified. *Journal of Human Evolution.*

Domínguez-Rodrigo, M., and R. Martí Lezana

1996 Estudio etnoarqueológico de un campamento temporal Ndorobo (Maasai) en Kulalu. *Kenia Trabajos de Prehistoria* 53:131-143.

Domínguez-Rodrigo, M., and T. R. Pickering

2003 Early hominids, hunting and scavenging: a summary of the discussion. *Evolutionary Anthropology* 12(6):275-282.

Domínguez-Rodrigo, M., C. Egeland, and R. Barba

2006 *Deconstructing Olduvai.* Cambridge, Cambridge University Press.

Domínguez-Rodrigo, M., J. Serrallonga, and V. Medina

1998 Food availability and social stress among captive baboons: referential data for early hominid food transport at sites. *Anthropologie* 36:225-230.

Domínguez-Rodrigo, M., J.A. López-Saez, A. Vincens, L. Alcalá, L. Luque, and J. Serrallonga

2001a Fossil pollen from the Upper Humbu Formation of Peninj (Tanzania): hominid adaptation to a dry open Plio-Pleistocene savanna environment. *Journal of Human Evolution* 40:151-157.

Domínguez-Rodrigo, M., J. Serrallonga, J. Juan-Treserras, L. Alcalá, and L. Luque

2001b Woodworking activities by early humans: a plant residue analysis on Acheulian stone tools from Peninj, Tanzania. *Journal of Human Evolution* 39:421-436.

Domínguez-Rodrigo, M., I. de la Torre Sainz, L. Luque, L., Alcalá, R. Mora, J. Serrallonga, and V. Medina

2002 The ST Site Complex at Peninj, West Lake Natron, Tanzania: implications for early hominid behavioral models. *Journal of Archaeological Science* 29: 639-665

Domínguez-Rodrigo, M, L. Alcalá, L. Luque, and J. Serrallonga

2005 Quelques aperçus sur les significations paléoécologique et comportamentale des sites oldowayens anciens et acheuléens du Peninj (Upper Humbu Formation, Ouest du Lac Natron, Tanzanie). In *Le Paléolithique en Afrique*, edited by M. Sahnouni. L'Histoire la plus longue, Paris, Editions Artcom, pp. 129-156.

Domínguez-Rodrigo, M., C. P. Egeland, and R. Barba

2007 *Deconstructing Olduvai.* Springer, New York.

Downey, C., and M. Domínguez-Rodrigo

2002 Palaeoecological reconstruction and hominid land use of the lake Natron basin during the early Pleistocene. *Before Farming* 4:1-53.

Ehleringer, J. R., and T. A Cooper

1988 Correlations between carbon isotope ratio and microhabitat in desert plants. *Oecologia* 76:562-566.

Ehleringer, J. R., C. B. Field, Z. F. Lin, and C. Y. Kuo

1986 Leaf carbon isotope and mineral composition in subtropical plants along an irradiance cline. *Oecologia* 70:520-526.

Eugster, J. F.

1967 Hydrous sodium silicate from lake Magadi, Kenya: precursors of bedded chert. *Science* 157:1177-1180.

Eugster, H. P.

1986 Lake Magadi, Kenya: a model for rift valley hydrochemistry and sedimentation? In L.E. Frostick et al. (Eds.), Sedimentation in the African Rifts. *Geol. Soc. Spec. Pub.* 25:177-190.

Evernden, J. F., and G. H. Curtis

1965 The Potassium-Argon Dating of Late Cenozoic Rocks in East Africa and Italy. *Current Anthropology* 6(4):343-385.

Fairhead, J. D.

1980 The structure of the cross-cutting volcanic chain of Northern Tanzania and its relation to the East African Rift System. *Tectonophysics* 65:193-208.

Fernández-Jalvo, Y., C. Denys, P. Andrews, T. Williams, Y. Dauphin, and L. Humphrey

1999 Taphonomy and palaeocology of Olduvai Bed I (Pleistocene, Tanzania). *Journal of Human Evolution* 34:137-172.

Fisher, J. W.

1995 Bone surface modification in zooarchaeology. *Journal of Archaeological Method and Theory* 1:7-65.

Foley, R. A.

1980 The spatial component of archaeological data: off-site methods and some preliminary results from the Amboseli Basin, Southern Kenya. *Proceedings of the VIII Pan-African Congress in Prehistory and Quaternary Studies*, pp. 39-40.

Foley, R. A.

1981 Off-site archaeology: an alternative approach for the short-sited. In *Patterns of the Past*, edited by I. Hodder, G. Isaac and N. Hammond, pp. 157-183. Cambridge University Press, Cambridge.

1987 *Another Unique Species*. Longmann, London.

Foster, J. B., and D. Kerney

1967 Nairobi National Park game census. *E. Afr. Wildl. J.* 5:112-120.

Foster, A. N., C. J. Ebinger, E. Mbede, and D. Rex

1997 Tectonic development of the Northern Tanzanian sector of the East African Rift System. *Journal of the Geological Society, London* 154:689-700.

Fritz, B., M. D. Zins-Paulas, and M. Gueddari

1987 Geochemistry of sislica-rich brines from lake Natron (Tanzania). *Sci. Geol. Bull.* 40(1-2):97-110.

Gamble, C.

1986 *The Paleolithic Settlement of Europe*. Cambridge University Press, Cambridge.

1999 *The Paleolithic Societies of Europe*. Cambridge World Archaeology, Cambridge.

Geneste, J. M.

1985 Analyse lithique d'industries mousteriennes du Périgord: un aproche technologique du comportement des groupes humains au Paléolithique Moyen. Unpublished Ph.D., Université de Bordeaux I.

1991 L'approvisionnement en matières premières dans les systèmes de production lithique: la dimension spatiale de la technologie. In *Tecnología y cadenas operativas líticas*, edited by R. Mora, X. Terradas, A. Parpal and C. Plana, pp. 1-36. Treballs d'Arqueologia, 1, Universidad Autónoma de Barcelona, Barcelona.

Geological Survey of Tanganyika

1961 Geological Map of Angata Salei, 1:125.000.

Geraads, D.

1987 La faune des dépôts pléistocénes de l'Ouest du Lac Natron (Tanzanie): interpretation biostratigraphique. *Sci. Geol. Bull.* 40(1-2):167-184.

Gifford, D.

1977 Observations of modern human settlements as an aid to archaeological interpretation. Ph.D. Thesis, University of California, Berkeley.

Gifford-Gonzalez, D. P.

1991 Bones are not enough: analogues, knowledge, and interpretive strategies in zooarcaheology. *Journal of Anthropological Archaeology* 10:215-254.

Gregory, J. W.

1921 *The Rift Valleys and the Geology of East Africa*. London Seeley Service, London.

Grimaud, P., J. P. Richert, J. Rolet, J. J. Tiercelin, J. P. Xavier, C. K. Morley, C. Coussement, S. W. Karanja, R. W. Renaut, G. Guérin, C. Le Turdu, and G. Michel-Noel

1994 Fault geometry and extension mechanisms in the central Kenya Rift. East África. A 3D remote sensing approach. *Bull. Cen. Rech. Explor. Prod. Elf-Aquit.* 18(1):59-92.

Guest, N. J.

1953 The geology and petrology of the Engaruka Oldonyo-Lengai Lake Natron Area of Northern Tanganika Territory. Ph.D. Thesis, University of Sheffield.

Hardy, B. L., and G. T. Garufi

1998 Identification of woodworking on stone tools through residue and use-wear analyses: experimental results. *J. Archeol. Sci.* 25:177-184.

Hay, R. L.

1966 Zeolites and zeolitic reactions in sedimentary rocks. *Geological Society of America Special Paper* 85:1-130.

1968 Chert and its precursors in sodium carbonate lakes of East Africa. *Contributions to Mineralogy and Petrology* 17:255-274.

1976 *The Geology of Olduvai Gorge*. Clarendon Press, Oxford.

1986 Role of Tephra in the preservation of fossils in Cenozoic deposits of East Africa. In Sedimentation in the African Rifts, edited by L.E. Frostick et al. *Geol. Soc. Spec. Pub.* 25:339-344.

Hillaire-Marcel, C.

1987 Hydrologie isotopique des lacs Magadi (Kenya) et Natron (Tanzania). *Sci. Geol. Bull.* 40(1-2):111-120.

Hillaire-Marcel, C., and J. Casanova

1987 Isotopic hydrology and paleohydrology of the Magadi (Kenya)-Natron (Tanzania) basin during the late Quaternary. *Palaeogeogr. Palaeoclimatol. Palaeoecol.* 58:155-181.

Hillaire-Marcel, C., J. Casanova, and M. Taieb

1987 Isotopic Age and Lacustrine Environment during Late Quaternary in the Tanzanian Rift (Lake Natron). In *Climate, History, Periodicity and Predictability*, edited by M. R. Rampino, pp. 117-123.

Howell, F. C.

1972 Pliocene/Pleistocene Hominidae in East Africa: absolute and relative ages. In *Calibration of Hominoid Evolution*, edited by W. W. Bishop and J. A. Miller, pp. 331-339.

Icole, M., M. Taieb, G. Perinet, P. Manega, and C. Robert

1987 Mineralogie des sediments du Groupe Peninj (Lac Natron, Tanzanie). Reconstitution des paleoenvironnements lacustres. *Sci. Geol. Bull.* 40(1-2):71-82.

Inizan, M. L., M. Reduron-Balliner, H. Roche, and J. Tixier

1995 *Technologie de la pierre taillée*. CREP, Meudon.

Isaac, G. L.

1965 The stratigraphy of the Peninj Beds and the provenance of the Natron Australopithecine mandible. *Cuaternaria* 7:101-130.

1967 The stratigraphy of the Peninj-Group; Early Middle Pleistocene Formations West of Lake Natron, Tanzania. In *Background to Evolution in Africa*, edited by W. W. Bishop et al., pp. 229-257,

1978 The food-sharing behavior of protohuman hominids. *Scientific American* 238:90-108.

1983 Bones in contention: competing explanations for the juxtaposition of Early Pleistocene artifacts and faunal remains. In *Animals and Archaeology 1. Hunters and Their Prey*, edited by J. Clutton-Brock and C. Grigson, pp. 3-19. B. A. R. International Series, 163, Oxford.

1984 The archaeology of human origins: studies of Lower Pleistocene in East Africa, 1971-1981. *Advances in World Archaeology* 3:1-87.

1997 *Plio-Pleistocene archaeology. Koobi Fora Research Project.* Volume V. Cambridge, Cambridge University Press.

Isaac, G. L., and G. H. Curtis

1974 Age of early Acheulian industries from the Peninj Group, Tanzania. *Nature* 249:624-627.

Isaac, G., and J. W. K. Harris

1975 *The scatters between the patches.* Kroeber Anthropological Society. University of California, Berkeley.

1980 *A method for determining the characteristics of artefacts between sites in the Upper Member of the Koobi Fora Formation, East Lake Turkana.* Proc. 8th Pan-African Congress on Prehistory and Quaternary Studies, Nairobi, 1977, pp. 19-22.

1997 The stone artefact aseemblages: a comparative study. In *Koobi Fora Research Project* 5, edited by G. Isaac and B. Isaac, pp. 262-362. Clarendon Press, Oxford.

Jaeger, J. J.

1976 Les rongeurs (Mammalia, Rodentia) du Pleistocene Inferieur d'Olduvai Bed I (Tanzanie), 1 partie: les muridés. In *Fossil Vertebrates of Africa*, edited by R. J. G. Savage, and S. C. Coryndon, pp. 4:57-120. London, Academic Press.

Jones, P.

1994 "Results of experimental work in relation to the stone industries of Olduvai Gorge". In *Olduvai Gorge, 5. Excavations in Beds II, IV and the Masek Beds, 1968-1971*, edited by M. Leakey, pp. 254-298. Cambridge University Press, Cambridge.

Juan, J., V. Medina, and J. Serrallonga

1996 Los frutos secos como recurso alimentario de los cazadores-recolectores. Estudios experimentales para la identificación de indicadores microscópicos y bioquímicos en el instrumental lítico. *XVII Reunió de Paleolitistes de l'Estat, Torroella de Montgrí, Girona*, pp. 99-107.

Kapitsa, A. P.

1968 *Preliminary Report of the Soviet East African Expedition of the Academy of Sciences of the USSR in 1967*. Moscow.

Katzenberg, M. A.

2000 Stable isotope analysis: a tool for studying past diet, demography, and life history. In *Biological Anthropology of the Human Skeleton*, edited by M. A. Katzenberg and S. R. Saunders, pp. 305-327. New York, Wiley-Liss.

Kaufulu, Z.

1983 The geological content of some early archaeological sites in Kenya, Malawi and Tanzania. Ph. D. Thesis, Department of Anthropology, University of California, Berkeley.

Kleindienst, M. R.

1962 Component of the East African Acheulean assemblage: an analytic approach. In *Actes du IV Congrès Panafricain de Préhistoire et de l'Étude du Quaternaire*, edited by G. Mortelmans and J. Nenquin, pp. 81-108. Leopoldville (1959) Tervuren.

Kruuk, H.

1972 *The spotted hyena*. Chicago, University of Chicago Press.

Kuman, K.

1998 The earliest South African industries. In Early human behaviour in global context, edited by M. Petraglia and R. Korisettar, pp. 151-186. Routledge, Londres.

Kutzbach, J. E., and F. Street-Perrot

1985 Milankovitch forcing of fluctuations in the level of tropical lakes from 18 to 0 Kyr. *Nature* 317:130-134.

Laplace, G.

1974 La typologie analytique et structurale: base rationelle d'étude des industries lithiques et osseuses. *Banques de données archéologiques* 932:91-143. CNRS.

Leakey, L.

1951 *Olduvai Gorge*, vol. 1. Cambridge, Cambridge University Press.

Leakey, L. S. B. and M. D. Leakey

1965 Recent Discoveries of Fossil Hominids in Tanganyika: at Olduvai and near Lake Natron. *Current Anthropology* 6(4):422-424.

Leakey, M.

1971 *Olduvai Gorge*, vol. 3. Excavations in Bed I and II, 1960-63. Cambridge, Cambridge University Press.

1976 A summary and discussion of the archaeological evidence from Bed I and Bed II, Olduvai Gorge, Tanzania. In Human origins. Louis Leakey and the East African evidence, edited by G. Isaac and E. McCown, pp. 431-459. W. A. Benjamin, Menlo Park.

Leakey. R.

1984 *One life. An Autobiography*. Salem House Publishers, London

Lee, P. C.

1983 Ecological influences on relationships and social structure. In *Primate Social Systems: An Integrated Approach*, pp. 225-230. Blackwell, Oxford,

Lee-Thorp, J. A., and N. J. van der Merwe

1987 Carbon isotope analysis of fossil bone apatite. *South African Journal of Science* 83:712-715.

1991 Aspects of the chemistry of fossil and modern biological apatites. *Journal of Archaeological Science* 18:343-354.

Lee-Thorp, J. A., L. Manning, and M. Sponheimer

1997 Exploring problems and opportunities offered by down-scaling sample sizes for carbon isotope analyses of fossils. *Bulletin de la Societè Geologique de France* 168:767-773.

Lee-Thorp, J. A., M. Sponheimer, and N. J. van der Merwe

2003 What do stable isotopes tell us about hominid dietary and ecological niches in the Pliocene? *International Journal of Osteoarchaeology* 13:104-113.

Lee-Thorp, J. A., J. F. Thackeray, and N. J. van der Merwe

2000 The hunters and the hunted revisited. *Journal of Human Evolution* 39:565-576.

Lenoir, M., and A. Turq

1995 Recurrent Centripetal Debitage (Levallois and Discoidal): Continuity or Discontinuity? In *The Definition and Interpretation of Levallois Technology*, edited by H. L. Dibble and O. Bar-Yosef, pp. 249-256. Prehistoric Press, Madison.

Lubala, R. T., and A. Rafoni

1987 Petrologie et signification géodynamique du volcanisme alcalin mio-pliocene de la région du Lac Natron (rift est-africain, *Tanzanie Sci. Geol. Bull.* 4(1-2):41-56.

Luque. L.

1996 Estratigrafia del grupo Peninj y geomorfología del margen occidental del lago Natron (Tanzania). M.A. Thesis, Complutense University, Madrid.

Lupo, K. D., and J. F. O'Connell

2002 Cut and tooth mark distributions on large animal bones: ethnoarchaeological data from the Hadza and their implications for current ideas about early human carnivory. *Journal of Archaeological Science* 29:85-109.

Luyt, J.

2001 Revisiting palaeoenvironments from the hominid-bearing Plio-Pleistocene sites: New isotopic evidence from Sterkfontein. University of Cape Town, unpublished master's thesis.

Mac Donald, R., L. A. J. Williams, and I. G. Gass

1994 Tectonomagmatic evolution of the Kenya rift valley: some geological perspectives. *J. Geol. Soc. London* 151:879-888.

Mac Intyre, R. M., J. G. Mitchell, and J. B. Dawson

1974 Age of Fault Movements in Tanzanian Sector of East African Rift System. *Nature*, 247:354-356.

Manega, P. C.

1993 Geochronology, geochemistry and isotopic study of the Plio-Pleistocene hominid sites and the Ngorongoro Volcanic Highland in Northern Tanzania. Ph. D. Thesis, University of Colorado.

Manega, P. C., and S. Bieda

1987 Modern sediments of Lake Natron. Tanzania. *Sci. Geol. Bull.* 40(1-2):83-96.

Margalef, R.

1968 *Perspectives in ecological theory*. Chicago, Chicago University Press.

1977 *Ecología*. Omega, Barcelona.

Marean, C. W.

1998 A critique of the evidence for scavenging by Neandertal and early modern humans: new data from Kobeh Cave (Zagros mountains, Iran) and Die Kelders Cave 1 Layer 10 (South Africa). *Journal of Human Evolution* 35:111-136.

Marean, C. W., and L. Bertino

1994 Intrasite spatial analysis of bone: Subtracting the effect of secondary carnivore consumers. *American Antiquity* 59(4):748-768.

Marean, C. W., and N. Cleghorn

2003 Large mammal skeletal element transport: applying foraging theory in a complex taphonomic system. *Journal of Taphonomy* 1:15-42.

Marean, C. W., and C. L. Ehrhardt

1995 Paleoanthropological and paleoecological implications of the taphonomy of a sabertooth's den. *Journal of Human Evolution* 29:515-547.

Marean, C. W., and C. Frey

1997 Animal bones from caves to cities: reverse utility curves as methodological artifacts. *American Antiquity* 62:698-716.

Marean, C. W., and S. Y. Kim

1998 Mousterian large mammal remains from Kobeh cave. *Current Anthropology* 39:79-113.

Marean, C. W., and L. M. Spencer

1991 Impact of carnivore ravaging of bone in archaeological assemblages. *Journal of Archaeological Science* 18:677-694.

Marean, C. W., M. Domínguez-Rodrigo, and T. R. Pickering

2004 Skeletal element equifinality begins with method. *Journal of Taphonomy* 2:69-98.

Marean, C. W., L. M. Spencer, R. J. Blumenschine, and S. Capaldo

1992 Captive hyaena bone choice and destruction, the Schlepp effect and Olduvai archaeofaunas. *Journal of Archaeological Science* 19:101-121.

Marshall, F.

1986 Implications of bone modification in a Neolithic faunal assemblage for the study of early hominid butchery and subsistence practices. *Journal of Human Evolution* 15:661-672.

1994 Food sharing and body part representation in Okiek faunal assemblages. *Journal of Archaeological Science* 21:65-77.

McDougall, I., F. H. Brown, T. E. Cerling, and J. W. Hillhouse

1992 A reppraisal of the Geomagnetic polarity time scale to 4 ma using data from the Turkana basin, East Africa. *Geophysical Research Letters* 19:2349-2352.

McNabb, J.

1998 On the Move. Theory, Time Averaging and Resource Transport at Olduvai Gorge. In *Stone Age Archaeology. Essays in honour of John Wymer*, edited by N. Ashton, F. Healy and P. Pettit, pp. 102:15-22. Oxford, Oxbow Monograph, 102.

McPherron, S.

2002 Technological and typological variability in the bifaces from Tabun Cave, Israel. In *Multiple approaches to the study of biracial technologies*, edited by M. Soresi, and H. Dibble, pp. 115:55-75. University of Pennsylvania Monograph.

Mechie, J., G. R. Keller, C. Prodehl, M. A. Khan and S. J. Graciri

1997 A model for the structure, composition and evolution of the Kenya rift. *Tectonophysics* 278:95-119.

Mineral Resources Division, Dodoma

1966 Geological Map, 1:125.000, numbers: 27, 28, 39, 40.

Monahan, C. M.

1996 New zooarchaeological data from Bed II, Olduvai Gorge, Tanzania: implications for hominid behavior in the Early Pleistocene. *Journal of Human Evolution* 31:93-128.

1998 The Hadza carcass transport debate revisited and its archaeological implications. *Journal of Archaeological Science* 25:405-524.

Mora, R., and I. de la Torre

2005 Percussion tools in Olduvai Beds I and II (Tanzania): Implications for early human activities. *Journal of Anthropological Archaeology* 24:179-192.

Mturi, A. A.

1987 The archaeological sites of Lake Natron (Tanzania). *Sci. Geol. Bull.* 40:1-2:209-215.

Novak, O., C. Prodehl, A. W. B. Jacob, and W. Okoth

1997 Crustal structure of the Southeastern flank of the Kenya Rift deduced from wide-angle P-wave data. *Tectonophysics* 278:171-186.

O'Connell, J. F.

1997 On Plio-Pleistocene archaeological sites and central places. *Current Anthropology* 38:86-88.

O'Connell, J. F., K. Hawkes, and N. Blurton Jones

1988 Hadza hunting, butchering and bone transport and their archaeological implications. *Journal of Anthropological Research* 44:113-61.

1990 Reanalysis of large mammal body part transport among the Hadza. *Journal of Archaeological Science* 17:301-316.

1991 Distribution of refuse-producing activities at Hadza residential base camps: implications for analysis of archaeological site structure. In *The Interpretation of Archaeological Spatial Patterning*, edited by E. M. Kroll and T. D. Price, pp. 61-76. Plenum Press, New York.

1992 Patterns in the distribution, site structure and assemblage composition of Hadza kill-butchering sites. *Journal of Archaeological Science* 19:319-45.

Oliver, J. S.

1994 Estimates of hominid and carnivore involvement in the FLK Zinjanthropus fossil assemblage: some socioecological implications. *Journal of Human Evolution* 27:267-294.

Pelegrin, J.

1990 Prehistoric Lithic Technology: Some Aspects of Research. *Archaeological Review from Cambridge* 9(1):116-125.

Peresani, M.

2003 *Discoid Lithic Technology. Advances and implications.* BAR International Series 1120, Oxford.

Peters, C. R., and R. J. Blumenschine

1995 Landscape perspectives on possible land use patterns for early hominids in the Olduvai Basin. *Journal of Human Evolution* 29:321-362.

Pickering, T. R., and M. Domínguez-Rodrigo

2003 *The acquisition and use of large mammal carcasses by Oldowan hominins in Eastern and Southern Africa: A selected review and assessment.* CRAFT publications, Indiana.

Pickering, T. R., and J. Wallis

1997 Bone modifications resulting from captive chimpanzee mastication: implication for the interpretation of Pliocene archaeological faunas. *Journal of Archaeological Science* 24:1115-1127.

Pickering, T. R., C. W. Marean, and M. Domínguez-Rodrigo

2003 Importance of limb bone shaft fragments in zooarchaeology: a response to "On *in situ* attrition and vertebrate body part profiles" (2002) by M.C. Stiner. *Journal of Archaeological Science* 30:1469-1482.

Pickering, T. R., M. Domínguez-Rodrigo, C. Egeland and C. K. Brain

2004 New data and ideas on the foraging behavior of early stone age hominids at Swartkrans cave. *South African Journal of Science* 100:215-220.

Pickford, M.

1986 Sedimentation and fossil preservation in the Nyanza Rift System, Kenya. In *Sedimentation in the African Rifts*, edited by L. E. Frostick et al. *Geol. Soc. Spec. Pub.* 25:345-362.

Plummer, T. W., and L. C. Bishop

1994 Hominid paleocology at Olduvai Gorge, Tanzania, as indicated by antelope remains. *Journal of Human Evolution* 27:47-76.

Potts, R.

1982 Lower Pleistocene Site Formation and Hominid Activities at Olduvai Gorge, Tanzania, Ph. D. Thesis. Harvard University, Massachussets.

Potts, R.

1988 *Early Hominid Activities at Olduvai*. Aldine, New York.

1984 Hominid hunters? Problems of identifying the earliest hunter/gatherers. In *Hominid evolution and community ecology*, edited by R. Foley, pp. 129-166. Academic Press, London.

1991 Why the Oldowan? Plio-Pleistocene tool making and the transport of resources. *Journal of Anthropological Research* 47:53-176.

1994 Variables versus models of early Pleistocene hominid land use. *Journal of Human Evolution* 27:7-24.

Potts, R., and P. Shipman

1981 Cutmarks made by stone tools from Olduvai Gorge, Tanzania. *Nature* 291:577-580.

Potts, R., A. K. Behrensmeyer, and P. Ditchfield

1999 Paleolandscape variation in early Pleistocene hominid activities. *Journal of Human Evolution* 37:747-788.

Quade, J., T. E. Cerling, J. C. Barry, M. E. Morgan, D. R. Pilbeam, A. R. Chivas, J. A. Lee-Thorp, and N. J. van der Merwe

1992 A 16-Ma record of paleodiet using carbon and oxygen isotopes in fossil teeth from Pakistan. *Chemical Geology* (Isotope Geoscience Section) 94:183-192.

Richards, M. P., P. B. Pettit, E. Trinkhaus, F. H. Smith, M. Paunovic, and I. Karavanic

2000 Neanderthal diet at Vindija and Neanderthal predation: the evidence from stable isotopes. *Proceedings of the National Academy of Science* 97:7663-7666.

Riou, G.

1995 *Savanes. L'herbe, l'arbre et l'homme.* Masson, Paris.

Roberts, N.

1990 Ups and downs of African Lakes. *Nature* 346:107.

Roberts, N., M. Taieb, P. Barker, B. Damnati, M. Icole, and D. Williamson

1993 Timing of the Younger Dryas event in East Africa from lake level changes. *Nature* 366:146-148.

Roche, H., and M. Kibunjia

1994 Les sites archéologiques plio-pléistocènes de la Formation de Nachukui, West Turkana, Kenya. *C.R.A.S.P* 318 II:1145-1151.

Roche, H., A. Delagnes, J. P.Brugal, C. Feibel, M. Kibunjia, V. Mourre, and P. J. Texier

1999 Early hominid stone tool production and technical skill 2.34 Myr ago in West Turkana, Kenya. *Nature* 399:57-60.

Rogers, M.

1997 A landscape archaeological study at East Turkana, Kenya. Ph. D. Thesis, Anthropology Department, Rutgers University, New Brunswick.

Rogers, M., C. S. Feibel, and J. W. K. Harris

1994 Changing patterns of land use by Plio-Pleistocene hominids in the Lake Turkana Basin. *Journal of Human Evolution* 27:139-158.

Rose, L,. and F. Marshall

1996 Meat eating, hominid sociality, and home bases revisited. *Current Anthropology* 37:307-338.

Rudnai, J. A.

1973 *The Social Life of the Lion.* Washington Sq. East, Willingford.

Schaller, G. B .

1972 *The Serengeti Lion.* University of Chicago Press, Chicago.

Schick, K. D.

1984 Processes of Palaeolithic Site Formation: An Experimental Study. Unpublished Ph.D. Thesis, University of California, Berkeley.

1998 A comparative perspective on Paleolithic cultural patterns". In *Neandertals and Modern Humans in Western Asia*, edited by T. Akazawa, K. Aoki and O. Bar-Yosef, pp. 449-460. Plenum Press, New York.

Schick, K., and Clark, J. D.

2000 Acheulean archaeology of the Western Middle Awash. In *The Acheulean and the Plio-Pleistocene deposits of the Middle Awash valley, Ethiopia*, edited by J. de Heinzelin, J. D. Clark, K. Schick, and H. Gilbert, pp. 125-181.

Schick, K., and N. Toth

1993 *Making Silent Stones Speak*. Simon and Schuster, New York.

Schoeninger, M., and K. Moore

1992 Bone stable isotope studies in archaeology. *Journal of World Prehistory* 6:247-296.

Schubel, K. A., and B. M. Simonson

1990 Petrography and diagenesis of cherts from Lake Magadi, Kenya. *Jour. Sed. Petrol.* 60(5):761-776.

Selvaggio, M. M.

1994 Identifying the Timing and Sequence of Hominid and Carnivore Involvement with Plio-Pleistocene Bone Assemblages from Carnivore Tooth Marks and Stone-Tool Butchery Marks on Bone Surfaces. Ph. D. Thesis, Rutgers University.

Selvaggio, M. M.

1998 Concerning the 3 stage model of carcass processing at the FLK Zinjanthropus. A reply to Capaldo. *Journal of Human Evolution* 35:319-321.

Sept, J. M.

1992 Was there no place like home? *Current Anthropology* 33:187-207.

Severtzov, A.

1947 *La dinámica de la población animal*. Lautaro, Buenos Aires.

Shipman, P., and J. M. Harris

1988 Habitat prefeence and paleoecology of Australopithecus boisei in Eastern Africa. In *Evolutionary History of the "Robust" Australopithecines*, edited by F. E. Grine, pp. 343-382. Aldine de Gruyter, New York.

Sikes, N.

1994 Early hominid habitat preferences in East Africa: paeosol carbonate isotopic evidence. *Journal of Human Evolution* 2725-45.

Simiyu, S. M., and G. R. Keller

1997 An integrated analysis of lithospheric structure across the East African plateau based on gravity anomalies and recent seismic studies. *Tectonophysics* 278(1-4):291-313.

Simiyu, S. M., and G. R. Keller

1998 Upper crustal structure in the vicinity of Lake Magadi in the Kenya Rift Valley region. *Journal of African Earth Science* 27(3-4):359-371.

Sinclair, A. R. E.

1979 The Serengeti environment. In *Serengeti: Dynamics of an Ecosystem*, edited by A.R.E. Sinclair and M. Norton-Griffiths (Eds.), pp. 31-45. Chicago University Press, Chicago.

Slimak, L.

1998-1999 La variabilité des débitages discoïdes au Paléolithique moyen: Diversité des méthodes et unité d´un concept. L´exemple des gisements de la Baume Néron (Soyons, Ardèche) et du Champ Grand (Saint-Maurice-sur-Loire, Loire). *Préhistoire-Anthropologie Méditerranéennes* 7-8:75-88.

Sponheimer, M.

1999 Isotopic ecology of the Makapansgat Limeworks fauna. Rutgers University, Ph. D. Thesis, Ann Arbor, MI, University Microfilms.

Sponheimer, M., and J. A. Lee-Thorp

1999 Oxygen isotope ratios in enamel carbonate and their ecological significance. *Journal of Archaeological Science* 26:723-728.

Stanley, S. M.

1992 An ecological theory for the origin of Homo. *Paleobiology* 3:237-257.

Stern, N.

1991 The scatter-between-the-patches: a study of early hominid land use patterns in the Turkana Basin, Kenya. Ph. D. Thesis, Harvard University, Cambridge, Mass.

1993 The structure of the Lower Pleistocene Archaeological Record. *Current Anthropology* 34:201-226.

Stiles, D.

1979 Early Acheulean and Developed Oldowan. *Current Anthropology* 20:126-129.

Stiner, M.

1994 *Honor Among Thieves: a Zooarchaeological Study of Neandertal Ecology*. Princeton University Press, Princeton.

Sullivan, C. H., and H. W. Krueger

1981 Carbon isotope analysis of separate phases in modern and fossil bone. *Nature* 292:333-335.

1983 Carbon isotope ratios of bone apatite and animal diet reconstruction. *Nature* 301:177.

Surdam, R. C., and R. A. Sheppard

1978 Zeolites in saline alkaline-like deposits. In *Natural Zeolites: Ocurrence, Properties, Use*, edited by L. B. Sand and F. A. Mumpton.

Taieb, M., and B. Fritz

1987 Introduction. *Sci. Geol. Bull.* 40:1-2:5-6.

Tappen, M.

1992 Taphonomy of a central African savanna: natural bone deposition in Parc National des Virunga, Zaire. Ph. D. Thesis, Department of Anthropology, Harvard University, Cambridge.

1995 Savanna ecology and natural bone deposition. implications for early hominid site formation, hunting and scavenging. *Current Anthropology* 36:223-260.

Terradas, X.

2003 Discoid flaking method: conception and technological variability. In *Discoid Lithic Technology*, edited by M. Peresani. BAR, International Series, 1120:19-31.

Texier, P-J., and Roche, H.

1995 "El impacto de la predeterminación en algunas cadenas operativas achelenses". In *Evolución human en Europa y los yacimientos de la Sierra de Atapuerca*, edited by J. M. Bermúdez, J. L. Arsuaga and E. Carbonell, pp. 403-420. Junta de Castilla y León, Valladolid.

Texier, P. J., and H. Roche

1995 Polyèdre, sub-sphéroïde, sphéroïde et bola: des segments plus ou moins longs d´une même chaîne opératoire. *Cahier Noir* 7:31-40.

Tixier, J.

1956 Le hachereau dans l'Acheuléen nord-africain. Notes typologiques. *Congrès Préhistorique de France, XVe session, Poitiers-Angoulême*, pp. 914-923.

Thouveny, N., and M. Taieb

1986 Preliminary Magnetostratigraphic record of Pleistocene deposits, Lake Natron basin, Tanzania. In L. E. Frostick et al. (Eds.), *Sedimentation in the African Rifts*. Geological Society Special Publication, 25:331-336.

1987 Étude paleomagnetique des formations du Plio-Pléistocène de la région de la Peninj (Ouest du lac Natron, Tanzanie). Limites de la interpretation magnetostratigraphique. *Sci. Geol. Bull.* 40(1-2):57-70.

de la Torre Sáinz, I.

2002 El Olduvayense de la Seccion Tipo de Peninj (Lago Natron, Tanzania). M Sc. Thesis. Complutense University, Madrid.

2005 Estrategias tecnológicas en el Pleistoceno inferior de África oriental (Olduvai y Peninj, norte de Tanzania). Unpublished Ph. D. Thesis, Universidad Complutense de Madrid, Madrid.

de la Torre, I., and R. Mora

2004 El Olduvayense de la Sección Tipo de Peninj (Lago Natron, Tanzania). *CEPAP*, vol. 1, Barcelona.

2005 Unmodified Lithic Material at Olduvai Bed I: Manuports or Ecofacts? *Journal of Archaeological Science* 32(2):273-285.

de la Torre, I., R. Mora, and J. Martínez-Moreno

2008 The early Acheulian in Pêninj (Lake Natron, Tanzania). *Journal of Anthropological Archaeology* 27:244-264

de la Torre, I. de la, R. Mora, and M. Domínguez-Rodrigo

2004 La tecnología lítica del "Complejo ST" de Peninj (Lago Natron, Tanzania): análisis de un conjunto del Olduvayense africano. *Trabajos de Prehistoria* 61(1):23-45.

de la Torre, I. de la, R. Mora, M. Domínguez-Rodrigo, L. Luque, and L. Alcalá

2003 The Oldowan industry of Peninj and its bearing on the reconstruction of the technological skills of Lower Pleistocene hominids. *Journal of Human Evolution* 44:203-224.

Tieszen, L. L.

1991 Natural variation in the carbon isotope values of plants: implications for archaeology, ecology and paleoecology. *Journal of Archaeological Science* 18:227-248.

Toth, N.

1982 The Stone Technologies of Early Hominids at Koobi Fora, Kenya: an Experimental Approach, Ph. D. Thesis, Department of Anthropology, University of California, Berkeley.

1987 Behavioral inferences from early stone artifact assemblages: an experimental model. *Journal of Human Evolution* 16:763-787.

Uhlig, C., and F. Jaeger

1942 Die Ostafrikanische Bruchstuffe und die Angrezende Gebiete zwischen den Seen Magad und Lawa ja Mweri sowie dem Westflus des Meru. *Wisenschaftliche Veroffenlichungen des Deutsches Institut fur Landerkunde Neu Folge*, 10.

Valet, J. P., and L. Meylander

1993 Geomagnetic Field intensity and reversals during the past four million years. *Nature* 366:234-238.

van der Merwe, N. J.

1982 Carbon isotopes, photosynthesis, and archaeology. *American Scientist* 70:596-606.

van der Merwe, N. J., and Medina, E.

1989 Photosynthesis and $^{13}C/^{12}C$ ratios in Amazonian rain forests. *Geochimica et Cosmochimica Acta* 53:1091-1094.

van der Merwe, N. J., and J. C. Vogel

1978 ^{13}C content of human collagen as a measure of prehistoric diet in woodland North America. *Nature* 276:815-816.

van der Merwe, N. J., A. C. Roosevelt, and J. C. Vogel

1981 Isotopic evidence for prehistoric subsistence change at Parmana, Venezuela. *Nature*, 292:536-538.

van Wyk de Vries, B., and O. Merle

1998 Extension induced by volcanic loading in regional strike-slip zones. *Geology* 26(11):983-986.

Vila, P.

1978 The stone artifact assemblage from Terra Amata: a contribution to the comparative study of Acheulian industries in southern Europe. Ph. D. Thesis. University of California, Berkeley.

Vincens, A., and J. Casanova

1987 Modern Background of Natron-Magadi basin (Tanzania-Kenya): physiography, climate, hydrology and vegetation. *Sci. Geol. Bull.* 40(1-2):9-22.

Vincens, A., J. Casanova, and J. J. Tiercelin

1986 Paleolimnology of Lake Bogoria (Kenya) during the 4,500 BP high lacustrine phase. In L.E. Frostick et al. (Eds.), Sedimentation in the African Rifts. *Geol. Soc. Spec. Pub.* 25:323-330.

Vincens, A., R. Bonnefille, and G. Buchet

1991 Étude palynologique du sondage Magadi NF1 (Kenya); implications paléoclimatiques. *Geobios* 24(5):549-558.

Vogel, J. C., and N. J. van der Merwe

1977 Isotopic evidence for early maize cultivation in New York State. *American Antiquity* 42:238-242.

Walter, R. C., P. C. Manega, R. L. Hay, R. E. Drake, and G. H. Curtis

1991 Laser fusion $^{40}Ar/^{39}Ar$ dating of Bed I, Olduvai Gorge, Tanzania. *Nature* 354:145-149.

Western, D.

1973 The structure, dynamics and changes of the Amboseli ecosystem. Ph. D. Thesis. University of Nairobi.

Williamson, D., M. Taieb, B. Damnati, M. Icole, and N. Thouveny

1993 Equatorial extension of the younger Dryas event: rock magnetic evidence from Lake Magadi (Kenya). *Global and Planetary Change* 7:235-242.

Wolpoff, M. H.

1999 *Paleoanthropology*. McGraw Hill, Madison, Wisconsin.

Woolley, A. R.

1989 The spatial and temporal distribution of carbonatites. In *Carbonatites: Genesis and Evolution*, edited by K. Bell, pp. 15-21.

Wynn. T.

1979 The intelligence of later Acheulean hominids. *Man* 14:371-391.

Appendix: Color Plates

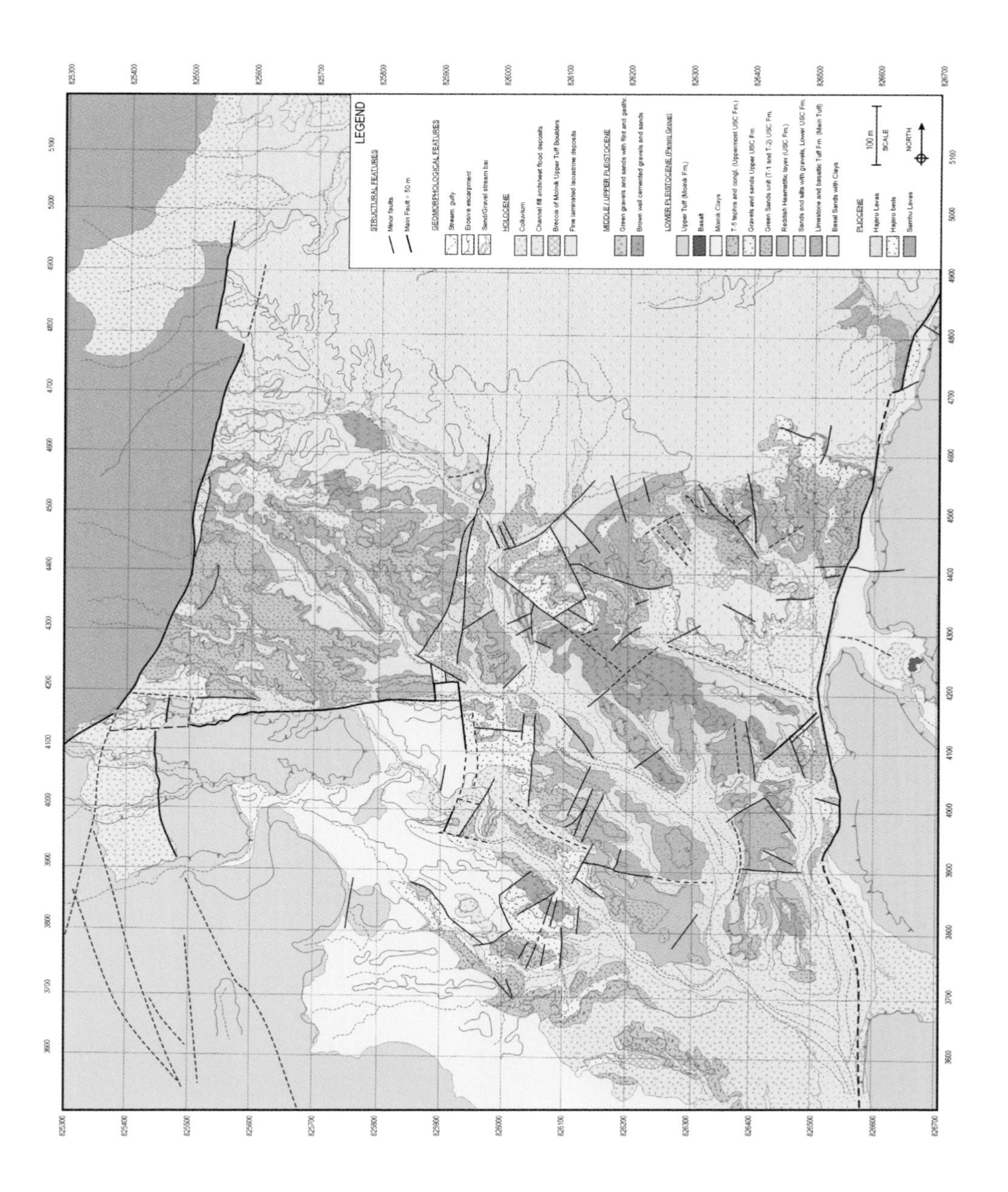

Figure A.1 (see also Figure 3.3) Detailed geological map of the Humbu and Moinik Formations.

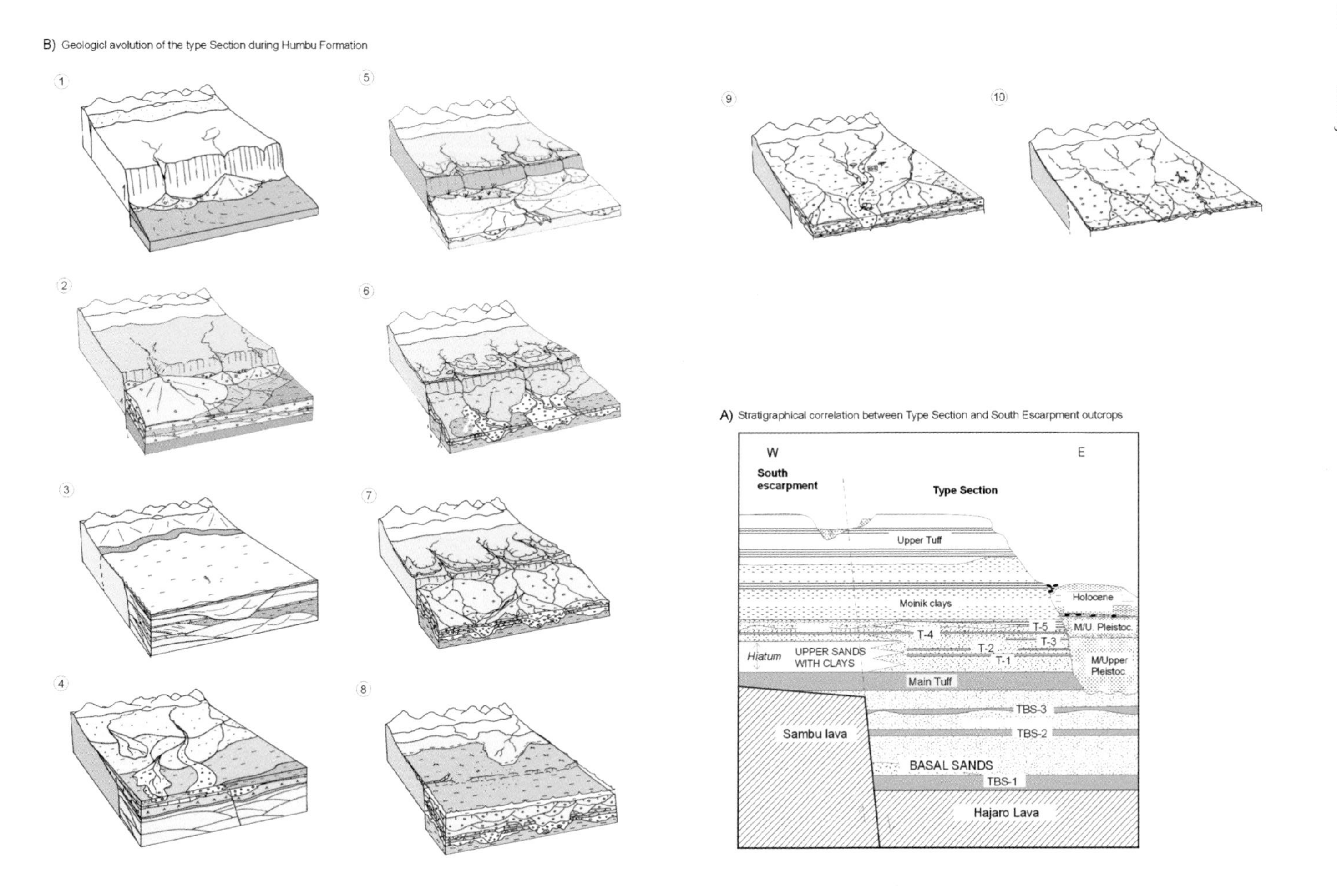

Figure A.2 (see also Figure 3.9) Landscape evolution of the Type Section during Humbu Formation.

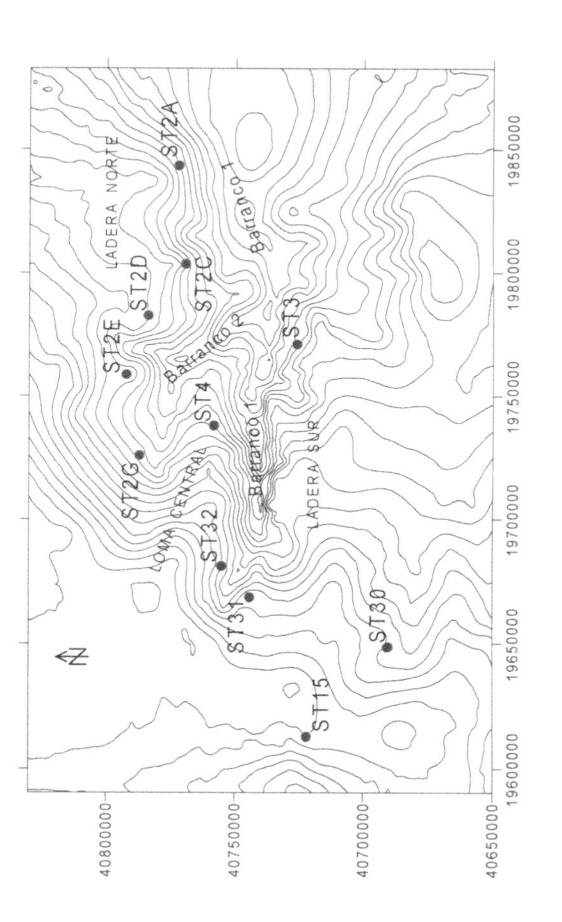

Figure A.3 (see also Figure 4.8) Distribution of the sites in the ST Site Complex.

Figure A.4 (see also Figure 6.6) Spatial distribution of bones (light circles) and stone tools (dark triangles) at ST4.

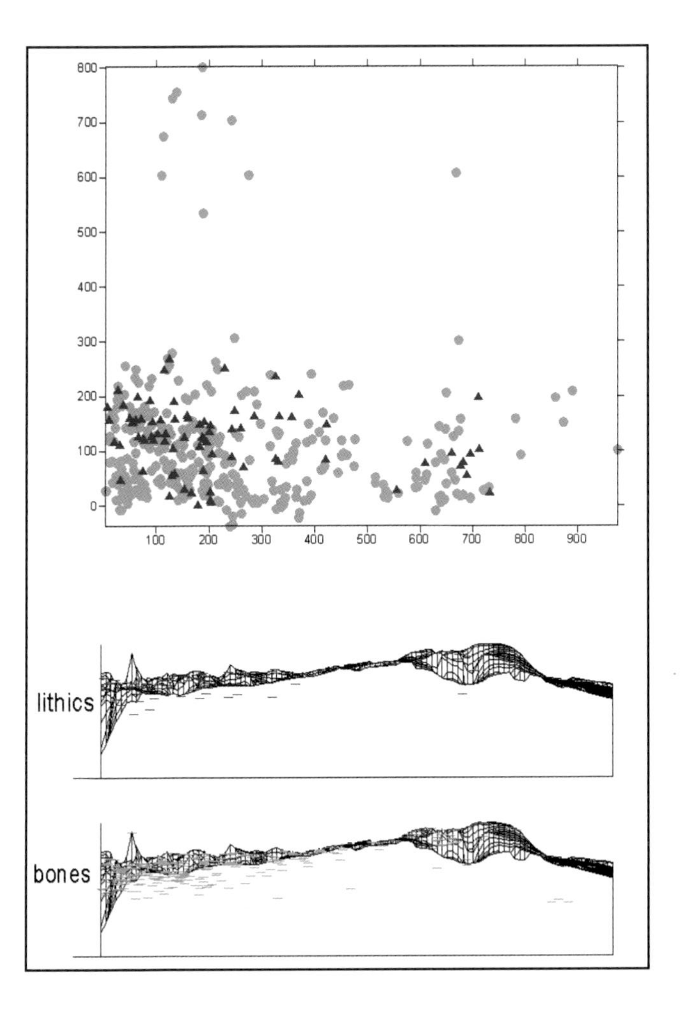

Figure A.5 (see also Figure 6.7) North-south view of the vertical and horizontal distribution of bones (light circles) and stone tools (dark triangles) at ST4.

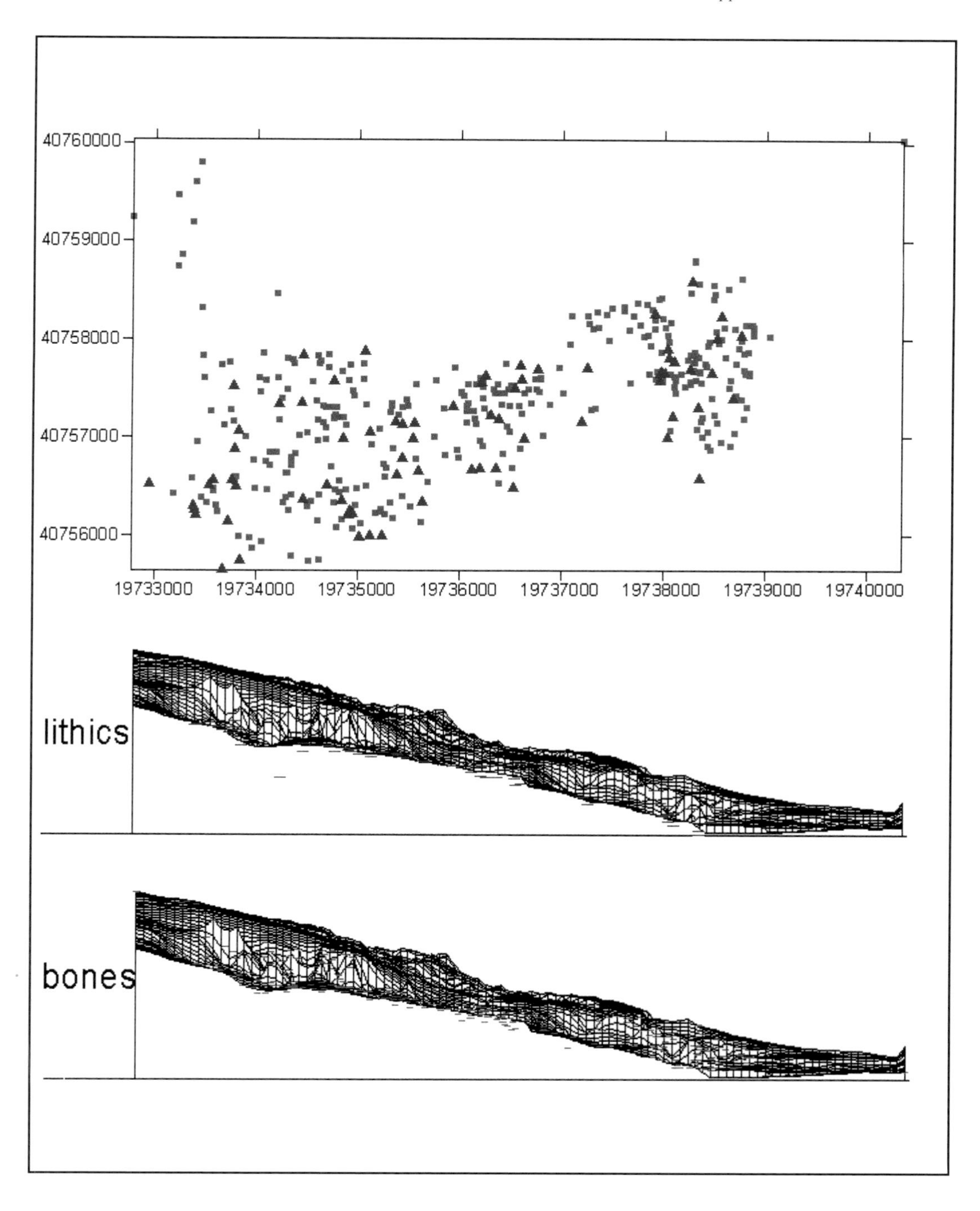

Figure A.6 (see also Figure 6.8) West-east view of the vertical and horizontal distribution of bones (light circles) and stone tools (dark triangles) at ST4.

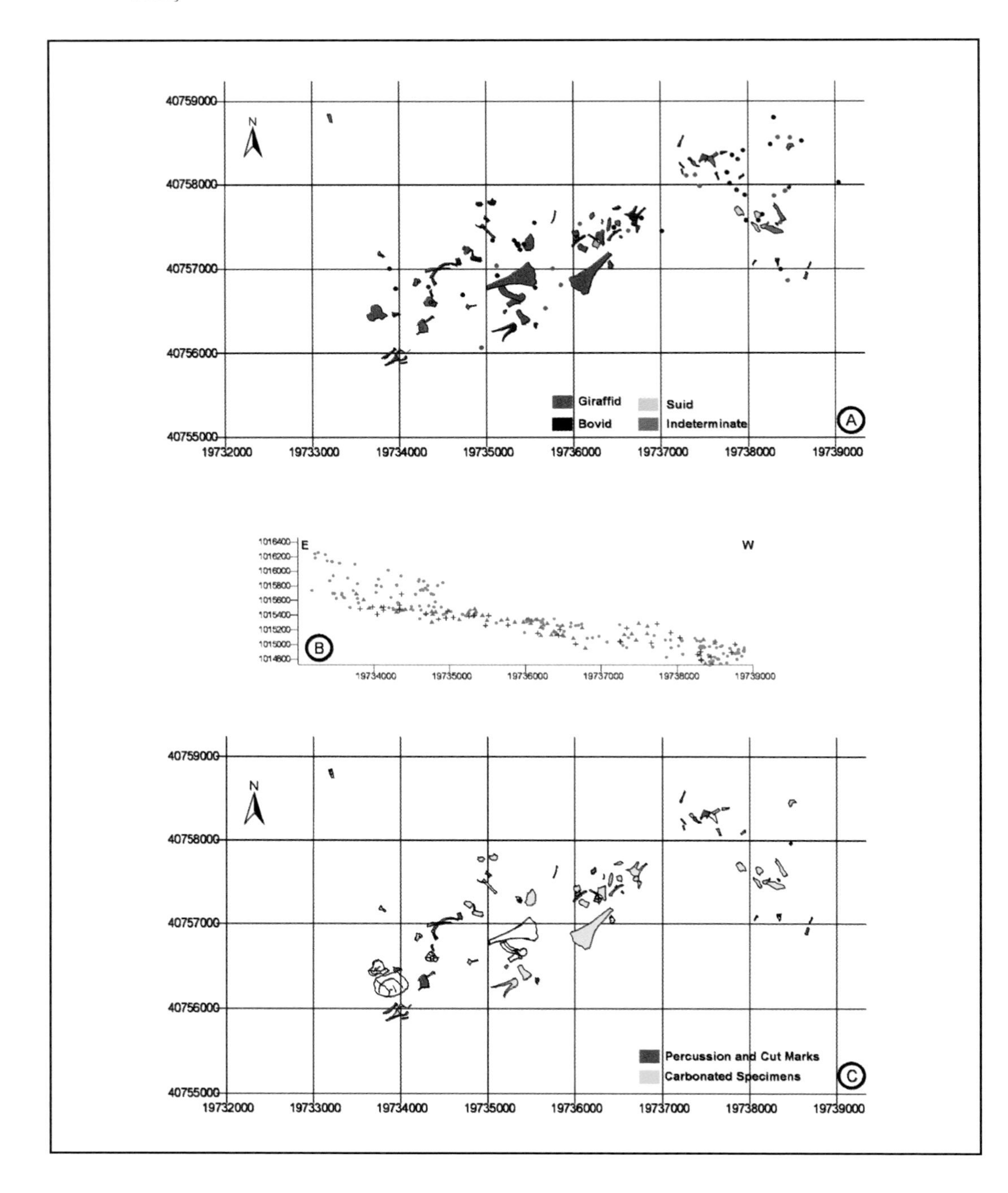

Figure A.7 (see also Figure 6.17) Spatial distribution of the bones from each of the four carcass types present in the bottom of ST4; A), including their vertical distribution B); C) Spatial distribution of carbonate-affected bones and specimens bearing cut marks and percussion marks. Figure courtesy of R. Mora.